Colloidal Quantum Dot Light Emitting Diodes

Colloidal Quantum Dot Light Emitting Diodes

Materials and Devices

Hong Meng

WILEY-VCH

Prof. Hong Meng
Peking University
Building G 306
Lishui Road, Nanshan Disctrict
Shenzhen
CH, 518055

Cover Image: © Prof. Hong Meng

■ All books published by **WILEY-VCH** are carefully produced. Nevertheless, authors, editors, and publisher do not warrant the information contained in these books, including this book, to be free of errors. Readers are advised to keep in mind that statements, data, illustrations, procedural details or other items may inadvertently be inaccurate.

Library of Congress Card No.: applied for

British Library Cataloguing-in-Publication Data
A catalogue record for this book is available from the British Library.

Bibliographic information published by the Deutsche Nationalbibliothek
The Deutsche Nationalbibliothek lists this publication in the Deutsche Nationalbibliografie; detailed bibliographic data are available on the Internet at <http://dnb.d-nb.de>.

© 2024 WILEY-VCH GmbH, Boschstr. 12, 69469 Weinheim, Germany

All rights reserved (including those of translation into other languages). No part of this book may be reproduced in any form – by photoprinting, microfilm, or any other means – nor transmitted or translated into a machine language without written permission from the publishers. Registered names, trademarks, etc. used in this book, even when not specifically marked as such, are not to be considered unprotected by law.

Print ISBN: 978-3-527-35327-9
ePDF ISBN: 978-3-527-84512-5
ePub ISBN: 978-3-527-84513-2
oBook ISBN: 978-3-527-84514-9

Typesetting   Straive, Chennai, India
Printing and Binding   CPI Group (UK) Ltd, Croydon, CR0 4YY

Contents

Preface *xi*

1   History and Introduction of QDs and QDLEDs *1*
1.1   Preparation Route of Quantum Dots *3*
1.2   Light-Emitting Characteristics of Quantum Dots *3*
1.2.1   Particle Size and Emission Color *3*
1.2.2   Quantum Dot Optical Property *4*
1.2.2.1   Quantum Surface Effect *4*
1.2.2.2   Quantum Size Effect *4*
1.2.2.3   Quantum Confinement Effect *5*
1.2.2.4   Quantum Tunnelling Effect *5*
1.2.2.5   Quantum Optical Properties *6*
1.2.3   Core–Shell Structure of QDs *8*
1.2.4   Continuously Gradated Core–Shell Structure of QDs (cg-QDs) *12*
1.2.5   Typical QDs Materials *14*
1.2.5.1   II–VI Semiconductor QDs *16*
1.2.5.2   IV–VI Semiconductor QDs *17*
1.2.5.3   II$_3$–V$_2$ Semiconductor QDs *17*
1.2.5.4   Ternary I–III–VI$_2$ Chalcopyrite Semiconductor QDs *17*
1.2.5.5   Single Element-Based Semiconductor QDs *17*
1.3   Application of Quantum Dots on Display Devices *18*
1.3.1   The Basic Structure of QDLED *18*
1.3.2   Main Factors Affecting QDLED Light Emission *19*
1.3.2.1   Auger Recombination (AR) *19*
1.3.2.2   Fluorescence Resonance Energy Transfer *21*
1.3.2.3   Surface Traps and Field Emission Burst *22*
1.3.3   History of QDLED Development *22*
1.4   Conclusion and Remarks *27*
    References *28*

2   Colloidal Semiconductor Quantum Dot LED Structure and Principles *33*
2.1   Basic Concepts *33*
2.1.1   Color Purity *33*

2.1.2	Solution Processability	34
2.1.3	Stability	35
2.1.4	Surface States of Quantum Dots	36
2.1.5	Energy Levels and Energy Bands	36
2.1.6	Metals, Semiconductors, and Insulators	37
2.1.7	Electrons and Holes	38
2.1.8	Fermi Distribution Function and Fermi Energy Level	39
2.1.9	Schottky Barrier	39
2.1.10	Energy Level Alignment	40
2.2	Colloidal Quantum Dot Light-Emitting Devices	40
2.2.1	The Basic Structure of QDLED	41
2.2.2	The Working Principle of QDLED	43
2.2.3	Operating Parameters of QDLED	44
2.2.3.1	Turn-on Voltage	44
2.2.3.2	Luminous Brightness	44
2.2.3.3	Luminous Efficiency	44
2.2.3.4	Luminescence Color	45
2.2.3.5	Luminous Lifetime	45
2.2.3.6	QDLED Device Fabrication Process	48
	References	48
3	**Synthesis and Characterization of Colloidal Semiconductor Quantum Dot Materials**	**51**
3.1	Background	51
3.2	Synthesis and Post-processing of Colloidal Quantum Dots	53
3.2.1	Direct Heating Method and Hot Injection Synthesis Method	53
3.2.1.1	Hot-Injection Method	54
3.2.1.2	Direct Heating Method	55
3.2.2	Precursor Chemistry	56
3.2.3	Ligating and Non-ligating Solvents	56
3.2.4	Mechanism of Nucleation and Growth of Colloidal Quantum Dots	58
3.2.5	Size Distribution Focus and Size Distribution Scatter	59
3.2.6	Crystalline Species-Mediated Growth and Orientation of Nanocrystals Attachment Growth	60
3.2.7	Synthesis Methods and Band Gap Regulation Engineering of Nuclear-Shell Quantum Dots	61
3.2.7.1	Non-alloyed Core–Shell Quantum Dots	63
3.2.7.2	Alloy Core–Shell Quantum Dots	64
3.2.8	Surface Chemistry of Colloidal Quantum Dots	65
3.2.8.1	Covalent Bond Classification Method	65
3.2.8.2	Entropic Ligands	66
3.3	Material Characterization	66
3.3.1	Ultraviolet–Visible (UV–Vis) Absorption and Fluorescence Spectra	67
3.3.2	Nuclear Magnetic Resonance Spectroscopy	69
3.3.3	Fourier Transform Infrared Spectroscopy (FTIR)	71

3.3.4	X-Ray Photoelectron Spectroscopy (XPS)	74
3.3.5	Transmission Electron Microscopy	76
3.3.6	Small-Angle X-Ray Scattering and Wide-Angle X-Ray Scattering	76
3.3.7	X-Ray Diffractometer	77
3.3.8	X-Ray Absorption Fine Structure Spectra	78
3.3.9	Measurement of Fluorescence Quantum Yield	79
3.4	Conclusion and Outlook	79
	References	81

4 Red Quantum Dot Light-Emitting Diodes 87

- 4.1 Background 87
- 4.2 Red Light Quantum Dot Materials 88
- 4.2.1 Materials 89
- 4.2.2 Quantum Dot Structure Design and Optimization 90
- 4.2.3 Surface Ligands 91
- 4.2.4 Core–Shell Structure 94
- 4.2.5 Alloy Core–Shell Structure 96
- 4.3 Red QDLED Devices 97
- 4.3.1 Red QDLED Device Architecture Development 97
- 4.3.2 Common Device Structures 99
- 4.4 Conclusion and Outlook 102
- References 104

5 Green Quantum Dot LED Materials and Devices 111

- 5.1 Background 111
- 5.2 Commonly Used Luminescent Layer Materials in Green QDLEDs 120
- 5.2.1 Discrete Core/Shell Quantum Dots 120
- 5.2.2 Alloyed Core/Shell Quantum Dots 121
- 5.2.3 Core/Multilayer Shell Quantum Dots 121
- 5.3 Development of Device Structures for Green QDLEDs 122
- 5.4 Factors Affecting the Performance of Green QDLEDs 125
- 5.4.1 QD Ligand Effect 126
- 5.4.2 QD Core/Shell Structure 129
- 5.4.3 Optimization of the Device Structure 130
- 5.4.4 Other Strategies to Improve Device Performance 132
- 5.5 Summary and Outlook 134
- References 135

6 Blue Quantum Dot Light-Emitting Diodes 141

- 6.1 Introduction 141
- 6.2 Blue Quantum Dot Luminescent Materials 143
- 6.2.1 Blue Quantum Dots Containing Cadmium 145
- 6.2.2 Cadmium-Free Quantum Dots 149
- 6.2.2.1 Quantum Dots Based on InP 149
- 6.2.2.2 Quantum Dots Based on ZnSe 151

6.2.2.3	Quantum Dots Based on Cu	*153*
6.2.2.4	Quantum Dots Based on AlSb	*155*
6.3	Optimization of Charge Transport Layer (CTL)	*155*
6.3.1	Hole Transport Layer	*156*
6.3.2	Electron Transport Layer	*161*
6.4	Device Structure	*164*
6.5	Summary	*166*
	References	*168*
7	**Near-Infrared Quantum Dots (NIR QDs)**	*173*
7.1	Introduction of Near-Infrared Quantum Dots	*173*
7.2	Near-Infrared Quantum Dot Materials	*174*
7.2.1	Chalcogenide Lead Quantum Dots	*176*
7.2.2	Chalcogenide Cadmium Quantum Dots	*177*
7.2.3	Silicon Quantum Dots	*178*
7.3	Optimization of Near-Infrared Quantum Dot Materials	*179*
7.3.1	Regulation of Near-Infrared Quantum Dots by Ligand Engineering	*179*
7.3.2	Control of Near-Infrared Quantum Dots by Core/Shell Structure	*180*
7.3.3	Quantum Dots in the Matrix	*181*
7.4	Summary and Prospect	*182*
	References	*183*
8	**White QDLED**	*187*
8.1	Generation of White Light	*187*
8.2	Quantum Dots for White LEDs	*188*
8.2.1	Yellow–Blue Composite White Light Quantum Dots	*189*
8.2.1.1	Cadmium-Containing Yellow Light Quantum Dots	*189*
8.2.1.2	Cadmium-Free Yellow Light Quantum Dots	*189*
8.2.2	Three-Base Color Quantum Dot Composite	*193*
8.2.3	Quantum Dots with Direct White Light Emission	*197*
8.3	Summary Outlook	*200*
	References	*203*
9	**Non-Cadmium Quantum Dot Light-Emitting Materials and Devices**	*207*
9.1	Introduction	*207*
9.2	Quantum Dots and QDLED	*208*
9.2.1	InP	*208*
9.2.2	ZnSe	*215*
9.2.3	I-III-VI	*218*
9.3	Methods for Optimizing QDLED Performance	*222*
9.3.1	Ligand Engineering	*223*
9.3.2	Shell Engineering	*224*
9.3.3	QDLED Device Structure Optimization	*225*
9.4	Summary and Outlook	*227*
	References	*230*

10   AC-Driven Quantum Dot Light-Emitting Diodes *235*
10.1   Principle of Luminescence of DC and AC-Driven QDLEDs *236*
10.2   Mechanism of Double-Emission Tandem Structure of AC QDLEDs *239*
10.2.1   Field-Generated AC QDLEDs *240*
10.2.2   Half-Field to Half-Injection AC QDLEDs *242*
10.2.3   AC/DC Dual Drive Mode QDLEDs *244*
10.3   Optimization Strategies for AC QDLEDs *245*
10.3.1   Optimization of the Field-Induced AC QDLED *247*
10.3.1.1 Dielectric Layer Optimization *248*
10.3.1.2 Quantum Dot Layer Optimization *250*
10.3.2   Optimization of Half-Field-Driven Half-Injected AC QDLEDs *251*
10.3.2.1 Charge Generation Layer Optimization *254*
10.3.2.2 Tandem Structure *254*
10.3.2.3 AC/DC Dual Drive Mode QDLED Optimization *255*
10.3.3   Conclusion and Future Direction of AC-QDLED *256*
          References *257*

11   Stability Study and Decay Mechanism of Quantum Dot Light-Emitting Diodes *259*
11.1   Quantum Dot Light-Emitting Diode Stability Research Status *259*
11.2   Factors Affecting the Stability of Quantum Dot Light-Emitting Diodes *261*
11.2.1   Quantum Dot Light-Emitting Layer *261*
11.2.2   Hole Transport Layer *263*
11.2.3   Electronic Transport Layer *265*
11.2.4   Other Functional Layers *267*
11.3   Quantum Dot Light-Emitting Diode Efficiency Decay Mechanism *268*
11.4   Aging Mechanisms of QDLEDs *271*
11.4.1   Positive Aging *272*
11.4.2   Negative Aging *273*
11.4.3   Electron Transport Layer *274*
11.4.4   Hole Transport Layer *275*
11.4.5   QDs Layer *276*
11.5   Characterization Technologies for QDLEDs *278*
11.5.1   Transient Electroluminescence *279*
11.5.2   Electro-Absorption (EA) Spectroscopy *281*
11.5.3   In-Situ EL–PL Measurement *282*
11.5.4   Differential Absorption Spectroscopy *283*
11.5.5   Displacement Current Measurement DCM Technology *285*
11.6   Outlook *286*
          References *287*

12   Electron/Hole Injection and Transport Materials in Quantum Dot Light-Emitting Diodes *291*
12.1   Introduction *291*
12.2   Charge-Transport Mechanisms *292*

12.3	Electron Transport Materials (ETMs) for QDLED	*293*
12.3.1	Metal-Doped ETMs	*293*
12.3.2	Metal Salt-Doped ETMs	*296*
12.3.3	Design of Composite Materials ETMs	*296*
12.3.4	Polymer-Modified ETMs	*296*
12.3.5	Inorganic Organic Hybrid ETMs	*296*
12.3.6	Double-Stacked ETMs	*297*
12.4	Electron Injection Materials for QDLED	*299*
12.5	Hole Transport Materials for QDLED	*301*
12.5.1	Doping of HTMs	*305*
12.5.2	Compositions of HTMS	*309*
12.5.3	New HTM Materials for QDLED	*311*
12.6	Hole Injection Materials for QDLED	*315*
12.7	Summary and Outlook	*321*
	References	*322*
13	**Quantum Dot Industrial Development and Patent Layout**	*327*
13.1	Introduction	*327*
13.2	Patent Layout	*330*
13.2.1	Nanosys	*330*
13.2.2	SAMSUNG	*332*
13.2.3	Nanoco	*335*
13.2.4	Najing Tech	*338*
13.2.5	CSOT	*344*
13.2.6	BOE	*347*
13.2.7	TCL	*351*
13.3	Summary and Outlook	*355*
	References	*355*
14	**Patterning Techniques for Quantum Dot Light-Emitting Diodes (QDLED)**	*361*
14.1	Introduction	*361*
14.2	Photolithography	*361*
14.3	Micro-Contact Transfer	*363*
14.4	Inkjet Printing	*366*
14.5	Other Patterning Techniques	*368*
14.6	Conclusion	*369*
	References	*370*

Index *373*

Preface

Light has played a significant role in human life since the beginning of civilization, from the discovery of fire to the invention of electric light. As technology has advanced, so has the need for more efficient, cost-effective, and environmentally friendly lighting solutions. The advent of quantum dot light-emitting diodes (QDLEDs) is a significant step forward in the development of new-generation display and lighting technologies. QDLEDs have shown tremendous potential in various fields, including flat-panel displays and solid-state lighting.

This book aims to provide an in-depth understanding of the recent developments in colloidal quantum dot light-emitting materials and devices. It comprises 14 chapters that discuss key materials and optimization schemes for QDLEDs, exploring the relationship between material synthesis and device performance. The first chapter provides a brief overview of the history and development of quantum dots and QDLEDs, while the second chapter introduces the structure and operating principles of QDLEDs. Chapters 3–9 review the latest research results in red, green, blue, near-infrared, white, and cadmium-free quantum dot light-emitting materials and devices. Chapter 10 discusses the luminescence principles of AC-driven QDLEDs, and Chapter 11 delves into the stability study and decay mechanism of QDLEDs. Chapter 12 covers electron/hole injection and transport materials in QDLEDs, and Chapter 13 summarizes the industrialization development and patent layout of quantum dots, while Chapter 14 illustrates the patterning and printing technology in QDLED development.

The author of this book has extensively researched the latest progress and results in academia and industry, citing relevant contents, diagrams, and data in the references. This book is a collaborative effort, with contributions from researchers at various levels, including PhD and MS students, who have made significant contributions to the formation and drafting of the book content in Chinese. The publication of this book would not have been possible without the support and help of the relevant staff of the publisher and the assistance of ChatGPT during the editing and finalization of the book in English.

The author of this book would like to express his sincere gratitude to everyone who has contributed to the creation and publication of this book. The team of authors is thankful to their students, postdocs, research assistants, and staff for their hard work and dedication in researching and writing this first-edition book in Chinese.

They also appreciate the editors, translators, proofreaders, and designers who helped to bring the book to life in multiple languages and formats. Lastly, the authors extend their heartfelt thanks to their families, colleagues, and friends for their unwavering support and encouragement throughout this project.

This book is a culmination of the authors' efforts to share their knowledge and expertise on colloidal quantum dot light-emitting materials and devices with readers. It is hoped that this book will serve as a useful reference for researchers, scientists, engineers, students, and anyone interested in the development and application of QDLEDs. The authors believe that the contents of this book will help promote the research and development of QDLEDs and contribute to the progress of the lighting and display industries.

May 2023

Hong Meng
School of Advanced Materials
Peking University Shenzhen Graduate School
Peking University
Shenzhen
518055 China
E-mail: menghong@pku.edu.cn

1

History and Introduction of QDs and QDLEDs

Semiconductor nanocrystals (NCs) are the most widely studied of the nanoscale semiconductors. In early 1981, Alexei Ekimov and Alexander Efros, working at the S.I. Vavilov State Optical Institute and A.F. Ioffe Institute, Russia, discovered nanocrystalline, semiconducting quantum dots (QDs) in a glass matrix and conducted pioneering studies of their electronic and optical properties. Simultaneously, in 1985, Louis Brus at Bell Laboratories in Murray Hill, NJ, discovered colloidal semiconductor NCs (QDs), for which he shared the 2008 Kavli Prize in Nanotechnology. Over the years, QDs have been established as a new type of semiconductor nanocrystalline material whose size is smaller than or close to the excitonic Bohr radius of its bulk material. Common semiconductor materials include Si, Ge, compounds of group II–VI (e.g. CdSe), and compounds of group III–V (e.g. indium phosphide [InP]). When the size of these bulk semiconductor materials is larger than their exciton Bohr radii, electrons and holes are able to move freely and independently in the bulk materials. However, when the size of QDs is smaller than their own exciton Bohr radius, after being excited by light, an electron in the valence band will leap to the conduction band, leaving a hole in the valence band, and the electron and hole form an exciton due to Coulomb effect, which is confined in a space smaller than the exciton Bohr radius, and the electron and hole will be quantized, which is called the "quantum size effect" of nanomaterials. This quantum size effect allows QDs to have discrete energy levels, thus giving them unique physicochemical properties [1]. Colloidal semiconductor NCs have size-dependent particle properties, while their surface ligands make them solution-processable, which gives them a "particle-solution" duality.

Figure 1.1a shows the energy level diagrams of molecular, QD, and bulk semiconductor materials. The molecular orbital energy level diagram is composed of the highest occupied molecular orbital (HOMO) and the lowest unoccupied molecular orbital (LUMO), while the energy level diagram of QDs consists of some discrete energy levels, and the bulk semiconductor material consists of conduction and valence bands. Figure 1.1b illustrates the spatial extent of the confined domains of electrons and holes and the respective energy as a function of the density of electronic states for bulk semiconductor materials, two-dimensional

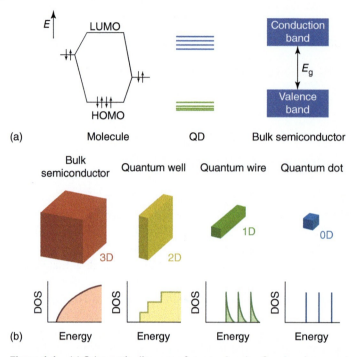

Figure 1.1 (a) Schematic diagram of energy levels of molecular, quantum dot, and bulk semiconductor materials; (b) spatial extent of the confined domain of electrons and holes and the respective energy as a function of the density of electronic states for bulk semiconductor materials, two-dimensional quantum sheets, one-dimensional quantum wires, and zero-dimensional quantum dots depending on the material size.

quantum sheets, one-dimensional quantum wires, and zero-dimensional QDs, depending on the size of the material. For a bulk semiconductor material, the dimensions in all three dimensions are larger than its own Bohr exciton radius, and electrons and holes are free to move independently in all three dimensions; while for a two-dimensional quantum sheet, the dimensions in two dimensions are larger than its own Bohr exciton radius, and electrons and holes are free to move independently in two dimensions; and for a one-dimensional quantum wire, the dimensions in one dimension are larger than its own Bohr exciton radius, while for a one-dimensional quantum wire, the dimensions in one dimension are larger than its own Bohr exciton radius, and electrons and holes are free to move independently in two dimensions. For a one-dimensional quantum wire, whose dimension in one dimension is larger than its own exciton radius, electrons and holes are free to move independently in one dimension; and for a zero-dimensional QD, whose dimension in all three dimensions is smaller than its own exciton radius, electrons and holes are restricted from moving freely and independently in all dimensions. In general, QD is a collective term for a two-dimensional quantum sheet, a one-dimensional quantum wire, and a zero-dimensional QD.

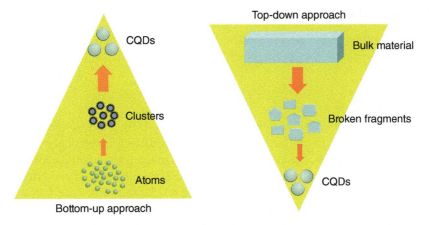

Figure 1.2 Preparation pathways for quantum dots: the "top-down" method and the "bottom-up" method.

1.1 Preparation Route of Quantum Dots

There are two completely different ways to prepare QDs, namely the "top-down method" and the "bottom-up method", as shown in Figure 1.2. The top-down method is to prepare QDs by reducing the dimensionality and size of the bulk semiconductor material; the bottom-up method is to combine atoms or molecules into QDs by chemical synthesis. The former approach is limited by the ultra-fine processing technology, which cannot produce QDs below 10 nm at present, and the morphological regulation of QDs is also limited to some extent. The latter is mainly achieved through colloidal chemical synthesis, which can produce colloidal QDs of different sizes and shapes.

1.2 Light-Emitting Characteristics of Quantum Dots

1.2.1 Particle Size and Emission Color

QDs are semiconductor particles having a few nanometers in size, their optical and electronic properties are quite different from those of larger particles as a result of quantum mechanics. When the QDs are illuminated by UV light, an electron in the QD can be excited from the transition of an electron valence band to the conductance band. The excited electron can drop back into the valence band releasing its energy as light. The color of that light depends on the energy difference between the conductance band and the valence band. The QD absorption and emission features correspond to transitions between discrete quantum mechanically allowed energy levels in the box, which are reminiscent of atomic spectra.

1.2.2 Quantum Dot Optical Property

The QDs are defined as the semiconductor NCs with the quantum confinement. Thus, the semiconductor nanoparticles with dimensions QDs have the following features:

1.2.2.1 Quantum Surface Effect

The surface effect refers to the fact that as the particle size of QDs decreases, most of the atoms are located on the surface of QDs, and the specific surface area of quantum dots increases with decreasing particle size. Due to the large specific surface area of QDs (nanoparticles), the increase in the number of atoms in the surface phase leads to the lack of coordination, unsaturated bonds, and suspension bonds of surface atoms. This makes these surface atoms highly reactive, extremely unstable, and easily bonded with other atoms. This surface effect will cause the large surface energy and high activity of nanoparticles. The activity of surface atoms not only causes changes in the surface atomic transport and structural type of nanoparticles but also causes changes in the surface electron spin conformation and electronic energy spectrum. This feature offers a route to manipulate QD interactions with their environment. QDs can be tethered to proteins, antibodies, or other biologic species and used as optically addressable bio-labels. On the other hand, passivation of QD surface can improve the QD stability and increase the photoluminescent quantum efficiency. Surface defects lead to trapped electrons or holes, which in turn affect the luminescent properties of QDs and cause nonlinear optical effects. Metallic materials show various characteristic colors through light reflection. Due to the surface effect and size effect, the light reflection coefficient of nanoparticles decreases significantly, usually less than 1%, so nanoparticles are generally black in color, and the smaller the particle size, the darker the color, i.e. the stronger the light absorption ability of nanoparticles, showing a broadband strong absorption spectrum. Surface effect or ligand modification offers an additional tool for manipulating energy levels and electronic and optical properties.

1.2.2.2 Quantum Size Effect

Quantum size effect refers to the phenomenon that the electron energy levels near the Fermi energy level change from quasi-continuous to discrete energy levels, that is, when the particle size drops to a certain value, the energy level splits or energy gap widens, in other words, the energy spectrum becomes discrete, and as a result, the bandgap becomes size-dependent. When the change in energy level is greater than the change in thermal, optical, and electromagnetic energy, it leads to the magnetic, optical, acoustic, thermal, electrical and superconducting properties of nanoparticles being significantly different from those of conventional materials. This feature of QDs is that the energy gap changes with the increase in the grain size, the larger the grain size, the smaller the energy gap, and *vice versa*, the larger the energy gap. That is, the smaller the QD, the shorter the wavelength of light (blueshift), and the larger the QD, the longer the wavelength of light (redshift). According to the size effect of QDs, we can use the method of changing the size of the grain to regulate

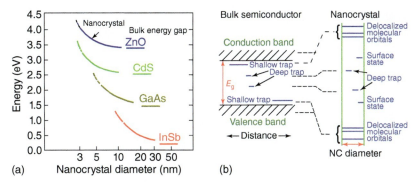

Figure 1.3 Theoretical ideas from the early 1980s. (a) Calculated size-dependent shift of the lowest exciton levels in strong confinement (b) Spatial electronic state correlation diagram for bulk semiconductors and NCs. The bulk valence and conduction bands, together with shallow trap states, evolve into the NC molecular orbitals. Deeply localized defect states in the bulk have essentially the same energy as those in the NC. New localized surface states exist in the NC. Source: Efros and Brus [2]/American Chemical society.

the tuning of the light spectrum of the material and no longer need to change the chemical composition of QDs.

1.2.2.3 Quantum Confinement Effect

Quantum confinement can be observed once the diameter of a material is of the same magnitude as the de Broglie wavelength of the electron wave function. When the QD size of the particle reaches the nanometer scale, the electronic energy level near the Fermi energy level splits from the continuum to the discrete energy level, and their electronic and optical properties deviate substantially from those of bulk materials (Figure 1.3). For semiconductor materials, the size of the bandgap can be adjusted by changing the scale of the particles, thus changing the reliance on certain very costly semiconductor materials (Figure 1.4). Quantum confinement effects in QDs can also result in fluorescence intermittency, called "blinking" [4].

1.2.2.4 Quantum Tunnelling Effect

Quantum tunning effect is one of the fundamental quantum phenomena, i.e. when the total energy of a microscopic particle is less than the height of the potential barrier, considering the motion of a particle encountering a potential barrier above the energy of the particle, the particle is still able to cross this barrier, which indicates that on the other side of the barrier, the particle has a certain probability that the particle penetrates the potential barrier. For QDs, electron movement in the nanoscale space, the carrier transport process will have obvious electronic fluctuations, the emergence of quantum tunneling effect, and the energy level of the electron is discrete [5]. To achieve the quantum effect, it requires the formation of nano-conducting domain in a few μm to tens of μm tiny area. When the voltage is low, the electrons are confined to the nanoscale range of motion, and increasing the voltage can make the electrons cross the nanopotential barrier to form a sea of Fermi electrons, making the system conductive. The

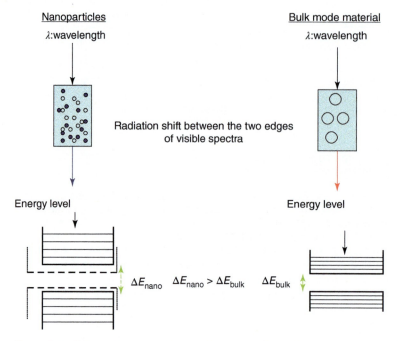

Figure 1.4 QDs Confinement effect: As the size of the particles decreases, the electrons and electron holes come closer, and the energy required to activate them increases, which ultimately results in a blueshift in light emission. Source: Adapted from Kang and Min [3].

quantum tunneling effect occurs when electrons cross the quantum barrier from one quantum well into another quantum well, and this insulating to conducting critical effect is a characteristic of the QDs in a nano-ordered array system. Most QD solids exhibit complex charge-carrier interactions between carrier confinement, interfacial properties, and quantum tunneling effects in the nature of electronic coupling [6].

1.2.2.5 Quantum Optical Properties

Owing to the above-mentioned effects of QDs, the QD absorption and emission features correspond to transitions between discrete quantum mechanically allowed energy levels in the box, which are reminiscent of atomic spectra. QDs have intermediate properties between bulk semiconductors and discrete atoms or molecules. Their optoelectronic properties change as a function of both size and shape. Larger QDs of 5–6 nm diameter emit longer wavelengths, with colors such as orange, or red. Smaller QDs (2–3 nm) emit shorter wavelengths, yielding colors like blue and green. Whereas, the specific colors vary depending on the exact composition of the QD. It turns out that QDs have broad absorption spectra, meaning that they can be excited across a pretty expansive range of light wavelengths. Figure 1.5 shows the emission color of QDs dependent on their respective sizes.

Recent studies have also shown that the shape of the QD may play a role in the band-level energy of the QD, thus affecting the frequency of fluorescence emission

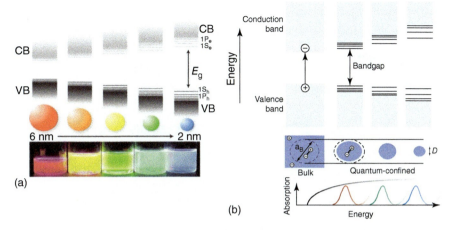

Figure 1.5 (a) Schematic representation of the quantum confinement effect on the energy level structure of a semiconductor material. The lower panel shows colloidal suspensions of CdSe NCs of different sizes under UV excitation. Source: Donegá [7]/Royal Society of Chemistry.; (b) Quantum confinement, leading to size-dependent optical and electrical properties that are distinct from those of parental bulk solids, occurs when the spatial extent of electronic wave functions is smaller than the Bohr exciton diameter. Source: García de Arquer et al. [8]/American Association for the Advancement of Science.

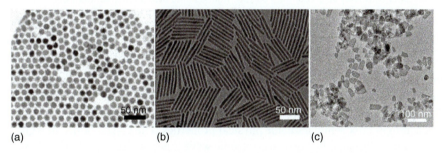

Figure 1.6 Pictures of quantum dots with various sizes and morphologies taken with transmission electron microscopy (a) quantum dots, (b) quantum rods, and (c) quantum sheets.

or absorption. The luminescence characteristics of QDs are closely related to their size and shape, the presence or absence of core–shell structure, and their surface chemistry. QDs of various sizes, morphologies, and core–shell structures can be synthesized by colloidal chemistry, as shown in the transmission electron microscope image in Figure 1.6.

The ligand type and ligand concentration on the surface of colloidal QDs also have an effect on their luminescence properties. For example, Peng et al. observed a shift of several nanometers in the peak position of the fluorescence emission peak of CdSe/CdS core–shell structured QDs ligated by fatty acid cadmium salt surface after ligand exchange treatment by aliphatic amines (Figure 1.7). The QDLED luminescence performance indexes such as external quantum yield and lifetime of CdSe/CdS core–shell QDs with different surface ligands are even more different when prepared

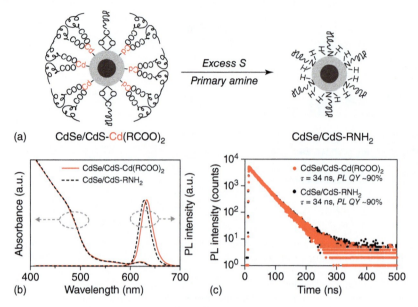

Figure 1.7 Photoluminescence and electroluminescence properties of CdSe/CdS-Cd(RCOO)$_2$ and CdSe/CdS-RNH$_2$ quantum dots. Schematic diagram of CdSe/CdS quantum dots with fatty acid cadmium surface ligands (and also a small amount of negatively charged carboxylates) undergoing ligand exchange to generate CdSe/CdS quantum dots with aliphatic amines. (b) Absorption and steady-state photoluminescence spectra. (c) Time-resolved photoluminescence spectrum with single-exponential decay lifetime (τ) and quantum yield (QY). Source: Pu et al. [9]/CC BY 4.0/Public Domain.

into QDLED devices [9]. Therefore, we should pay special attention to the surface chemical state when studying the luminescence properties of QDs.

The effect of crystal structure can also affect the emission color. Table 1.1 illustrates common bulk semiconductor physical properties. The crystal lattice of a QD semiconductor has an effect on the electronic wave function. As a result, QDs have a specific energy spectrum equal to the bandgap and a specific density of electronic states outside the crystal. Compared to conventional fluorophores, QDs have unique optical and electronic properties. Examples include high quantum yields and molar extinction coefficients, large effective Stokes shifts, broad excitation spectra, narrow emission spectra, a high resistance to reactive oxygen species, protection against material degradation, and nearly impervious to photobleaching.

In highly monodisperse colloidal QD samples, due to the quantum size effect, electrons and holes are subjected to quantum confinement effect, atomic-like structure of electronic states of QDs leads to the formation of discrete energy levels of narrow full-width at half maximum (FWHM) width of 20–80 meV at room temperature and symmetrical fluorescence emission peaks [10] (Figure 1.8).

1.2.3 Core–Shell Structure of QDs

For most core QDs, due to their low PLQYs and poor stabilities, they tend to exhibit a broad red-shifted emission, owing to the surface defects. The issues can be

Table 1.1 Common bulk semiconductor physical parameters.

Material name	Crystal structure type (300 K)	Category	E_{gap} (eV)	Lattice parameters (Å)	Density (g cm^{-3})
ZnS	Sphalerite	II–VI	3.61	5.41	4.09
ZnSe	Sphalerite	II–VI	2.69	5.67	5.27
ZnTe	Sphalerite	II–VI	2.39	6.10	5.64
CdS	Wurtzite	II–VI	2.49	4.14/6.71	4.82
CdSe	Wurtzite	II–VI	1.74	4.3/7.01	5.81
CdTe	Sphalerite	II–VI	1.43	6.48	5.87
InP	Sphalerite	III–V	1.35	5.87	4.79
GaAs	Sphalerite	III–V	1.42	5.65	5.32
PbS	Rock salt	IV–VI	0.41	5.94	7.60
PbSe	Rock salt	IV–VI	0.28	6.12	8.26
PbTe	Rock salt	IV–VI	0.31	6.46	8.22

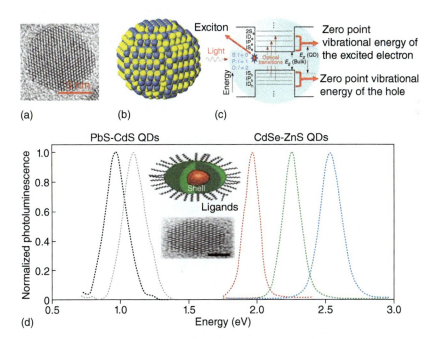

Figure 1.8 (a) TEM image of CdSe nanocrystal; (b) nanocrystal atomic structure; (c) energy level discreteness of the excited electron and the hole in an exciton entity. (d) PL spectra of CdSe–ZnS and PbS–CdS core/shell colloidal QDs. Source: Efros and Brus [2]/American Chemical Society.

addressed to improve efficiency and brightness of semiconductor NCs by growing shells of another higher-bandgap semiconducting material around them, resulting in core–shell QDs. The improvement is due to the reduced access of electrons and holes to non-radiative surface recombination pathways, and in other cases, due to the reduced Auger recombination (AR). Core–shell QDs (core@shell QDs) hold the promise of being emissive components through the precise control of shade and an improved color-rendering index. They exhibit improved optical properties over pure core-only QDs due to the growth of the shell around the QD core, which improves stability and photoluminescence efficiency. A fundamental feature of QDs is the tunability of their emission color through precise control of their size and composition, giving access to UV, visible, and near-infrared wavelengths. Continuing improvements in engineering core–shell QD structures, where a 1–10 nm binary, ternary, or alloyed semiconductor core particle is surrounded by a shell composed of one or more semiconductors of a wider bandgap, have resulted in materials with fluorescence quantum yields that approach unity, narrow symmetric spectral line shapes, and remarkable stabilities, as shown in Figure 1.9, for CdSe/ZnS QDs [11], CdSe/CdS quantum rods [12], CdSe quantum sheets [13]. Interestingly, the fluorescence emission peaks of QDs and quantum rods have a large Stocks shift with the peak position of their first exciton absorption peaks, while the fluorescence emission peaks of quantum sheets have almost no Stocks shift with the peak position of their first exciton absorption peaks. In addition, in terms of fluorescence lifetime, the fluorescence lifetime of rare-earth luminescent materials is at millisecond or microsecond level, while the fluorescence lifetime of QDs is usually in the range of milliseconds [14–16]. While the fluorescence lifetime of QDs is usually below 100 ns [17–19]. It has been found that the fluorescence emission of single QDs has severe blinking behavior, ranging from a few milliseconds to a few minutes, which is mainly due to the non-radiative compounding process caused by the "surface defects" of QDs [20–22].

Although the PL of II–VI QDs can be bright and stable under a reasonable range of excitation light intensities, core–shell QDs universally showed significant "blinking" under the high fluxes used in single QD fluorescence spectroscopy, whereby the PL of single QDs turns "on" and "off" under continuous excitation. This single QD "blinking" not only limits the use of QDs as single photon sources but also

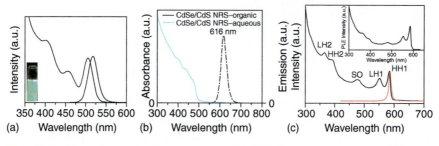

Figure 1.9 Absorption and emission spectra of (a) CdSe/ZnS quantum dots, (b) CdSe/CdS quantum rods, and (b) CdSe quantum sheets.

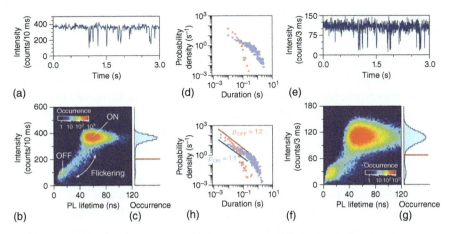

Figure 1.10 (a) Emission intensity from a single CdSe/CdS/ZnS core/shell/shell quantum dot with a temporal resolution of 10 ms under pulsed excitation (405 nm; 10 MHz; 1 µJ cm^{-2}). The diameter of the quantum dot core is 3.2 nm and the shell has 8(2) CdS(ZnS) monolayers. (b) The corresponding "fluorescence lifetime intensity distribution" is a two-dimensional histogram. The plot is based on a 300 seconds experiment, which was divided into 30 000 10 ms time intervals. The effect of scintillation is highlighted. (c) Corresponding 1D intensity histogram with thresholds used for statistical analysis indicated by red lines. (d) Flicker periods (blue) and non-flicker periods (red) extracted from the 10 ms merged and thresholded data in (a) plot. (e–h) Same as (a–d), but using the same single photon data on which there are 3 ms time bins for statistics. Source: Rabouw et al. [22]/American Chemical Society.

potentially limits QDs from being a stable photoluminescent output source under relatively high fluxes. A different approach to blinking suppression was explored by growing a thick CdS or CdS/CdZnS/ZnS shell (>5 nm in shell thickness) onto CdSe core QDs with the idea of fully isolating the excited carriers from the QD surface and the surface environment [22]. However, these QDs generally do not have a very good size distribution, exhibit broad PL spectra, and display moderate PL QYs. More recently, synthesis of CdSe/CdS QDs at a high reaction temperature (310 °C) using octanethiol as a sulfur precursor resolved many of these issues (Figure 1.10).

The terms Type I and Type II QDs are used to classify QDs based on their band structure and electron–hole recombination dynamics (Figure 1.11). Type I and Type

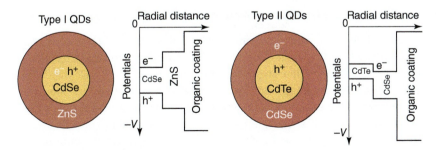

Figure 1.11 Comparison of Type I and Type II CdSe/ZnS core–shell quantum dots. Source: Vaneski [23]/Aleksandar Vaneski.

II QD materials are two different categories of semiconductor nanostructures that exhibit unique electronic and optical properties due to their quantum confinement effects. In a Type I QD, the electrons and holes are confined in the same region of the QD, resulting in a strong overlap of their wave functions. This leads to efficient radiative recombination of the electron–hole pairs, resulting in a high quantum yield and bright fluorescence. Examples of Type I QDs include CdSe, CdTe, and InP QDs. In contrast, in a Type II QD, the electrons and holes are spatially separated between two different regions of the QD due to a band offset at the interface. As a result, radiative recombination of the electron–hole pairs is less efficient than in a Type I QD, leading to lower quantum yields and longer-lived excitons. However, Type II QDs exhibit unique properties such as multiple exciton generation and efficient charge separation, which make them attractive for applications such as solar cells and photocatalysis. Examples of Type II QDs include CdS/CdTe, CdSe/CdTe, and CdSe/ZnTe QDs. Both Type I and Type II QDs have found numerous applications in areas such as optoelectronics, bioimaging, and sensing due to their unique properties and tunability.

Giant QDs are a type of QD structure that is characterized by a larger size and a more complex core–shell structure than traditional QDs. While traditional QDs typically have dimensions on the order of a few nanometers, giant QDs can have dimensions up to tens of nanometers. The larger size of giant QDs offers several advantages over traditional QDs. For example, giant QDs have a higher absorption cross-section, which allows them to absorb more light and generate more charge carriers per photon. They also have a higher quantum yield and longer emission lifetime due to reduced surface recombination and enhanced confinement of excitons. Giant QDs can be synthesized using various methods, including the core–shell approach, which involves the growth of a large core particle, followed by the deposition of multiple layers of shell material. The resulting core–shell structure can have a complex morphology, such as a core–shell–shell or a core–shell–shell–shell structure. Applications of giant QDs include bioimaging, sensing, and photovoltaics. In bioimaging, giant QDs can be used as contrast agents due to their bright and stable emission. In sensing, they can be used as probes for detecting biomolecules due to their high sensitivity and specificity. In photovoltaics, giant QDs can be used as absorbers in thin-film solar cells due to their high absorption efficiency and tunable bandgap.

1.2.4 Continuously Gradated Core–Shell Structure of QDs (cg-QDs)

For most core/shell QDs, such as giant CdSe/CdS core–shell QDs (denoted CdSe/CdS g-QDs, where the small CdSe core is passivated by the large CdS shell) synthesized by the successive ion layer adsorption and reaction (SILAR) method, exhibit reduced surface trapping and AR. Notably, this core/shell QD shows a significant redshift of the emission peak, which indicates that the CdSe core wave function extends into the CdS shell region, i.e. the effective size of the core increases. In addition, the first absorption peaks of CdSe/CdS g-QDs are relatively suppressed, which is due to the fact that the bandgap of CdS is larger than that of CdSe, so the absorption mainly comes from the thick CdS shell. However, due to the large bandgap of CdS (e.g. 2.42 eV; corresponding to an absorption onset of 512 nm), the emission position of CdSe/CdS g-QDs achieving reabsorption suppression is limited throughout the visible region compared to CdSe. In addition, g-QDs are typically prepared by the SILAR method, which requires multiple time-consuming steps to epitaxially deposit the desired shell material. It is clearly desirable to develop alternative, simple, robust, and convenient synthetic routes to prepare g-QDs with enhanced photostability and quantum yields. It has been recently demonstrated that the implementation of larger CdSe QDs and CdSe QDs with hierarchical shell structures can reduce the AR rate. In this context, cg-QDs with smooth confinement potentials hold promise for effectively reducing nonradiative AR, such as suppressing intraband conversion of additional carriers during Auger recombination and greatly balancing the charge injection in QDLEDs brought about by the fine nanostructure of cg-QDs. These QDs with chemical composition gradients possess excellent photostability due to the judicious incorporation of CdSe/Cd$_{1-x}$Zn$_x$Se$_{1-y}$S$_y$ graded shells, which mitigate the lattice strain between CdSe and ZnS as shown in Figure 1.12 [24]. Furthermore, the Stokes shift (i.e. the difference between absorption and emission maxima) of these QDs with graded shell structures can be easily engineered by simply further adjusting the thickness of the outermost ZnS shell (i.e. redshift with increasing ZnS shell thickness), which is not observed in conventional CdSe/ZnS QDs because of their energy level mismatch. It is also noteworthy that CdSe/Cd$_{1-x}$Zn$_x$Se$_{1-y}$S$_y$/ZnS QDs are more advantageous than giant CdSe/CdS QDs because the bandgap of ZnS is larger than that of CdS and thus highly tunable due to the suppression of reabsorption when choosing the emission position in the visible region. All CdSe/Cd$_{1-x}$Zn$_x$Se$_{1-y}$S$_y$/ZnS

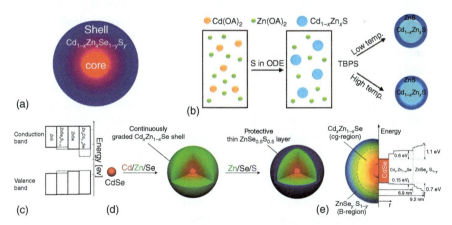

Figure 1.12 Fabrication methods of cg-QDs. (a) One-step synthesis of Cd$_{1-x}$Zn$_x$Se$_{1-y}$S$_y$ cg-QDs with continuously graded nanocompositions along the whole radial direction. (b) Schematic illustration of continuously graded Cd$_{1-x}$Zn$_x$S cores with ZnS shell. The smoothness of the core/shell interface can be adjusted by changing the reaction temperatures during shell growth. (c) The energetic band alignments of Zn$_{1-x}$Cd$_x$Se/ZnSe/ZnSe$_x$S$_{1-x}$/ZnS QDs. (d) Reaction schematics of CdSe/Cd$_x$Zn$_{1-x}$Se/ZnSe$_{0.5}$S$_{0.5}$ QDs with multistep synthesis. (e) Structures and band alignments of CdSe/Cd$_x$Zn$_{1-x}$Se/ZnSe$_y$S$_{1-y}$ QDs. Source: Shen et al. [24]/American Chemical Society.

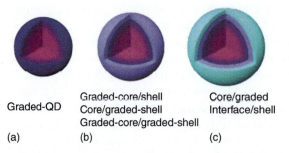

Graded-QD	Graded-core/shell Core/graded-shell Graded-core/graded-shell	Core/graded Interface/shell
(a)	(b)	(c)

Figure 1.13 Schematics of three types of cg-QDs. (a) cg-QDs with no distinct boundary between the core and shell layers; (b) cg-QDs with one or both of the core and shell are continuously graded along the radial direction; (c) cg-QDs with continuously graded intermediate layers between the core and shell of the QD. Source: Cheng et al. [25]/American Chemical Society.

QDs with different emission positions exhibit longer lifetimes compared to ordinary CdSe QDs, which implies a reduced AR rate as a direct result of electron wave function delocalization over the whole QD.

The CdS- and CdS-based cg-QDs reported so far can be divided into three main categories: (i) QDs that are continuously graded along the entire radial direction, i.e. no distinct boundary between the core and shell layers (Figure 1.13a); (ii) QDs in which one or both of the core and shell are continuously graded along the radial direction (Figure 1.13b); (iii) continuously graded intermediate layers between the core and shell of the QD (Figure 1.13c) [25]. It should be emphasized that although unintentional alloying of the core/shell interface has been achieved in the early stages of QD synthesis, it is now possible to precisely and quantitatively control the elemental composition ratio and the thickness of the cg-QD layer. In contrast to the shell layer growth method based on continuously graded cores, another widely used technique is to grow shell layers directly on discontinuous cores, which also exhibit good optical properties. Continuously graded CdSe/Cd$_x$Zn$_{1-x}$Se/ZnSe$_{0.5}$S$_{0.5}$ QDs were synthesized by Lim et al. and are expected to have great potential in DC-pumped lasers [26]. After dispersing the CdSe core in a mixed solution and controlling the constant injection of reactants, they obtained a continuously graded Cd$_x$Zn$_{1-x}$Se shell on the CdSe core. In addition, they coated these CdSe/Cd$_x$Zn$_{1-x}$Se particles with a thin ZnSe$_{0.5}$S$_{0.5}$ shell to protect the QDs from degradation. Lim et al. also proposed a novel multishell QD consisting of a CdSe core coated with a continuously graded Cd$_x$Zn$_{1-x}$Se inner shell followed by a ZnSe$_y$S$_{1-y}$ barrier shell. The Cd$_x$Zn$_{1-x}$Se inner shell is able to suppress the AR process, while the ZnSe$_y$S$_{1-y}$ shell with a wide gap is able to optimize electron and hole injection in QDLEDs [25].

1.2.5 Typical QDs Materials

QDs are nanoscale semiconducting materials with unique optical and electronic properties. QDs can be made from a variety of materials, including metals, semiconductors, and organic molecules. Here are some common QD materials and their properties:

Cadmium selenide (CdSe): CdSe is one of the most commonly used QD materials. It has a high quantum yield and narrow emission spectra, making it useful for a variety of applications, including biological imaging and photovoltaics;

Cadmium telluride (CdTe): CdTe has a similar bandgap to CdSe but a larger Bohr exciton radius, which makes it useful for photovoltaic applications. It is also more environmentally friendly than CdSe;

Indium arsenide (InAs): InAs QDs have a narrow bandgap, which makes them useful for infrared applications. They are also good for quantum computing and single-photon sources;

Lead sulfide (PbS): PbS QDs have a large Bohr exciton radius and can absorb light in the near-infrared region, making them useful for imaging and sensing applications;

Silicon (Si): Silicon QDs have a tunable bandgap and good biocompatibility, which makes them useful for biological imaging and sensing;

Perovskite QDs: Perovskite QDs are a new class of QDs that have recently gained attention due to their high quantum yield, tunable bandgap, and low toxicity. They have potential for use in solar cells, LEDs, and other optoelectronic devices.

These are just a few examples of many QD materials that have been developed. The choice of QD material depends on the specific application and desired properties.

Similar to the bulk semiconductor materials, typical QDs can also be classified as single-element QDs such as C and Si-based QDs (denoted as 1-D QDs), binary QDs (denoted as 2-D QDs), ternary QDs, and alloy QDs (denoted as 3-D QDs), as shown in Figure 1.14.

According to the specific structures and properties, the types of QDs can also be divided into core–shell QDs and Janus QDs. The former refers to a central core surrounded by a shell, while the latter refers specifically to a core with two different material "halves."

Core–shell QDs are nanoscale semiconductor materials that have a core of one material surrounded by a shell of another material. The core of the QD typically consists of a semiconductor material with a small bandgap, such as CdSe or PbS, while the shell is made up of a material with a wider bandgap, such as ZnS or CdS.

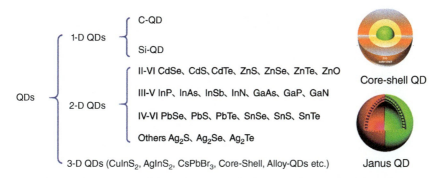

Figure 1.14 Classification of QDs based on composition type and structure type.

The core–shell structure of QDs provides several advantages over simple core-only QDs. First, the shell material can act as a barrier to prevent the diffusion of impurities into the core, which can degrade the optical properties of the QD. Second, the shell can provide additional stability to the QD by preventing oxidation and protecting the core from environmental degradation. Third, the shell can modify the electronic properties of the QD, such as the exciton energy and electron–hole recombination rate, which can improve the photoluminescence efficiency of QD and make it more suitable for various applications in optoelectronics and biological imaging. Core–shell QDs have shown great potential in a wide range of applications, including bioimaging, light-emitting diodes, solar cells, and single-electron transistors.

Janus QDs are a type of QD with a core–shell structure, where one half of the shell is made of one material and the other half is made of a different material. The name "Janus" refers to the two-faced Roman god, who is often depicted with two different faces looking in opposite directions. Janus QDs can have a number of interesting properties, such as asymmetric surface charge distributions, anisotropic shapes, and tunable surface properties. These properties make Janus QDs attractive for a range of applications, including catalysis, sensing, and imaging. Janus QDs can be synthesized using a number of methods, including co-precipitation, SILAR, and reverse micelle methods. The choice of synthesis method can impact the properties of the resulting Janus QDs, such as their size, shape, and composition. Janus QDs are an exciting area of research with potential for a wide range of applications due to their unique properties and versatility in synthesis.

1.2.5.1 II–VI Semiconductor QDs

Semiconductor QDs of the II–VI family, especially those based on CdSe, exhibit a tunable band-edge emission covering the visible spectrum (480–650 nm). They have been the most extensively investigated QDs and are recognized as a model system. The full range of visible colors from emissive II–VI QDs has given rise to a series of images as shown in Figure 1.15a, highlighting the color tunability of core–shell QDs. Figure 1.15b illustrates the bandgap of the bulk semiconductors.

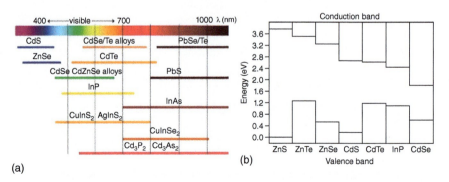

Figure 1.15 (a) Reported spectral range of emission for different semiconductor quantum dots. (b) Bandgap of the bulk semiconductors.

1.2.5.2 IV–VI Semiconductor QDs

Lead chalcogenide IV–VI semiconductor QDs are characterized by their tunable NIR emission from 600 to 2200 nm. PbSe and PbS NCs are widely explored including, the development of core–shell PbSe/CdSe and PbS/CdS QDs. These QDs exhibit intrinsically strong quantum confinement and have seen deep interest in various applications such as photodetectors, LEDs, and photovoltaic solar cells.

1.2.5.3 II$_3$–V$_2$ Semiconductor QDs

II$_3$–V$_2$ semiconductor QDs (e.g. Cd$_3$P$_2$ and Cd$_3$As$_2$) with efficient PL have only been reported very recently. Cd$_3$As$_2$ QDs with emission in the NIR range have the highest QY of 85%, an average size of 2.5 nm, and an emission wavelength of ~900 nm. Good-quality core–shell QDs for this class of NIR- and IR-emitting QDs would enhance their properties and are lacking thus far.

1.2.5.4 Ternary I–III–VI$_2$ Chalcopyrite Semiconductor QDs

Ternary semiconductor QDs such as the I–III–VI$_2$ chalcopyrites CuInSe$_2$ (CISe) and CuInS$_2$ (CIS) have emerged in the past few years as further alternative materials to cadmium-based systems without toxic elements. They are direct semiconductors and exhibit a relatively narrow bandgap (1.05 eV for CISe, 1.5 eV for CIS). Ternary CISe and CIS QDs were mainly studied because of their potential use in photovoltaics.

1.2.5.5 Single Element-Based Semiconductor QDs

In recent years, many studies on single-element QD materials in nonlinear optics and ultrafast lasers have been reported, such as graphene QDs, carbon QDs, black phosphorus (BP) QDs, sulfur QDs, silicon QDs, selenium QDs, boron QDs, and metal elemental QDs. In recent years, the interest has been in C-QDs and Si-QDs [27]. Carbon quantum dots, also commonly referred to as carbon dots (abbreviated as CQD), are carbon nanoparticles less than 10 nm in size with surface passivation possibility. As a new class of fluorescent carbon nanomaterials, CQD has high stability, good electrical conductivity, low toxicity, environmental friendliness, simple synthesis routes, and optical properties comparable to QDs. Most CQD research applications are in the fields of chemical sensing, biosensing, bioimaging, nanomedicine, photocatalysis, and electrocatalysis [28].

Silicon QDs are metal-free, biocompatible quantum dots with photoluminescence emission peaks that can be modulated from the visible to near-infrared spectral regions. These QDs have unique properties due to their indirect bandgap, including long-lived luminescent excited states and large Stokes shifts. Silicon quantum dots (SiQDs) have size-tunable photoluminescence similar to that observed for conventional QDs. By varying the size of the Si QDs, the LED emission can be tuned from deep red (680 nm) to orange/yellow (625 nm), although Si QDs-LEDs exhibit low efficiency and broad luminescence emission that can be improved by further studies [29].

1.3 Application of Quantum Dots on Display Devices

As QDs have adjustable emission peaks, high color purity, and high photoluminescence quantum yield, they have attracted more and more attention in academia and industry, and have now been commercially used in LCD backlight products [30, 31]. For white LED products in traditional LCD backlighting, yellow phosphor is used for down conversion, and its color gamut range only reaches 70% of the National Television Standards Committee (NTSC) standard [32]. While products with QDs as backlight can achieve high saturation and a color gamut greater than 100% of the NTSC standard. At present, Samsung, TCL, BOE, and other domestic and foreign display panel manufacturers have adopted QDs as the backlight technology solution for high-end display panels.

Compared to other display technologies such as LED-backlit LCDs and OLEDs, the design and manufacture of quantum dot light-emitting diodes (QDLEDs), which are directly driven by voltage, has greater appeal and development potential in terms of display technology specifications such as contrast, color gamut, response time, and viewing angle. In addition, QDLEDs have better temperature and moisture resistance than OLEDs and have better application prospects in the field of flexible devices.

1.3.1 The Basic Structure of QDLED

The schematic diagram of QDLED device structure is shown in Figure 1.16, which is a typical sandwich structure with multiple functional layers stacked together, in which indium tin oxide (ITO) material is used as the anode, s-NiO material as the hole transport layer, Al_2O_3 material as the electron blocking layer, QD light-emitting

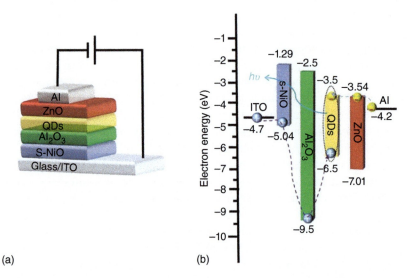

Figure 1.16 Basic structure of QDLED: (a) schematic diagram of QDLED device structure, (b) schematic diagram of energy level of each functional layer of QDLED device.

material as the light-emitting layer, ZnO material as the electron transport layer, and Al material as the cathode. In QDLED, after passing voltage, electrons and holes are transferred from the cathode and anode to the QD light-emitting layer, respectively, and combine in the light-emitting layer to form electron–hole pairs (excitons) to produce photons. In addition, the electron-blocking layer composed of Al_2O_3 material is also required to assume the charge-blocking role to improve the luminescence efficiency of the QDLED device. The LUMO/conducting band bottom of the hole transport layer needs to be shallow enough for hole transport; the HOMO/valence band top of the electron transport layer needs to be deep enough for electron transport.

1.3.2 Main Factors Affecting QDLED Light Emission

The main challenge for QDLED commercialization is that active matrix QDLED (AM-QDLED) devices are difficult to achieve relatively high luminous efficiency and up-to-standard lifetime [33]. The main factors affecting their performance include charge carrier combination, fluorescence resonance energy transfer, and field effect quenching. The prerequisites for high external quantum efficiencies (EQEs) are a high PLQY and good balance between electron and hole injection currents to avoid CQD charging because the formation of charged excitons promotes nonradiative AR. During Auger decay, the electron–hole recombination energy is not released as a photon but instead transferred to the resident charge carrier. AR has been identified as at least one of the reasons for EQE droop – a decrease in device efficiency with increasing current density (Figure 1.17). Compositionally graded QD multishell heterostructures have been shown to impede AR because of creation of a "smooth" confinement potential that suppresses the intragap transition involving the energy-accepting carrier and thus minimizes the efficiency roll-off [35].

1.3.2.1 Auger Recombination (AR)

When an electron is excited to a higher energy level by a photon, a hole is created at the same time, forming an electron–hole pair; and when the electron–hole pair recombines, a photon is emitted. The recombination dynamics of these single

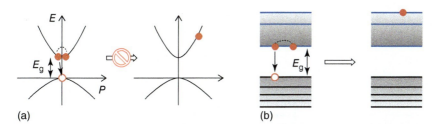

Figure 1.17 (a) In a bulk semiconductor, at $T = 0$ when the kinetic energies of all three charge carriers are zero, Auger recombination is not possible, as it violates translational momentum conservation. (b) In QDs, translational momentum conservation is relaxed, resulting in efficient Auger recombination even at $T = 0$. Source: Pietryga et al. [34]/American Chemical Society.

excitons in well-passivated QDs are dominated by intrinsic radiative recombination. The situation dramatically changes in the case when one or more extra carriers are introduced into the NC (for example, at higher current densities), which opens a new nonradiative pathway associated with AR. However, the energy can be directly transferred to the third carrier in this process, which is called AR, if a third carrier is present [36, 37]. In this process, the recombination energy of the e–h pair is not converted into radiation but instead is used to excite a third charge carrier (an electron or a hole) into a higher energy level [34]. In bulk materials, the AR is hindered because the conservation of energy and momentum leads to a threshold that limits the rate of AR (Figure 1.17a). In QDs, however, momentum conservation is relaxed, especially in strongly confined regions such as interfaces or defect sites. Therefore, QDs usually have efficient AR (Figure 1.17b).

In bulk semiconductors, due to the requirement of simultaneous conservation of energy and flat mobile quantities, AR is a temperature-activated process whose rate can be expressed as $r_A \propto \exp(-\gamma_A E_g/k_B T)$, where γ_A is a constant dependent on the electronic structure of a particular semiconductor. Based on this expression, the rate of Auger decay quickly decreases with increasing bandgap. As a result, AR is considerably less efficient in bulk wide-gap (e.g. CdSe and CdS) compared to narrow-gap (e.g. PbSe and PbS) materials. However, in QDs subject to strong spatial constraints, the translational momentum conservation is relaxed and replaced by a less stringent angular momentum conservation, making the AR in QDs unusually strong. The AR process is highly dependent on the size of the QDs [38]. Although there are different paths for the AR process in direct and indirect bandgap semiconductors, the increase or decrease of the volume can be applied to QDs with direct and indirect bandgap structures. In the former case, AR is a three-particle process, while in the latter case, photons require additional emission or absorption to satisfy momentum conservation. Relevant calculations also predict this size-dependent AR [39]. Auger decay rates of multiexciton states are expected to progressively increase with the number of charge carriers in a QD due to the increase in both the number of recombination pathways shown in Figure 1.18. The simplest form of multicarrier states in a QD is

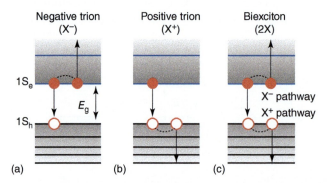

Figure 1.18 Auger recombination of (a) a negative trion (X^-), (b) a positive trion (X^+), and (c) a biexciton (2X). The energy released during e–h recombination is transferred to an electron for X^- or to a hole for X^+. In the case of biexciton Auger recombination, the energy of the e–h pair can be transferred to either an electron (X^- pathway) or a hole (X^+ pathway). Source: Pietryga et al. [34]/American Chemical Society.

a charged exciton composed of one exciton plus an extra electron or hole, which is referred to as a negative (X^-) or a positive (X^+) trion, respectively. The nonradiative Auger process is also the dominant recombination mechanism of a biexciton, which is a neutral state consisting of two e–h pairs.

AR is also related to the scintillation of QDs. The scintillation phenomenon of QDs, also known as fluorescence intermittency, is a random switch between the strong emitting state (ON) and the dark emitting state (OFF). This phenomenon has been studied by many research groups, different theoretical models have been developed and attempts have been made to explain this physical phenomenon using theoretical models [20, 40, 41]. One of the most widely accepted theories is the charge/discharge model, which attributes the scintillation to excess carriers that cause a radiation-free oscillatory process that affects the overall emission [42]. However, this model has also been challenged in some studies where size effects were not found to exist [43] and cannot explain the ultrafast nonradiative combination [44, 45]. The physical mechanism of the oscillation process and scintillation needs to be further investigated, and therefore the oscillation mechanism of the device efficiency decrease needs to be further explored.

Since the charge imbalance is expected to increase with the applied voltage, the deleterious effect of Auger decay should also increase. Therefore, AR of charged QDs has been used to explain the so-called efficiency "drop" or "roll-off" phenomenon, i.e. the decrease in EQE at higher currents typically observed in QDLEDs. These considerations suggest that the optimization of QDs applied in QDLEDs may involve not only the improvement of their single-exciton PLQYs but also the yield of charged and multi-exciton states, the latter of which can be achieved by developing methods to reduce AR.

1.3.2.2 Fluorescence Resonance Energy Transfer

The exciton energy is transferred to the defect state and the third carrier during the OSI combination process, while the exciton energy is transferred to the other radiative state during the fluorescence resonance energy transition [45]. In hybrid organic/colloidal QDLEDs, excitons are formed in organic molecules around the QD film, and exciton energy is transferred to the QDs by resonance. In addition to this layer-to-layer exciton energy transfer, there is another phenomenon of fluorescence resonance energy transfer between dots, which leads to the so-called "self-quenching" [46].

The fluorescence resonance energy transfer is influenced by the distance, so that its effective range is at the nanometer level. By assuming that the QDs are uniformly distributed, the distance between the dots can be calculated. Whereas in the solid state, the luminescent layer is usually a closed film of QDs. Since the average face-to-face distance between dots, in a QD film is usually within an energy transfer window, this structure facilitates fluorescence resonance energy transfer. Exciton diffusion among QDs within EMLs can also be responsible for efficiency losses in QDLEDs. This means that excitons can diffuse in QD films by: (i) nonradiative Forster resonant energy transfer (ET), with an efficiency that varies with dot-to-dot distance, d, as $\sim d^{-6}$; or (ii) reabsorption of emitted photons by neighboring QDs with a rate varying as $\sim d^{-2}$. Exciton diffusion intensifies the contribution to emission efficiency loss of fast nonradiative trapping, even if this occurs only in a minority of QDs.

Traditional fluorescence electrophoresis studies using organic, biological, or inorganic materials as donors or acceptors in solution are also referred to as homogeneous transfer. In homogeneous transfer, the energy transfer process occurs between the same materials [47]. Primitive QD energy transfer studies have shown that a recognizable redshift is produced in the emission spectrum, which implies a transfer of electronic energy [48]. Due to the inhomogeneous size distribution of the QDs, the non-radiative energy transfer leads to a burst of blue luminescence and enhanced red luminescence; this results in an overall redshift in the emission spectrum. Related theoretical calculations investigated the effect of size distribution on the spectral shape and found that the increase in size nonuniformity leads to an increase in spectral shift and spectral narrowing. It has also been reported that the fluorescence resonance energy transfer process may contribute to the self-bursting of fluorescence in QDLEDs [49]. Although the so-called reduction of self-burst is associated with an increase in the inter-dot spacing, the potential mechanism of the grating contribution to the reduction of quantum efficiency remains uncertain.

1.3.2.3 Surface Traps and Field Emission Burst

For a singly excited QD, nonradiative losses are dominated by recombination at structural defects that are most often associated with the QD surface. Typically, a continuous decrease in EQE is observed in many types of QDLEDs as the current density increases, a phenomenon known as efficiency roll-off or efficiency degradation [50]. Some studies have specifically measured the efficiency roll-off of the EQE at current density. Longitudinal studies of efficiency roll-off typically quantify a range of devices by comparing parameters such as critical current density or critical luminance, and the fitted trends show that it is difficult to achieve the desired relationship between efficiency and luminance.

To understand the cause of efficiency roll-off in QDLEDs, Shirasaki et al. utilized an intelligent device design [51]. It was shown that electric fields alone can contribute to efficiency roll-off and that the drop in EQE can be predicted using a quantitative approach. Their idea builds on the relationship between the offset and intensity of the field-related spectrum. After considering the contributions of charge leakage and charge-induced Osher combinations, they propose that high field strength is a major factor in the drop in QDLED luminescence efficiency. By applying an anti-deflection field while other factors are kept constant. They observed the emission spectra at different electric field strengths and then measured the energy transfer of the emitted photons and compared them with the luminescence intensity at different bias voltages. By analyzing transient fluorescence emission spectra, they concluded that the reduced emissivity could be the cause of the efficiency roll-off.

1.3.3 History of QDLED Development

At the end of the twentieth century, few people were optimistic about the application prospects of QDLEDs because they showed only very low EQE in the early days of

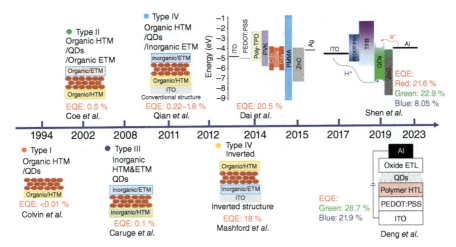

Figure 1.19 Outlines the technological development of QDLEDs in recent years (1994–2023).

electroluminescent displays. However, with the gradual maturation of OLED technology after 2000, the development of QDLEDs was inspired by the optimization of OLED structures and the working mechanisms. Since then, QDLED technology has developed rapidly, and its performance has been improving, approaching the requirements for commercial applications. In order to have an intuitive understanding of the development process of QDLED display technology, we will then briefly introduce some representative technological breakthroughs and innovative ideas in the development of QDLED display technology. It is worth noting that the continuous innovation and performance improvement of QDs and charge transport layer materials play a crucial role in the development of QDLED display technology.

The technology road map development of QDLED is illustrated in Figure 1.19. The first time reported QDLED (Type I QDLED: Organic HTM) based on a bilayer structure consisting of hole transport polymer PPV and CdSe NC active layers was reported in *Nature* by Alivisatos et al. in 1994 [52]. In 1995, the research group at MIT reported a single-layer CdSe-QDLED with the NCs incorporated into thin films ~1000 Å of polyvinylcarbazole PVK and an oxadiazole derivative *t*-Bu-PBD and sandwiched between ITO and Al electrodes. These devices exhibited very small values of EQEs of 0.001–0.01% and 0.0005%, though the room-temperature electroluminescence (EL) is nearly identical to the PL, with the same peak positions and similar linewidths. Because of the low conductivity of the organic matrix, and the poor conduction through QD multilayers led to an injected charge imbalance, the driving voltage is very high starting at 17 V, and consequently, the luminescence efficiency of these early devices never exceeded 0.10 cd A^{-1}, and considerable parasitic polymer-related emission observed in addition to QD emission [53].

The deficiencies of these devices likely originated from the charge imbalance, which led to the formation of charged excitons decaying not by radiative processes but by nonradiative AR (see Section 11.3). In the early twenty-first century, inspired by the design of OLED device structures, Coe et al. demonstrated a QDLED device with an OLED-like device structure, in which some organic materials used in OLED devices were employed as the electron transport layer and hole transport layer materials for QDLED devices (Type II QDLED: Organic HTM and ETM), where a single monolayer of QDs was sandwiched between two organic thin films via phase segregate between the QD aliphatic capping groups and the aromatic organic materials during the spinning process, which allowed them to produce a well-defined EML/HTL bilayer and helped regulate the electron flow into the QDs. This hybrid light-emitting diode QDLED showed a 25-fold improvement in luminescence efficiency (1.6 cd A^{-1} at 2000 cd m^{-2}) over the best previously reported QDLED at that time. The application of organic materials as charge transport layer materials and the formation of QD monolayers are believed to be responsible for the increased device efficiency. In this type of QDLED devices, exciton formation is mainly achieved by the Forster resonance energy transfer (FRET) process, which is very different from direct charge injection [10]. For the FRET process, excitons are first formed in the donor electron transport layer, and then the exciton energy is transferred to the QDs via non-radiative dipole-dipole coupling. Due to the decoupling of the emission process from the charge transport, such QDLED devices can obtain an EQE of 0.5–5% [3]. Despite the considerable progress in device performance achieved with all-organic CTLs, peak EQE was still far below the theoretical limit of ~20%, determined by optical outcoupling. Imbalanced charge injection remained a significant factor; the band positions of traditionally used organic materials are not ideally suited for CdSe-based QDs. Specifically, the large offset from the LUMO of typical organic ETLs and the QD conduction band edge means that electron injection is strongly energetically favorable, even without an applied bias. At the same time, holes in the same all-organic device structure face an energetic barrier in transiting from the HTL to the QDs, requiring a significant overpotential. The prevalence of the electron-over-hole injection current leads to an excess number of electrons in the emitting layer. The resulting negatively charged dots are then susceptible to nonradiative AR, thus lowering the overall EQE. The reason for the difficulty in improving the efficiency of such devices is the difficulty in achieving a tightly packed pinhole-free monolayer to prevent carrier leakage through the QDs. In addition, the low conductivity of organic materials limits the carrier's injection [54].

A drawback to the above-mentioned device is the utilization of the organic charge support layers in QDLEDs, which creates an unwanted contribution to the light emission of the device. To overcome this, the use of inorganic charge transport materials as both hole and electron transport materials comprise an inorganic metal oxide semiconductor such as the ETL and an inorganic semiconductor such as the HTL. This new design idea for QDLED structures is realized by replacing organic charge transport layer materials with inorganic materials. According to Caruge's study, sputtered zinc oxide, tin oxide, and nickel oxide can be used as n-type and p-type charge transport layer materials, respectively. Due to the superior conductivity of metal oxides over organic transport materials, these inorganic QDLED devices (Type III QDLED: inorganic HTM and ETM) all exhibit high current densities up to 4 A cm^{-2}. However, the luminescence efficiency is low (EQE < 0.1%) due to damage to the QD layer during the sputtering of the upper layer of ZnO:SnO$_2$ and insufficient hole injection due to the excessive barrier between NiO_x and QDs. In addition, exciton dynamics studies show that charge transport between QDs and adjacent metal oxides tends to occur spontaneously, leading to exciton bursts and lower device efficiency [55]. However, such all-inorganic QDLED devices are still attractive because the excellent inherent stability of metal oxides contributes to the device's lifetime. In addition, with the development of sol–gel method and NC synthesis, solution-processed metal oxides can reduce the damage to the underlying QDs. The brightness of the device was comparable to that of a Type-II device, albeit suffering from the relatively low efficiency attributed to additional quenching by the free carriers in the $ZnO:SnO_2$ ETL, yet with the benefit of improved shelf-life robustness inherent in the environmental stability of metal oxide charge transport layers [56]. The great progress was achieved using all-solution-processed QDLED device of Type IV structure consisting of a quantum dot emissive layer sandwiched between an organic hole transport layer and an electron transport layer of ZnO nanoparticles. The design of hybridized structures (Type IV QDLED: Organic HTM and Inorganic ETM) of inorganic electron transport material layers and organic hole transport material layers has been a hot research topic in the field of QDLED devices in the last decade in order to exploit the high conductivity of n-type metal oxides and the superior hole transport capability of organic materials [57–59]. In 2011, Qian et al. introduced full solution-processed QDLEDs based on ZnO nanoparticle electron transport layer materials. the resulting red, green, and blue (R/G/B) QDLEDs showed good performance with EQE peaks of 1.7%, 1.8%, and 0.22% and maximum luminance of 31 000, 68 000, and 4200 cd m^{-2} [60]. Moreover, with the incorporation of the ZnO nanoparticles as ETM, these unencapsulated devices have operating lifetimes exceeding 250 hours in low vacuum with an initial brightness of 600 cd m^{-2}. Since then, due to the advantages of high mobility, suitable electronic structure, and simple synthesis process, ZnO nanoparticles have been widely used as electron transport layer materials in QDLEDs, which has led to a leap forward in the performance of QDLED devices [4]. Under this hybrid structure, the inverted QDLED devices with QDs using ITO cathodes at the bottom reached a record of 18% EQE, which greatly exceeded the previous research results [3]. In 2014, Peng et al. achieved the first high-efficiency hybrid QDLED device (ITO/PEDOT:PSS/poly-TPD/PVK/QDs/ZnO/Ag) with an EQE of more than 20% by using a poly(methyl methacrylate) (PMMA) insulating electron blocking layer between the QD emitting layer and the ZnO electron transport layer. Since then, the improvement of the charge transport layer has received increasing attention and is considered as one of the effective strategies to achieve high-performance QDLED devices [61]. The improvement of the charge transport layer has since received increasing attention and is considered as one of the effective strategies to achieve high-performance QDLED devices.

In addition to the development of charge transport materials and modulation of device structures, it is necessary to prepare high-quality QDs by carefully controlling their nanostructure and composition in order to achieve high-performance QDLEDs. Through the fine-tuning of nanostructure of QDs, especially the composition of the graded intermediate shell and the thickness of the outer shell,

Qian et al. designed and synthesized QDs with a graded, alloyed intermediate shell ($Cd_{1-x}Zn_xSe_{1-y}S_y$) sandwiched between the Cd- and Se-rich core and the Zn- and S-rich outer shell QDs. They have successfully fabricated a full series of blue, green, and red quantum dot-based light-emitting devices (QDLEDs) with solution process, all with high external quantum efficiencies over 10%. The devices have maximum current and external quantum efficiencies of 63 cd A^{-1} and 14.5% for green QDLEDs, 15 cd A^{-1} and 12.0% for red devices, and 4.4 cd A^{-1} and 10.7% for blue devices, all of which are well maintained over a wide range of luminance from 100 to 10 000 cd m^{-2}. More importantly, the device half-lifetimes for the green and red QDLEDs showed over 90 000 and 300 000 hours, respectively, for a brightness of 100 cd m^{-2}, although the blue QDLEDs still has a relatively short lifetime of only 1000 hours. These QDLEDs also feature extremely low turn-on voltages (1.5–2.6 V), narrow full-width at half maximum (FWHM of <30 nm) of the EL peaks, highly saturated pure colors, and high brightness [62].

By incorporating ZnO NPs as the electron transport layer, and highly controlled QD synthesis, Manders et al. reported in SID 2015 the colloidal quantum dot-based hybrid light-emitting diodes (QDLEDs), which exhibited high maximum current and power efficiency of 6.1 cd A^{-1} and 5.0 lm W^{-1} for blue, 70 cd A^{-1} and 58 lm W^{-1} for green, and 12.3 cd A^{-1} and 17.2 lm W^{-1} for red emissions. It was the first time to achieve a green QDLED with EQE over 20% (21%) and 82 cd A^{-1}. These peak efficiencies occurred at a desirable luminance of 1000 cd m^{-2} and low voltage of 3.5 V [63]. High-efficiency blue CdSe/ZnS QDs were reported in 2017. Using small-size ZnO NPs, the authors have obtained a maximum current efficiency (CE) of 14.1 cd A^{-1} and a maximum EQE of 19.8 % for QDLEDs with an EL peak at 468 nm, with the CIE 1931 color coordinates (0.136, 0.078) [64]. Most previous work has focused on CdSe-based QDs, which present severe toxicity and environmental issues. To improve the operating stability of the devices and to replace their toxic cadmium composition with a more environmentally benign alternative QD, a uniform indium phosphide (InP)-based materials core and a highly symmetrical core/shell QD exhibit narrow FWHM; 35 nm at 630 nm with a quantum yield of approximately 100% was developed by Samsung Group. The device based on InP/ZnSe/ZnS QDs showed a theoretical maximum EQE of 21.4%, a maximum brightness of 100 000 cd m^{-2}, and an extremely long lifetime of a million hours at 100 cd m^{-2}. The InP-based QDLEDs will aid in fabricating Cd-free QDLEDs for next-generation displays [65].

Efforts have been made to improve the efficiency and lifetime of QDLEDs. In 2019, Shen et al. used a "low-temperature core and high-temperature shell growth" synthesis method to synthesize CdSe/ZnSe core/shell QDs and applied Se throughout the core/shell region in the presence of an alloy bridging layer at the core/shell interface, resulting in balanced charge injection and high current density at low voltage. As a result, QDLEDs based on CdSe/ZnSe core/shell structures for red, green, and blue quantum dot light-emitting diodes (ITO/PEDOT-PSS/TFB/QDs/ZnO/Al) showed maximum external quantum efficiencies of 21.6%, 22.9%, and 8.05% with corresponding luminances of 13 300, 52 500, and 10 100 cd m^{-2}. The peak luminance of these devices was also 356 000, 614 000, and 62 600 cd m^{-2}. This work represents a significant step forward in the realization of QDLEDs for display and potential lighting applications [66].

Near-infrared NIR–QD light-emitting diodes with an EQE of 16.98% and a power conversion efficiency of 11.28% at wavelength 1397 nm have recently been reported. The authors employed a binary emissive layer consisting of silica-encapsulated silver sulfide ($Ag_2S@SiO_2$) QDs dispersed in a cesium-containing triple cation perovskite matrix that serves as an additional passivation medium and a carrier supplier to the emitting QDs. Assisted by the hole-injection thin porphyrin interlayer, which balances the device current and enhances carrier radiative recombination, The IR–QDLEDs deliver an enhanced device performance. The present approach paves the way for the development of all-solution-processable, low-cost, high performance, and large-area NIR-II light sources for biomedical and imaging applications [67].

By using deep HOMO hole transport polymers with both low electron affinity and reduced energy disorder to eliminate electron leakage at the organic/inorganic interface, the authors demonstrate green and blue QDLEDs with approximately 100% conversion of the injected charge carriers into luminescent excitons, resulting in devices that exhibit high EQE over a wide range of luminance values (green with a peak EQE of 28.7% and blue with 21.9%) and excellent stability (inferred T_{95} lifetime of 580 000 hours for green and 4400 hours for blue QDLED). The elimination of charge leakage channels may inspire other designs of solution-processed QDLEDs with organic/inorganic interfaces [68].

1.4 Conclusion and Remarks

QDLEDs offer several promising features, such as size-dependent emission wavelengths, narrow emission spectrum, high efficiency, flexibility, and low-processing cost of organic light-emitting devices. QDLEDs not only reduce the consumption of energy but also show high color purity. It exhibits the ability to be more than twice as power efficient as OLEDs at the same color purity and has also presented a 30–40% luminance efficiency advantage over OLEDs for the same color point.

Semiconductor QDs offer a great opportunity in optical electronics because of their nanometer scale size in all three dimensions, the restricted electron motion and quantum confinement effects lead to a discrete atom-like electronic structure and size-dependent energy levels. These features enable us to design nanomaterials with widely tunable light absorption, bright emission, and narrow-band pure colors. Because of the quantum size and surface effects, which provide much opportunity for control over electronic transport, and a wide tuning of chemical and physical functions benefits the applications in optical electronic devices. Specifically, the bright and narrowband light emission feature of semiconductor QDs, with tunable capability across the visible and near-infrared spectrum is attractive to realize more efficient displays with purer colors in future. In addition, the advent of colloidal QDs, which can be fabricated and processed in solution at mild conditions, enabled large-area manufacturing and widened the scope of QD applications to QDLED display and lighting markets because QDLEDs present an ideal blend of high brightness, efficiency with long lifetime, flexibility, and low-processing cost of solution process.

Although much progress has been made, several factors limit the performance of QDLEDs, including AR, FRET, and field quenching (FIQ). In addition, most of the research and development efforts have been focused on cadmium-based QDs, which also limits their further commercialization. Therefore, the development of cadmium-free QDLEDs has high demands for their wide and practical implementation. In Chapter 2, the basic principle of QDLED, the materials and device development of QDLEDs, and a summary and perspective concerning the issues and limitations of the applicability of QDLEDs are presented.

References

1 Sun, Y., Jiang, Y., Sun, X.W. et al. (2019). Beyond OLED: efficient quantum dot light-emitting diodes for display and lighting application. *Chemical Record* 19 (8): 1729–1752.
2 Efros, A.L. and Brus, L.E. (2021). Nanocrystal quantum dots: from discovery to modern development. *ACS Nano* 15 (4): 6192–6210.
3 Kang, K. and Min, B.I. (1997). Effect of quantum confinement on electron tunneling through a quantum dot. *Physical Review B* 55 (23): 15412–15415.
4 Cordones, A.A. and Leone, S.R. (2013). Mechanisms for charge trapping in single semiconductor nanocrystals probed by fluorescence blinking. *Chemical Society Reviews* 42 (8): 3209–3221.
5 Joughi, Y.D.G. and Sahrai, M. (2022). Spatial-dependent quantum dot-photon entanglement via tunneling effect. *Scientific Reports* 12 (1): 7984.
6 Yeyati, A.L., Cuevas, J.C., López-Dávalos, A., and Martín-Rodero, A. (1997). Resonant tunneling through a small quantum dot coupled to superconducting leads. *Physical Review B* 55 (10): R6137–R6140.
7 Donegá, C.M. (2011). Synthesis and properties of colloidal heteronanocrystals. *Chemical Society Reviews* 40 (3): 1512–1546.
8 García de Arquer, F.P., Talapin, D.V., Klimov, V.I. et al. (2021). Semiconductor quantum dots: technological progress and future challenges. *Science* 373 (6555): eaaz8541.
9 Pu, C., Dai, X., Shu, Y. et al. (2020). Electrochemically-stable ligands bridge the photoluminescence-electroluminescence gap of quantum dots. *Nature Communications* 11 (1): 937.
10 Shirasaki, Y., Supran, G.J., Bawendi, M.G., and Bulović, V. (2012). Emergence of colloidal quantum-dot light-emitting technologies. *Nature Photonics* 7 (1): 13–23.
11 Shen, H., Wang, H., Tang, Z. et al. (2009). High quality synthesis of monodisperse zinc-blende CdSe and CdSe/ZnS nanocrystals with a phosphine-free method. *CrystEngComm* 11 (8): 1733–1738.
12 Lübkemann, F., Rusch, P., Getschmann, S. et al. (2020). Reversible cation exchange on macroscopic CdSe/CdS and CdS nanorod based gel networks. *Nanoscale* 12 (8): 5038–5047.

13 Cho, W., Kim, S., Coropceanu, I. et al. (2018). Direct synthesis of six-monolayer (1.9 nm) thick zinc-blende CdSe nanoplatelets emitting at 585 nm. *Chemistry of Materials* 30 (20): 6957–6960.

14 Wang, F., Chen, B., Pun, E.Y.B., and Lin, H. (2015). Alkaline aluminum phosphate glasses for thermal ion-exchanged optical waveguide. *Optical Materials* 42: 484–490.

15 Wang, F., Chen, B., Pun, E.Y.-B., and Lin, H. (2014). Dy^{3+} doped sodium–magnesium–aluminum–phosphate glasses for greenish–yellow waveguide light sources. *Journal of Non-Crystalline Solids* 391: 17–22.

16 Wang, F., Chen, B.J., Lin, H., and Pun, E.Y.B. (2014). Spectroscopic properties and external quantum yield of Sm^{3+} doped germanotellurite glasses. *Journal of Quantitative Spectroscopy and Radiative Transfer* 147: 63–70.

17 Pu, C., Qin, H., Gao, Y. et al. (2017). Synthetic control of exciton behavior in colloidal quantum dots. *Journal of the American Chemical Society* 139 (9): 3302–3311.

18 Zhang, A., Dong, C., Liu, H., and Ren, J. (2013). Blinking behavior of CdSe/CdS quantum dots controlled by alkylthiols as surface trap modifiers. *The Journal of Physical Chemistry C* 117 (46): 24592–24600.

19 Omogo, B., Aldana, J.F., and Heyes, C.D. (2013). Radiative and nonradiative lifetime engineering of quantum dots in multiple solvents by surface atom stoichiometry and ligands. *The Journal of Physical Chemistry C* 117 (5): 2317–2327.

20 Efros, A.L. and Nesbitt, D.J. (2016). Origin and control of blinking in quantum dots. *Nature Nanotechnology* 11 (8): 661–671.

21 Hohng, S. and Ha, T. (2004). Near-complete suppression of quantum dot blinking in ambient conditions. *Journal of the American Chemical Society* 126 (5): 1324–1325.

22 Rabouw, F.T., Antolinez, F.V., Brechbühler, R., and Norris, D.J. (2019). Microsecond blinking events in the fluorescence of colloidal quantum dots revealed by correlation analysis on preselected photons. *The Journal of Physical Chemistry Letters* 10 (13): 3732–3738.

23 Vaneski, A. (2013). Synthesis and design of hybrid nanoastructures [sic] based on II–VI semiconductor nanocrystals. Doctoral thesis. City University of Hong Kong.

24 Shen, H., Zhou, C., Xu, S. et al. (2011). Phosphine-free synthesis of $Zn_{1-x}Cd_xSe/ZnSe/ZnSe_xS_{1-x}/ZnS$ core/multishell structures with bright and stable blue–green photoluminescence. *Journal of Materials Chemistry* 21 (16): 6046–6053.

25 Cheng, Y., Wan, H., Liang, T. et al. (2021). Continuously graded quantum dots: synthesis, applications in quantum dot light-emitting diodes, and perspectives. *The Journal of Physical Chemistry Letters* 12 (25): 5967–5978.

26 Lim, J., Park, Y.-S., and Klimov, V.I. (2018). Optical gain in colloidal quantum dots achieved with direct-current electrical pumping. *Nature Materials* 17 (1): 42–49.

27 Michler, P. (2003). *Single Quantum Dots: Fundamentals, Applications and New Concepts*. Springer Science & Business Media.

28 Lim, S.Y., Shen, W., and Gao, Z. (2015). Carbon quantum dots and their applications. *Chemical Society Reviews* 44 (1): 362–381.
29 Morozova, S., Alikina, M., Vinogradov, A., and Pagliaro, M. (2020). Silicon quantum dots: synthesis, encapsulation, and application in light-emitting diodes. *Frontiers in Chemistry* 8: 00191.
30 Jang, E., Jun, S., Jang, H. et al. (2010). White-light-emitting diodes with quantum dot color converters for display backlights. *Advanced Materials* 22 (28): 3076–3080.
31 Ziegler, J., Xu, S., Kucur, E. et al. (2008). Silica-coated InP/ZnS nanocrystals as converter material in white LEDs. *Advanced Materials* 20 (21): 4068–4073.
32 Anandan, M. (2008). Progress of LED backlights for LCDs. *Journal of the Society for Information Display* 16 (2): 287–310.
33 Bae, W.K., Park, Y.-S., Lim, J. et al. (2013). Controlling the influence of Auger recombination on the performance of quantum-dot light-emitting diodes. *Nature Communications* 4 (1): 2661.
34 Pietryga, J.M., Park, Y.-S., Lim, J. et al. (2016). Spectroscopic and device aspects of nanocrystal quantum dots. *Chemical Reviews* 116 (18): 10513–10622.
35 Bi, C., Yao, Z., Hu, J. et al. (2023). Suppressing Auger recombination of perovskite quantum dots for efficient pure-blue-light-emitting diodes. *ACS Energy Letters* 8 (1): 731–739.
36 Klimov, V.I. (2014). Multicarrier interactions in semiconductor nanocrystals in relation to the phenomena of Auger recombination and carrier multiplication. *Annual Review of Condensed Matter Physics* 5 (1): 285–316.
37 Klimov, V.I. (2007). Spectral and dynamical properties of multiexcitons in semiconductor nanocrystals. *Annual Review of Physical Chemistry* 58: 635–673.
38 Robel, I., Gresback, R., Kortshagen, U. et al. (2009). Universal size-dependent trend in Auger recombination in direct-gap and indirect-gap semiconductor nanocrystals. *Physical Review Letters* 102 (17): 177404.
39 Chepic, D.I., Efros, A.L., Ekimov, A.I. et al. (1990). Auger ionization of semiconductor quantum drops in a glass matrix. *Journal of Luminescence* 47 (3): 113–127.
40 Tang, J. and Marcus, R.A. (2005). Diffusion-controlled electron transfer processes and power-law statistics of fluorescence intermittency of nanoparticles. *Physical Review Letters* 95 (10): 107401.
41 Frantsuzov, P.A., Volkan-Kacso, S., and Janko, B. (2009). Model of fluorescence intermittency of single colloidal semiconductor quantum dots using multiple recombination centers. *Physical Review Letters* 103 (20): 207402.
42 Shimizu, K.T., Neuhauser, R.G., Leatherdale, C.A. et al. (2001). Blinking statistics in single semiconductor nanocrystal quantum dots. *Physical Review B* 63 (20): 205316.
43 Califano, M. (2011). Off-state quantum yields in the presence of surface trap states in CdSe nanocrystals: the inadequacy of the charging model to explain blinking. *The Journal of Physical Chemistry C* 115 (37): 18051–18054.
44 Rosen, S., Schwartz, O., and Oron, D. (2010). Transient fluorescence of the off state in blinking CdSe/CdS/ZnS semiconductor nanocrystals is not governed by Auger recombination. *Physical Review Letters* 104 (15): 157404.

45 Kagan, C.R., Lifshitz, E., Sargent, E.H., and Talapin, D.V. (2016). Building devices from colloidal quantum dots. *Science* 353 (6302): 5523.

46 Anikeeva, P.O., Madigan, C.F., Halpert, J.E. et al. (2008). Electronic and excitonic processes in light-emitting devices based on organic materials and colloidal quantum dots. *Physical Review B* 78 (8): 085434.

47 Chou, K.F. and Dennis, A.M. (2015). Forster resonance energy transfer between quantum dot donors and quantum dot acceptors. *Sensors (Basel)* 15 (6): 13288–13325.

48 Spanhel, L. and Anderson, M.A. (1990). Synthesis of porous quantum-size cadmium sulfide membranes: photoluminescence phase shift and demodulation measurements. *Journal of the American Chemical Society* 112 (6): 2278–2284.

49 Michalet, X., Pinaud, F.F., Bentolila, L.A. et al. (2005). Quantum dots for live cells, in vivo imaging, and diagnostics. *Science* 307 (5709): 538–544.

50 Lingley, Z., Lu, S., and Madhukar, A. (2011). A high quantum efficiency preserving approach to ligand exchange on lead sulfide quantum dots and interdot resonant energy transfer. *Nano Letters* 11 (7): 2887–2891.

51 Shirasaki, Y., Supran, G.J., Tisdale, W.A., and Bulovic, V. (2013). Origin of efficiency roll-off in colloidal quantum-dot light-emitting diodes. *Physical Review Letters* 110 (21): 217403.

52 Colvin, V.L., Schlamp, M.C., and Alivisatos, A.P. (1994). Light-emitting diodes made from cadmium selenide nanocrystals and a semiconducting polymer. *Nature* 370 (6488): 354–357.

53 Dabbousi, B.O., Bawendi, M.G., Onitsuka, O., and Rubner, M.F. (1995). Electroluminescence from CdSe quantum-dot/polymer composites. *Applied Physics Letters* 66 (11): 1316–1318.

54 Coe, S., Woo, W.-K., Bawendi, M., and Bulović, V. (2002). Electroluminescence from single monolayers of nanocrystals in molecular organic devices. *Nature* 420 (6917): 800–803.

55 Kim, S.H., Man, M.T., Lee, J.W. et al. (2020). Influence of size and shape anisotropy on optical properties of CdSe quantum dots. *Nanomaterials* 10 (8): 1589.

56 Caruge, J.M., Halpert, J.E., Wood, V. et al. (2008). Colloidal quantum-dot light-emitting diodes with metal-oxide charge transport layers. *Nature Photonics* 2 (4): 247–250.

57 Kim, H.Y., Park, Y.J., Kim, J. et al. (2016). Transparent InP quantum dot light-emitting diodes with ZrO_2 electron transport layer and indium zinc oxide top electrode. *Advanced Functional Materials* 26 (20): 3454–3461.

58 Stouwdam, J.W. and Janssen, R.A.J. (2008). Red, green, and blue quantum dot LEDs with solution processable ZnO nanocrystal electron injection layers. *Journal of Materials Chemistry* 18 (16): 1889.

59 Cho, K.-S., Lee, E.K., Joo, W.-J. et al. (2009). High-performance crosslinked colloidal quantum-dot light-emitting diodes. *Nature Photonics* 3 (6): 341–345.

60 Qian, L., Zheng, Y., Xue, J., and Holloway, P.H. (2011). Stable and efficient quantum-dot light-emitting diodes based on solution-processed multilayer structures. *Nature Photonics* 5 (9): 543–548.

61 Dai, X., Zhang, Z., Jin, Y. et al. (2014). Solution-processed, high-performance light-emitting diodes based on quantum dots. *Nature* 515 (7525): 96–99.

62 Yang, Y., Zheng, Y., Cao, W. et al. (2015). High-efficiency light-emitting devices based on quantum dots with tailored nanostructures. *Nature Photonics* 9 (4): 259–266.

63 Manders, J.R., Qian, L., Titov, A. et al. (2015). High efficiency and ultra-wide color gamut quantum dot LEDs for next generation displays. *Journal of the Society for Information Display* 23 (11): 523–528.

64 Wang, L., Lin, J., Hu, Y. et al. (2017). Blue quantum dot light-emitting diodes with high electroluminescent efficiency. *ACS Applied Materials and Interfaces* 9 (44): 38755–38760.

65 Won, Y.-H., Cho, O., Kim, T. et al. (2019). Highly efficient and stable InP/ZnSe/ZnS quantum dot light-emitting diodes. *Nature* 575 (7784): 634–638.

66 Shen, H., Gao, Q., Zhang, Y. et al. (2019). Visible quantum dot light-emitting diodes with simultaneous high brightness and efficiency. *Nature Photonics* 13 (3): 192–197.

67 Vasilopoulou, M., Kim, H.P., Kim, B.S. et al. (2020). Efficient colloidal quantum dot light-emitting diodes operating in the second near-infrared biological window. *Nature Photonics* 14 (1): 50–56.

68 Deng, Y., Peng, F., Lu, Y. et al. (2022). Solution-processed green and blue quantum-dot light-emitting diodes with eliminated charge leakage. *Nature Photonics* 16 (7): 505–511.

2

Colloidal Semiconductor Quantum Dot LED Structure and Principles

2.1 Basic Concepts

Colloidal quantum dot light-emitting diodes (QDLEDs) have attracted a lot of attention and have good application prospects due to their excellent luminescence properties (e.g. high color purity) and good stability and solution processability. In this chapter, we will review the advantages of colloidal quantum dots for display and lighting applications, including color purity, solution processability, and stability. In addition, in this chapter, I will introduce the application of semiconductor physics theory to colloidal semiconductor quantum dots from the most basic semiconductor physics concepts. This chapter will prepare the knowledge involved in the structure (Chapter 2) and performance parameters (Chapter 11) of QDLEDs in subsequent chapters, including device engineering applications (Chapters 4–10).

2.1.1 Color Purity

Since the electronic structure of colloidal quantum dots depends on the quantum size effect, colloidal quantum dots have narrow band emission, which can be spectrally localized by controlling the size of the nanocrystals during chemical synthesis. For example, CdSe quantum dots of different sizes can cover fluorescence emission from blue to red, while smaller bandgap materials such as PbS or CdTe quantum dots can cover fluorescence emission in the near-infrared spectral region. Generally, colloidal chemically synthesized quantum dots have a size distribution of less than 5%, making the full width at half maximum of their emission peaks in the range of 30–40 nm. The narrow band emission of colloidal quantum dots can show their hue and saturation advantages in the CIE chromaticity diagram, as shown in Figure 2.1.

The boundary of the CIE diagram is defined by the different saturated hues perceptible to the human eye, which range from a wavelength-to-wave ratio of 380 nm to a wavelength-to-wave ratio of 780 nm. The purer the color, the closer it is to the boundary of the CIE diagram. The chromaticity coordinates define the position of the emitter on the CIE diagram. The color gamut covered by the red, green, and blue pixel displays is a triangle defined by the coordinates of individual pixels. The triangle surrounded by black dots is the gamut range that red, green, and blue emitting

Colloidal Quantum Dot Light Emitting Diodes: Materials and Devices, First Edition. Hong Meng.
© 2024 WILEY-VCH GmbH. Published 2024 by WILEY-VCH GmbH.

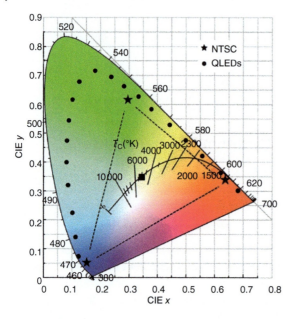

Figure 2.1 The color gamut of QDLEDs on the CIE chromaticity diagram. The position of the square black dot is where the color coordinates of sunlight are located.

quantum dots can display, which is larger than the gamut range of the International Telecommunication Union HDTV standard (dashed triangle), demonstrating the great advantage of quantum dot emitters in gamut range.

2.1.2 Solution Processability

The ligand passivation layer retained on the surface of quantum dots after synthesis prevents their agglomeration in solution and allows them to be processed in solution for various processes. The ligand usually has a polar group that fits the surface of the quantum dot and a hydrocarbon chain that ensures the long-term dispersion of the quantum dot in solution (Figure 2.2). After synthesis, typical hydrophobic ligands such as oleic acid and trioctylphosphine can be exchanged for hydrophilic ligands with amine or sulfhydryl groups to make the quantum dots compatible with aqueous solutions. The choice of ligand plays an important role in the conductivity of quantum dots. Metal-sulfur complexes have been shown to be a good choice of ligands for improving the interparticle interactions of quantum dots. Non-metallic inorganic ligands such as S^{2-}, HS^-, and HSe^- can improve carrier transport in QDLEDs [1]. "Entropic ligands" can greatly increase the solubility of quantum dots in solution [2]. The solution processability of quantum dots is necessary for various low-cost, large-area deposition processes, which have been successfully used to prepare QDLED devices by phase separation, inkjet printing, mist deposition, and microcontact printing. Suitable ligands can be selected to deposit quantum dot films of different colors in sequential solutions in orthogonal solvents or to allow post-deposition cross-linking to generate quantum dot films that can withstand subsequent solvent-based deposition steps.

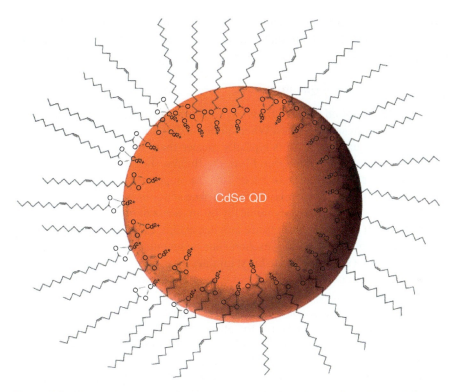

Figure 2.2 Schematic diagram of CdSe quantum dots and their surface oleic acid ligand layers.

2.1.3 Stability

Compared to organic light-emitting molecules, quantum dots are composed of inorganic semiconductors that are more resistant to degradation and photobleaching and have superior stability in display and lighting applications. A single CdSe quantum dot surface may undergo photooxidation resulting in $CdSeO_x$, and by covering the CdSe surface with a ZnS shell layer, it can provide a barrier to hinder the diffusion of oxygen, thus improving the photostability of quantum dot nanocrystals [3]. In addition, the core–shell structure can also provide physical separation of excitons from defect states on the surface of quantum dots, which may lead to nonradiative complexation. This is highlighted by thick-shell quantum dots, which have been reported to maintain a high degree of thermal stress and maintain luminescence even when passivating ligands are removed [4]. These thick-shell quantum dots can also suppress scintillation, which is associated with the suppression of oscillatory complexation and is thought to be a source of luminescence bursts. A problem in integrating colloidal quantum dot light-emitting devices into solid-state devices is the presence of organic ligands; however, the replacement of organic aliphatic ligands on quantum dots with metal-sulfur-based ligands makes the quantum dot films completely inorganic and exhibit excellent electron transport properties [5]. The core–shell structure of quantum dots has very good stability, which lays the foundation for the preparation of high-performance QDLED devices.

2.1.4 Surface States of Quantum Dots

Quantum dots in solution and thin films have different fluorescence quantum yields due to their different surface states. When quantum dots are suspended in solution, the fluorescence quantum yield is typically greater than 50%, while when quantum dots are deposited in dense thin films, the fluorescence quantum yield drops by about an order of magnitude to 5% or 10%. An important reason for this phenomenon is that the dark or emission-free surface states on the quantum dots in solid films cannot be passivated dynamically by an excess of ligands as in solution. In addition, when the quantum dots are in a dense film, excitons on one quantum dot can transfer energy to the dark exciton state on any neighboring quantum dot. Thus, a single defect state can lead to burst luminescence of 5–10 surrounding quantum dots. This Forster energy transfer can be easily observed by observing the red shift between the fluorescence emission spectrum of the dilute solution and the fluorescence emission spectrum of the dense film. Embedding quantum dots into an insulating polymer matrix can mimic the effect of dilute solutions and reduce the amount of quantum dot luminescence bursts observed in the closed quantum dot structure. However, the low conductivity of the wide bandgap polymers hinders the DC conductivity through these polymer composites, which makes them unsuitable for the preparation of QDLED structures with similar P–N junctions. In this case, quantum dot shell layers that can spatially separate excitons from the surface or the selection of ligands that are tightly bound to the quantum dot surface play an important role in maintaining quantum dot fluorescence quantum efficiency in the solid state. In addition, it is shown that field-driven QDLEDs can electrically excite quantum dot clusters embedded in insulating polymer composites.

2.1.5 Energy Levels and Energy Bands

Interface energetics are inherent to all organic electronic devices. In particular, the energy-level alignment at these organic–inorganic and organic–organic interfaces is decisive for the device's performance. As shown in Figure 2.3a, for an isolated atom,

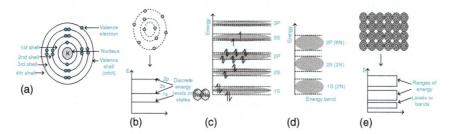

Figure 2.3 (a) Atom; (b) isolated atomic levels; (c) shared motion of electrons; (d) shared motion of n number of atoms; and (e) formation of energy bands. Source: Kare [6]/Testbook Edu Solution Pvt. Ltd.

the electron outside the nucleus will only be subjected to the potential field of the nucleus and other electrons outside the nucleus and thus move, and the energy level of this electron is the discrete energy level as shown in Figure 2.3b. And in the process of forming a crystal, the atoms come closer and closer together. When the distance between the atoms is large, their interactions can be neglected, and each atom can still be seen as isolated, even though they have the same electron energy level. If these atoms are considered as a system, then the energy levels of these electrons are concise. For example, a system of two atoms is a double degeneracy, as shown in Figure 2.3c, and a system of N atoms is an N degeneracy, as shown in Figure 2.3d. And when atoms approach each other to form a crystal, different crossover phenomena are formed in both the inner and outer orbitals of the atoms, as shown in Figure 2.3e. Due to the overlapping of the orbitals, these electrons are not confined to their own energy level orbitals but can be transferred to the same orbitals of neighboring electrons, and the transfer of electrons between neighboring orbitals is called the electron communalization movement motion. The shared electron motion is stronger in the outer electrons, and the electrons can only be transferred to orbitals with the same energy. For example, after the formation of a crystal, electrons on the 2s energy level can only move in the 2s energy level of the neighboring atoms, i.e. the individual energy levels will form a shared chemistry motion corresponding to them, as shown in Figure 2.3e. At the same time, the interaction between atoms is enhanced during the process of approaching each other, so that the original simplicity is eliminated and the energy levels with the same energy originally split into energy bands composed of energy levels of different energies; the smaller the distance between atoms, the stronger the interaction and the larger the energy bandwidth. For inorganic materials, the interatomic forces and the bandwidths are large, and the energy levels are quasi-continuous. For organic materials, on the other hand, the van der Waals forces are too weak, and the energy levels between individual molecules can be seen as discrete. In each energy band, the state in which each electron communalization movement motion may exist is called an allowed band, and the transfer of electrons between the allowed bands is forbidden and is called a forbidden band. The allowed bands can be divided into conduction bands, valence bands, and empty bands. The empty band refers to the band that is not occupied by electrons; the conduction band refers to the band that is occupied by electrons; and the valence band refers to the band that is occupied by valence electrons under low-temperature conditions.

2.1.6 Metals, Semiconductors, and Insulators

Materials can be classified as conductors, semiconductors, and insulators according to the strength of their electrical conductivity. The essence that distinguishes them is the difference in the structure of the energy bands. In general, the ability of a material to conduct electricity lies in the presence or absence of a dissatisfaction band. The energy band structure of a conductor is shown in Figure 2.4a. For a conductor at low temperatures, the highest energy level occupied by electrons is a discontent band, which means that under the action of an electric field, the electrons

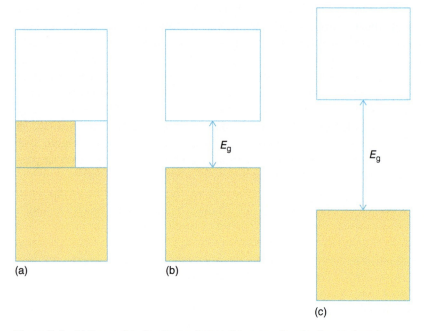

Figure 2.4 (a) Energy bands of a conductor; (b) energy bands of a semiconductor; (c) energy bands of an insulator.

in the discontent band will migrate to the empty state, causing a change in the density of states that will contribute to the conductivity. As shown in Figure 2.4b, for semiconductors at low temperatures, electrons in the valence band cannot leap to the conduction band to form a dissatisfied band due to the presence of the forbidden band, so the conduction band is empty and the valence band is full. Therefore, under the action of electric field, although the electrons in the valence band will move continuously with the electric field, it does not cause a change in the density of states, so there is no conductivity phenomenon. In contrast, at room temperature, due to the narrow width of the forbidden band of a semiconductor, it is possible to make electrons jump to the conduction band by intrinsic excitation, thus forming two unhappy bands for conducting electricity. For insulators, the bandwidth is too large (greater than or equal to 5 eV), so even at room temperature, electrons cannot jump from the valence band to the conduction band to form a dissatisfaction band, and therefore cannot conduct electricity, as shown in Figure 2.4. In organic semiconductors, although the forbidden bandwidth is large, through reasonable energy level deployment, electrons can be injected from the external electrode, which can also form a dissatisfied band to produce electrical conductivity.

2.1.7 Electrons and Holes

In semiconductors, there are two types of conducting carriers, namely free electrons and holes. In the low-temperature state, the electrons in the valence band cannot leap to the conduction band due to the existence of the forbidden band, so the conduction band is empty and the valence band is full. At room temperature, due to the narrow width of the forbidden band, the semiconductor can leap to the conduction band by intrinsic excitation, thus forming a hole in the valence band and an electron in the conduction band. Obviously, the number of holes formed by intrinsic excitation is the same as the number of electrons, and the number of electrons in the valence band is too large, so it is very complicated to analyze the motion of electrons. In order to simplify the problem, holes are introduced, which are considered to be positively charged charges. Overall, the hole is a hypothetical particle that replaces the contribution of the other electrons in the valence band to the current density. Generally, electrons are denoted by e and holes by h.

2.1.8 Fermi Distribution Function and Fermi Energy Level

When we use the Fermi distribution function, we need to pay attention to the conditions of use, and the Fermi Dirac statistics have certain applicability. The conditions of applicability are: (i) the interaction between electrons in a semiconductor is weak and can be regarded as an independent system; (ii) the motion of electrons obeys the laws of quantum mechanics, i.e. the energy of electrons is quantized, and a quantum state is occupied by one electron and does not affect the other quantum states; (iii) the electrons in the same system are interchangeable, that is, the all-same electron system; (iv) the distribution of electrons is restricted by the bubbly incompatibility principle.

$$f_E = \frac{1}{\exp\left(\frac{E-E_F}{KT}\right) + 1} \tag{2.1}$$

Under conditions of thermal equilibrium, the odds of a single electron of energy E being occupied by an electron are shown in Eq. (2.1). where f_E is known as the Fermi distribution function, describing the odds of an electron occupying an eigenstate of energy E with values from 0 to 1. Where K is the Boltzmann function, T is the absolute temperature, and E_F is the Fermi energy level. The Fermi energy level reflects the distribution of electrons in each energy level as a function of the level at which the electron fills the energy level. In the electron energy level diagram, electrons jump from low to high energy levels and holes jump from high to low energy levels, so the hole energy at the higher electron energy levels is lower. As for metals, the highest energy level occupied by electrons at absolute zero is the Fermi energy level.

2.1.9 Schottky Barrier

Figure 2.5a shows a schematic diagram of the formation of a Schottky barrier between a metal and an n-type semiconductor. In this figure, we assume that there is no effect of interfacial and surface states. In the figure, $q\phi_m$ is the work function of the metal and $q\phi_S$ is the work function of the semiconductor. As shown in Figure 2.5a, $E_{FS} > E_{FM}$ means that the electrons in the semiconductor occupy a higher energy level than those in the metal, so the electrons will leap from the semiconductor to the metal, leveling the Fermi energy levels of both. At this time,

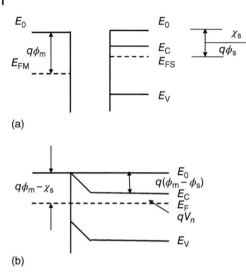

Figure 2.5 (a) Energy band diagram before metal, semiconductor contact; (b) Schottky contact.

the ionized sender ions with positive charge are left in the semiconductor due to electron migration, so a space charge layer is formed. While the metal side will accumulate electrons, because there are a large number of free electrons in the metal, the space charge region in the metal is very thin and even negligible. And the positively charged sender ions in the space charge region generate an internal electric field, which organizes the injection of electrons. And after reaching thermal equilibrium, a stable electric field and an internal potential difference are formed, the internal potential difference is $\psi_0 = \phi_m - \phi_s$ and the electron flow from the metal to the semiconductor needs to cross the potential barrier $q\phi_b = q\phi_m - \chi_s$. Here, the potential barrier $q\phi_b$ is the so-called Schottky barrier as shown in Figure 2.5b. In QDLED devices, carriers need to overcome the Schottky barrier in order to migrate.

2.1.10 Energy Level Alignment

The most common inorganic–organic interface consists of a conducting electrode and an organic hole (electron) transport layer, usually denoted as HTL (ETL). The size of the barrier to charge injection at such an interface depends on the arrangement of energy levels between the Fermi level of the electrode and the highest (lowest) occupied (unoccupied) molecular orbital HOMO (LUMO) of the transport layer. Under the assumption of an ordinary vacuum level (Schottky–Mott limit), an estimate of the hole injection barrier E^F_V can be made when the ionization potential (IP) of the HTL and the work function (ϕ_s) of the anode are known (Figure 2.6) [7].

2.2 Colloidal Quantum Dot Light-Emitting Devices

In 1994, Colvin et al. invented and prepared the first colloidal quantum dot light-emitting device [8]. Subsequently, colloidal quantum dots have become a hot

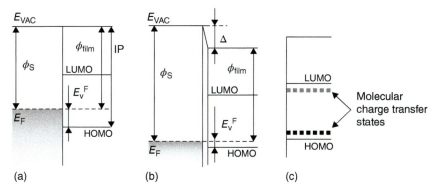

Figure 2.6 Energy level diagram for interfaces of organic semiconductor and conducting substrate: (a) vacuum level alignment, (b) vacuum level shift, and (c) molecular charge transfer states. Source: Braun et al. [7]/with permission of Elsevier.

research topic in recent years due to their high luminescence efficiency, high color purity, narrow line width, and large area flexibility [9–11]. The structure design of QDLED is based on the structure design and preparation process of organic light-emitting diodes (OLED), so the device structure of QDLED is similar to that of OLED [12]. Therefore, the device structure of QDLED is similar to that of OLED.

2.2.1 The Basic Structure of QDLED

The structure of QDLED devices is sandwich laminated structure, the light-emitting layer is generally located in the center of the device, the common structure of QDLED, as shown in Figure 2.7. Depending on the structure, it can be divided into single-layer devices, as shown in Figure 2.7a, and multilayer devices, as shown in Figure 2.7b. Although the single-layer devices are simple and low-cost to prepare, the efficiency of the devices is low and the performance is poor due to the Schottky barrier, the imbalance of the injected carriers, and the nailing effect of the interfacial state energy level. In contrast, multilayer QDLEDs can achieve carrier

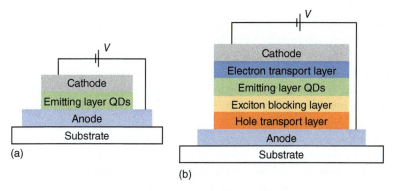

Figure 2.7 (a) Single-layer quantum dot light-emitting device; (b) multilayer quantum dot light-emitting device. Source: Neda et al. [13]/CC BY 3.0/Public domain.

injection balance through carrier buffer layers to improve device performance. The DC-driven QDLED device structure would include:

(1) *Anode*: The most widely used anode is indium tin oxide (ITO), because the principle of QDLED is injection luminescence, so the anode will provide holes into the HOMO energy level of organic semiconductors, providing the conditions for carrier combination;

(2) The hole buffer layer includes the hole injection layer as well as the HTL. In multilayer devices, the hole injection layer plays the role of reducing the Schottky barrier between the anode and the active layer, enhancing the carrier injection, and avoiding the energy loss caused by excessive charge accumulation at the interface as well as the roll-off of efficiency. Common QDLED hole injection layers include MoO_3, HAT-CN, and CuPc. The existence of the HTL not only enhances the hole transport and balances the combination of carriers in the active region, but also the general p-type HTL can have the effect of an electron blocking layer, which avoids the diffusion of excitons at the interface and effectively reduces the roll-off of efficiency. Common hole transport materials include TFB, TAPC, TCTA, NPB, and mCP;

(3) *Light-emitting layer*: The light-emitting layer is also known as the active layer. In QDLED, the light-emitting layer is composed of quantum dot nanocrystals with a core–shell structure. Unlike OLED, there are four ways of exciton formation inside the quantum dot device, which are shown in Figure 2.8 [14]. Figure 2.8a shows the process of exciton formation under the stimulation of external high-energy photons; Figure 2.8b shows the process of carrier combination injection through the buffer material; Figure 2.8c shows the formation of excitons by organic molecules and then transferring energy to the quantum dot material by Forster energy resonance transfer; and Figure 2.8d shows the process of electron transfer from a quantum electron–hole pair inside a quantum dot combining to form an exciton.

(4) The electron buffer layer includes an ETL as well as an electron injection layer. The existence of the ETL can not only enhance the electron transport and balance the carrier combination in the active region, but the general n-type ETL can also play the effect of the hole blocking layer, avoiding the diffusion of excitons at the interface and effectively reducing the efficiency roll-off. Common electron transport materials include TPBi, TmPyPb, and Bphen. The electron injection layer in multilayer devices plays the role of reducing the Schottky barrier between the cathode and the active layer, enhancing the carrier injection, and avoiding the energy loss and efficiency roll-off caused by excessive charge accumulation at the interface. Common QDLED electron injection layers include Liq and LiF.

(5) *Cathode*: The cathode will provide electrons, which will subsequently enter the LUMO energy level of the organic semiconductor, providing the basis for the combination of excitons. The cathode material generally requires a low-work-function metal material, and the materials usually used include Al, Mg:Ag.

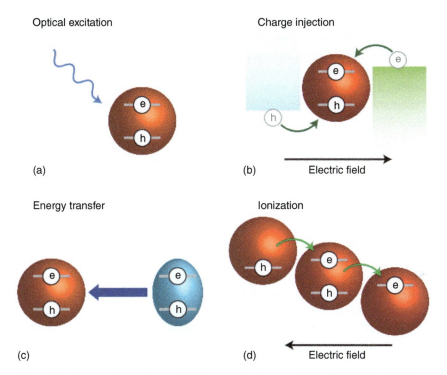

Figure 2.8 Exciton formation in QDLEDs by (a) photoexcitation; (b) carrier injection; (c) energy transfer; and (d) ionization. Source: Shirasaki et al. [14]/Macmillan Publishers Limited.

2.2.2 The Working Principle of QDLED

The current explanation of the QDLED luminescence mechanism is generally categorized as an injection carrier combination and energy transfer mechanism, where the device is divided into four processes: carrier injection, transport, exciton formation, and combination luminescence under the action of an electric field.

(1) *Carrier injection*: Under the action of the applied electric field, holes enter the HOMO of the hole injection layer from the anode by overcoming the Schottky barrier, while electrons enter the LUMO of the electron injection layer from the cathode by overcoming the potential barrier, and this process is called carrier injection.
(2) *Carrier transport*: The hole injected under the action of electric field moves to the active layer through the HTL, and the electron also moves to the active layer through the ETL. In organic semiconductors, the mobility of holes is much greater than that of electrons, so it is very important to synthesize high-mobility electron transport materials with reasonable deployment of energy levels. Charge transport layers (CTLs) include HTL and ETL.
(3) *Exciton formation*: When holes and electrons meet in the luminescent layer, they form electron–hole pairs and subsequently excitons, which are unstable,

high-energy excited states. The exciton is then transferred to the adjacent QD material via Forster energy transfer.

(4) *Radiative recombination of excitons*: Excitons recombine to emit light under the action of an electric field, but in the process of exciton recombination, they are recombined in two ways, including radiative recombination and non-radiative recombination; only after radiative recombination can QDLED devices emit light.

The performance and stability of QDLEDs are largely dependent on the choice of CTL materials. Good CTL materials should have high carrier mobility and balance the electron/hole injections well.

2.2.3 Operating Parameters of QDLED

2.2.3.1 Turn-on Voltage

Turn-on voltage refers to the voltage required when the QDLED luminance is $1\,\text{cd}\,\text{m}^{-2}$ under the action of an applied electric field. Its magnitude is related to the Schottky barrier between the gold half contacts, the potential barrier between different organic functional materials, and the intrinsic potential barrier of the light-emitting material. In general, the turn-on voltage of a QDLED device is not lower than the bandgap of the device's light-emitting material.

2.2.3.2 Luminous Brightness

In a specific voltage QDLED device, luminous intensity is called luminous brightness. Luminous brightness is proportional to the current density of the device in optics and refers to the flux of light emitted in the unit stereo angle. The unit is candela per square meter ($\text{cd}\,\text{m}^{-2}$).

2.2.3.3 Luminous Efficiency

In order to achieve a more energy-efficient and environmentally friendly use, more efficient QDLED has become a hot product in the market, so luminous efficiency has become an essential factor to measure the performance of QDLED. Efficiency mainly includes quantum efficiency, power efficiency, and current efficiency.

Quantum Efficiency
It includes both external quantum efficiency (EQE) and internal quantum efficiency (IQE). EQE refers to the ratio of the number of photons emitted by the QDLED device to the number of carriers injected, while IQE refers to the ratio of the number of photons formed by exciton combinations to the number of carriers injected. Due to the limitations of measurement instruments, the quantum efficiency of QDLED is generally expressed by EQE. The relationship between EQE and IQE is

$$\text{EQE} = \text{IQE} \times \eta_{\text{out}} = (\gamma_{\text{e-h}} \times \eta_{\text{r}} \times \phi_{\text{p}}) \times \eta_{\text{out}} \tag{2.2}$$

where η_{out} is the light extraction rate, which is generally 20%, and $\gamma_{\text{e-h}}$ refers to the ratio of the number of holes injected to the number of electrons under the action of an electric field, which is generally calculated as 1. η_{r} refers to the number of radiation-emitting excitons as a proportion of the total number of excitons; ϕ_p refers to the fluorescence quantum yield of luminescent materials, the maximum is 100%.

Current Efficiency (CE)
Current efficiency (CE) refers to the luminous intensity of the device under the condition of unit current density. The unit is candela per ampere (cd A^{-1}).

Power Efficiency (PE)
Power efficiency (PE) refers to the luminance of QDLED luminescence per unit power. The unit is lumens per watt (lm W^{-1})

$$\text{CE} = B/j \tag{2.3}$$

$$\text{PE} = \pi \times B \times S/I \times U \tag{2.4}$$

where B is the luminous brightness with unit of cd m^{-2}; S is the effective luminous area with unit of cm^2; I is the current with unit of A; and U is the device-added voltage with unit of V. From Eqs. (2.2) and (2.3), we can deduce the relationship between PE and CE as follows:

$$\text{PE} = \pi \times \text{CE}/U \tag{2.5}$$

PE is a measure of the competitiveness of the device in the market. On the one hand, the potential barrier can be lowered by matching the energy level structure of the device to enhance carrier injection and migration, thus reducing the operating voltage of the device; on the other hand, the fluorescence quantum yield and EQE of the luminescent layer can be improved by synthesizing new materials, thus increasing the luminescence efficiency of QDLED.

2.2.3.4 Luminescence Color
In the testing process of QDLED devices, the peak of luminescence can be determined from electroluminescence spectrogram (EL) and photoluminescence spectrogram (PL), thus determining the luminescence color of the device; in addition, it can also be measured by the chromaticity coordinate standard provided by the International Commission on Illumination. As shown in Figure 2.9, CIE is a method of measuring color by the three primary colors of light, the method uses x, y, z to indicate the percentage of red, green, and blue, respectively, and $\text{CIE}_x + \text{CIE}_y + \text{CIE}_z = 1$. The quality of color in general can be used (x, y) to determine the color of light (Table 2.1). For white organic light-emitting diodes (WOLED), the color coordinates are used to measure the color, and the color rendering index (CRI) and correlated color temperature (CCT) are used to determine the purity of white light.

2.2.3.5 Luminous Lifetime
Luminous lifetime is currently a key factor in determining whether QDLEDs can be widely used in the market. Luminous life is defined as the time required to reduce the luminous brightness of QDLED to half of the initial brightness under constant voltage and current conditions, and the factors affecting the device life include

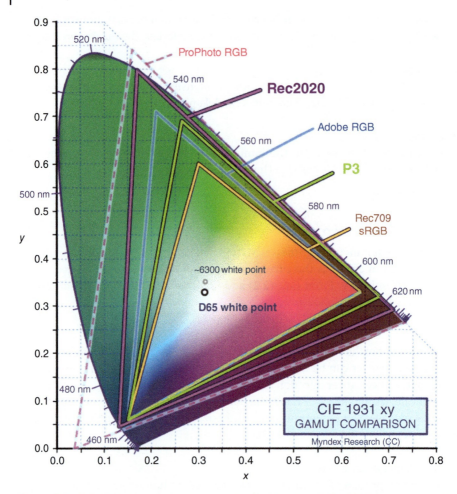

Figure 2.9 CIE 1931 chromaticity diagram showing the Rec. 2020 (UHDTV) color space in the triangle and the location of the primary colors. Rec. 2020 uses Illuminant D65 for the white point. Source: Myndex [15]/Wikipedia Commons/CC BY-SA 4.0.

Table 2.1 RGB color space parameters.

Color space	White point		Primaries					
	x_W	y_W	x_R	y_R	x_G	y_G	x_B	y_B
ITU-R BT. 2020	0.3127	0.3290	0.708	0.292	0.170	0.797	0.131	0.046

electrochemical corrosion between metal and organic materials, the effect of water and oxygen on the stability of the device interface, injection carrier balance problems, and the intrinsic stability of QDLED organic materials. Although QDLEDs have been greatly improved after nearly 30 years of development, the roll-off of device efficiency (roll-off) still needs a lot of research. Under the conditions of

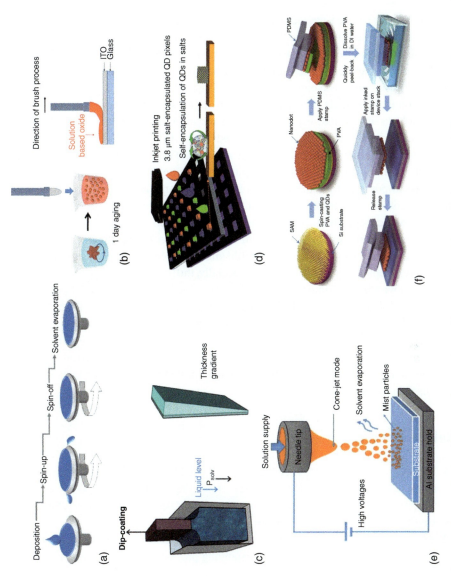

Figure 2.10 QDs layer printing methods: (a) spin-coating process; (b) brush process; (c) dip coating process; (d) ink jet printing process; (e) electrohydrodynamic jet-sprayed process; and (f) transfer process.

QDLED operation, due to the low mobility and potential barriers, a large number of carriers will accumulate at the QDLED interface, and these carriers will form an internal electric field, which will cause the quenching of excitons under the action of the internal electric field.

2.2.3.6 QDLED Device Fabrication Process

Since QD materials are normally in colloidal solution, the active layer of QDs is usually fabricated by solution-processed method. The active layer of thin film forming technology for QDLED fabrication becomes a crucial factor and needs to be largely developed to be compatible with industry techniques and to satisfy mass production requirements. There are several ways for solution process. Spin-coating is a widely used method, especially in the research lab to prepare QDLEDs, as shown in Figure 2.10a. QDs are dispersed in an ink-formulated solvent such as toluene and other high boiling point solvents. The drawback of spin-coating method is that most of the QDs ink is spined off during the spin-coating processes. It is also impossible to achieve RGB-pixel pattern and large-area deposition by this spin-coating processing. Other film-forming techniques such as dip coating, electrophoretic deposition, and Chinese brush are also developed to form QDs films and to save materials, as shown in Figure 2.10b–e. Currently, inkjet printing and micro-contact transfer are two reliable and available patterning techniques, as respectively depicted in Figure 2.10d,f. Especially, the inkjet printing technology has been employed in the industry production. However, high-quality RGB-pixel pattern is still difficult to gain with these technologies mentioned above. There is also another alternative patterning process through photolithography technology. Very recently, the electrohydrodynamic jet-sprayed technique, as shown in Figure 2.10e, has been applied to all solution-processed QDLEDs except for the electrodes. More feasible techniques for high-resolution and large-scale RGB-patterning QD films on various substrates are necessary.

References

1 Nag, A., Kovalenko, M.V., Lee, J.-S. et al. (2011). Metal-free inorganic ligands for colloidal nanocrystals: S^{2-}, HS^-, Se^{2-}, HSe^-, Te^{2-}, HTe^-, TeS_3^{2-}, OH^-, and NH^{2-} as surface ligands. *Journal of the American Chemical Society* 133 (27): 10612–10620.
2 Yang, Y., Qin, H., Jiang, M. et al. (2016). Entropic ligands for nanocrystals: from unexpected solution properties to outstanding processability. *Nano Letters* 16 (4): 2133–2138.
3 van Sark, W.G.J.H.M., Frederix, P.L.T.M., Bol, A.A. et al. (2002). Blueing, bleaching, and blinking of single CdSe/ZnS quantum dots. *ChemPhysChem* 3 (10): 871–879.
4 Pal, B.N., Ghosh, Y., Brovelli, S. et al. (2012). 'Giant' CdSe/CdS core/shell nanocrystal quantum dots as efficient electroluminescent materials: strong

influence of shell thickness on light-emitting diode performance. *Nano Letters* 12 (1): 331–336.

5 Liu, W., Lee, J.-S., and Talapin, D.V. (2013). III–V Nanocrystals capped with molecular metal chalcogenide ligands: high electron mobility and ambipolar photoresponse. *Journal of the American Chemical Society* 135 (4): 1349–1357.

6 Kare, D. (2022). Band theory of solids: learn various energy bands and their importance. https://testbook.com/learn/physics-band-theory-of-solids (accessed 24 May 2023).

7 Braun, S., Osikowicz, W., Wang, Y., and Salaneck, W.R. (2007). Energy level alignment regimes at hybrid organic–organic and inorganic–organic interfaces. *Organic Electronics* 8 (1): 14–20.

8 Colvin, V.L., Schlamp, M.C., and Alivisatos, A.P. (1994). Light-emitting-diodes made from cadmium selenide nanocrystals and a semiconducting polymer. *Nature* 370 (6488): 354–357.

9 Alexandrov, A., Zvaigzne, M., Lypenko, D. et al. (2020). Al-, Ga-, Mg-, or Li-doped zinc oxide nanoparticles as electron transport layers for quantum dot light-emitting diodes. *Scientific Reports* 10 (1): 7496.

10 Jiang, X.H., Liu, G., Tang, L.P. et al. (2020). Quantum dot light-emitting diodes with an Al-doped ZnO anode. *Nanotechnology* 31 (25): 8.

11 Lee, C.-Y., Chen, Y.-M., Deng, Y.-Z. et al. (2020). Yb:MoO_3/Ag/MoO_3 multi-layer transparent top cathode for top-emitting green quantum dot light-emitting diodes. *Nanomaterials (Basel, Switzerland)* 10 (4): 663.

12 Tang, C.W. and VanSlyke, S.A. (1987). Organic electroluminescent diodes. *Applied Physics Letters* 51 (12): 913–915.

13 Neda, H., Ghorashi, S.M.B., Wooje, H., and Park, H.-H. (2017). Quantum dot-based light emitting diodes (QDLEDs): new progress, Chapter 3. In: *Quantum-Dot Based Light-Emitting Diodes* (ed. M.S. Ghamsari), 25. Rijeka: IntechOpen.

14 Shirasaki, Y., Supran, G.J., Bawendi, M.G., and Bulovic, V. (2013). Emergence of colloidal quantum-dot light-emitting technologies. *Nature Photonics* 7 (1): 13–23.

15 Myndex (2022). CIE1931xy gamut comparison of sRGB P3 Rec2020. https://commons.wikimedia.org/w/index.php?curid=116654642 (accessed 4 April 2022).

3

Synthesis and Characterization of Colloidal Semiconductor Quantum Dot Materials

3.1 Background

The use of quantum dots (QDs) dates back more than 2000 years to Greco-Roman times, when PbS nanocrystals were used as hair dye pigments [1]. However, it was not until 1981, when Ekimov and Onushchenko discovered the blue-shift phenomenon of the absorption peak of semiconductor nanocrystals in glass, that the scientific study of QDs officially began [2]. A year later, Efros suggested that the optical properties of semiconductor nanocrystals were influenced by their size [3]. Then, in 1984, Brus and coworkers reported the variation of optical properties of colloidal CdS nanocrystals with their size [4]. In 1988, the term "quantum dots" was used for the first time to refer to semiconductor nanocrystals [5]. QDs are nanoscale fragments of their corresponding bulk materials. Due to the quantum size effect, the energy bandgap of QDs can be continuously tuned when their size is smaller than their exciton Bohr radius, and thus their fluorescence emission peaks can be tuned [6, 7]. Taking CdSe/CdS/ZnS colloidal QDs as an example (Figure 3.1), the tunable photoluminescence emission peaks can cover the major part of the visible spectrum [8, 9]. In addition, QDs have attracted great interest and attention from the scientific and industrial communities in the past three decades due to their unique optical properties, such as a wide absorption band, narrow emission peaks, high fluorescence quantum yields (QYs), and excellent stability.

The QD synthesis methodologies have progressed substantially in the past two decades, and various synthetic approaches have been developed for the synthesis of QDs, ranging from liquid-phase methods to vapor-phase epitaxial growth.

With extensive research on the synthetic chemistry of colloidal QDs, precise control of the size and shape of QDs has been successfully achieved, as shown in Figure 3.2 [10]. Monodisperse QDs with high fluorescence QYs and narrow emission peaks can now be produced and used to manufacture a variety of commercial products such as QD light-emitting diode televisions (QD TVs), bio-imaging agents, luminescent greenhouse films, solar windows, and security inks. International display giants such as Samsung and TCL have invested significant intellectual and financial resources to refine the manufacturing process, commercialize it, and improve the all-around performance of QD TV sets. Since the first release of quantum dot TV sets, ultra HD quantum dot TV sets with wide color gamut and

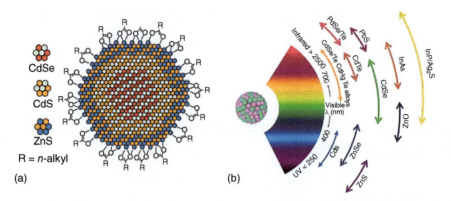

Figure 3.1 (a) Archetypal quantum dot heterostructure. An emissive CdSe core is shelled in increasingly electrically insulating CdS and ZnS layers that improve the luminescence efficiency. Surface-bound ligands (e.g. carboxylates) are bound to a metal-enriched surface. Source: Owen and Brus [8]/American Chemical Society; (b) representative QDs core materials scaled as a function of their emission wavelength superimposed over the spectrum. Source: Pu et al. [9]/American Chemical Society.

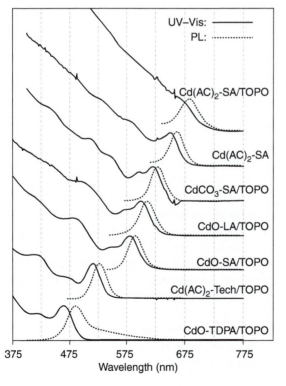

Figure 3.2 Ultraviolet-visible (UV–Vis) absorption spectra and photoluminescence (PL) spectra of different sizes of CdSe nanocrystals. TDPA: tetradecylphosphonic acid. Source: Qu et al. [10]/American Chemical Society.

high color purity have attracted a lot of attention at the Consumer Electronics Show every year and are becoming increasingly popular among consumers. We believe that QD TVs will gradually become a mainstream TV product in the next decade. In addition, Nanoco, UbiQD, and IQDEMY are expanding the commercial applications of colloidal QDs in biology, agriculture, solar energy harvesting, and security printing.

In this chapter, the colloidal synthesis methods of QD and the characterization methods of related materials will be outlined and discussed. We aim to give the readers a full understanding and fundamental knowledge about colloidal QDs, thus providing them with a solid foundation for further research. In addition, current challenges and bottlenecks for the future development of colloidal QDs will be presented to stimulate and broaden the readers' research ideas and approaches.

3.2 Synthesis and Post-processing of Colloidal Quantum Dots

The synthesis of QDs has developed relatively rapidly over the last two decades, and various methods for synthesizing QDs have been investigated and reported, essentially based on two synthetic methods: liquid-phase methods and physical vapor phase epitaxial growth synthesis methods. Although vapor phase epitaxial growth methods have been successfully used to prepare QDs with tunable dimensions, there are some inherent disadvantages, such as the use of expensive instrumentation and the difficulty of separation from the substrate. Liquid-phase methods have the potential to produce highly dispersible QDs in a variety of solutions and are inexpensive for easy production at scale and energy savings. In general, liquid-phase synthetic routes involve the preparation of colloidal QDs via hot-injection organometallic synthesis, direct heating organometallic synthesis, aqueous solution synthesis, and biosynthesis methods in batch reactors. From a scale-up perspective, industry has begun to develop techniques for various types of scale-up such as continuous reactors, including microchannels, high-gravity reactors, and spray-based technologies.

3.2.1 Direct Heating Method and Hot Injection Synthesis Method

Colloidal synthesis techniques have gradually evolved into a largely separate branch of synthetic chemistry in their own merit with their distinct set of thermodynamic and kinetic considerations. As depicted schematically in Figure 3.3a, it shows a schematic of a glassware setup widely used for colloidal synthesis of nanocrystals (NCs) in solutions, and the so-called LaMer diagram is depicted in Figure 3.3b, showing the three NC growth regimes on the temporal evolution of the monomer concentration curve. There are two main methods for the synthesis of colloidal QDs, namely the "direct heating method" and the "hot injection synthesis method."

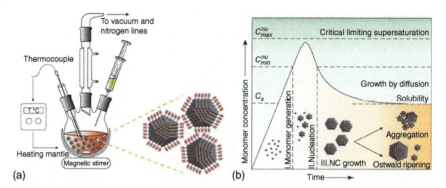

Figure 3.3 Aspects of the colloidal technique for solution growth of NCs. (a) Schematic of a glassware setup widely used for colloidal synthesis of NCs, generating surfactant-capped NCs soluble in a variety of solvents. (b) LaMer growth curve depicting the various stages of monomer generation (I), nucleation (II), and growth (III) involved in the solution synthesis. Source: Ghosh and Manna [11]/American Chemical Society.

3.2.1.1 Hot-Injection Method

Conventional methods for the synthesis of QDs rely on the heating of specific organic solvents and the injection of semiconductor precursors. In a typical preparation of CdSe QDs, for example, the Cd precursor and Se precursor solutions are first prepared by mixing $Cd(CH_3)_2$ and elemental Se in a liquid tri-n-octylphosphine (TOP) solution, respectively, which is then rapidly injected into a heated tri-n-octylphosphine oxide (TOPO) solution in a batch reactor (usually a three-necked round-bottom flask) under an inert atmosphere. TOPO acts as stabilizer and high boiling point solvent, bringing the mixture to high reaction temperatures (up to 320 °C). By controlling the growth temperature, a series of QDs with different sizes from 1.5 to 11.5 nm could be obtained within a few hours. This synthetic scheme is considered an inspiring example and a model system for the synthesis of high-quality CdSe QDs as well as other CdS QDs and CdTe QDs. However, the reactant $Cd(CH_3)_2$ is extremely toxic, expensive, and explosive, which makes this route impractical for the large-scale preparation of QDs. A milestone in the synthesis of QDs by thermal injection is the more environmentally friendly route proposed by Peng and coworkers, who replaced $Cd(CH_3)_2$ with cheap and less toxic cadmium oxide to produce CdSe NCs, with no degradation in the quality of the final QDs [10]. Much effort has been made by researchers to develop "green" thermal injection methods. The use of the non-coordinated solvent octadecene (ODE) instead of the coordination solvent TOPO has been well developed. Other green solvents, such as paraffinic liquids and oleic acid instead of TOP/TOPO as reaction media, have been introduced and have greatly simplified the green and low-cost synthesis of CdSe QDs [12]. By appropriate selection of source and synthesis parameters, the thermal injection method has been extended to the preparation of many types of QDs, such as PbS, PbSe, InP, Ag_2S, and Ag_2Se.

For the hot-injection methods, one of the cation or anion precursors is injected into the other at an appropriate temperature to trigger the rapid formation of a large number of nuclei, and then a suitable reaction temperature is set to further complete the growth of nanocrystals. These processes appear to be the best way to produce high-quality QDs with optimum physical and chemical properties at laboratory scale by precisely controlling reaction parameters such as precursor injection rate, stirring rate, and temperature. However, the synthesis of colloidal QDs by hot injection in large-scale production may face the following challenges: First, the inherent mixing time between cationic and anionic precursors can be longer and unpredictable for scaled-up batch synthesis, which may severely affect the final synthesis quality of colloidal QDs. Second, the cooling rate of the reaction mixture after hot injection of precursors is not linearly proportional to the reaction volume, which may severely interfere with the reaction process. Third, it is not very feasible to inject a large amount of one precursor into another precursor in a very short period of time. Possible solutions to these challenges are a smart choice of cationic and anionic precursors and a controlled yield of colloidal QDs produced in a single batch.

3.2.1.2 Direct Heating Method

The greatest advantage of the "direct heating method" is the ease of scaled-up synthesis, since all reaction chemicals are mixed into the reaction vessel prior to the reaction and further heated in a controlled manner to complete the nucleation and crystal growth of colloidal QDs, where the only reaction control variable is the reaction temperature. In order to synthesize monodisperse colloidal QDs, controlled reaction precursor chemistry, ligands, and heating conditions play a key role.

A number of "direct heating method" synthesis strategies have been developed with the aim of producing QDs on large scales. Efrima and coworkers report a simple and versatile method for the controlled production of high-quality tunable metal sulfide nanoparticles in a one-pot, low-temperature process using a universal precursor [13]. The metal sulfide particles were made to form at temperatures as low as 70 °C by heating metal xanthates in a strong electron-donating solvent such as hexadecylamine (HDA). The synthesis process follows the classical colloidal LaMer behavior, and QDs of various sizes can be obtained by adjusting the reaction temperature. Later, a one-pot colloidal synthesis method for high-quality CdS nanocrystals was investigated and developed by Cao and Wang by introducing a nucleation initiator into the reaction system, the separation of nucleation and growth was automatically achieved, and the quality of the produced CdS QDs was obtained comparable to that of the thermal injection method [14]. The one-pot "direct heating method" has been extended to other types of QDs such as CdSe, CdTe, Cu_7S_4, PbS, and Ag_2S. Liu et al. developed a non-injection and low-temperature method for small PbS QD assemblies with bandgaps at wavelengths shorter than 900 nm and with narrow bandwidths [15]. Du et al. reported the first QD synthesis by the thermal reaction of oleic acid, octadecylamine, and ODE [16]. In their approach, different combinations of capping ligands and solvents were found to play a key role in obtaining monodisperse Ag_2S QDs. Monodisperse Ag_2S QDs were formed in a mixture of oleic acid, octadecylamine, and ODE solvents, whereas using only oleic acid as capping ligand and solvent produced soluble but aggregated Ag_2S nanoparticles. The grain size of the obtained Ag_2S QDs could be adjusted mainly by changing the composition of the solution.

However, the design of reactions for the synthesis of high-quality QDs by the "direct heating method" is very challenging, although the basic principles of QD modulation and crystal growth are similar to those of the "hot injection synthesis method," since reaction precursors and ligands are required to ensure the generation of QDs in a short time at a suitable reaction temperature. The reaction precursors and ligands are required to ensure the generation of a large number of nuclei in a short period of time at the right reaction temperature, thus properly separating the temporal phases of nucleation and crystal growth. The synthesis of multi-element ternary, quaternary, and alloyed colloidal QDs by the "direct heating method" is difficult because this method requires matching the reactivity of each reaction precursor without guaranteeing the desired elemental composition of the multi-element QDs. Also, the requirement to synthesize high-purity crystalline phases of QDs has hindered the popularity of using the "direct heating method." This method requires the reaction precursor to remain stable at low temperatures but decompose or react rapidly once the reaction temperature reaches the threshold temperature. Usually, the ligand will participate in the reaction and affect the reactivity of the reaction precursor. In other words, the strength of the interaction between the ligand and the reaction precursor profoundly affects the colloidal QD nucleation and crystal growth stages. The strength of the interaction between the ligand and the reaction precursor can be used to qualitatively guide our judgment using the soft and hard and soft acid and base theory (HSAB).

3.2.2 Precursor Chemistry

Precursor chemistry of colloidal QDs is one of the key factors determining QD nucleation and crystal growth, as well as achieving control over QD shape and size. Usually, a balanced chemical reactivity between cationic and anionic precursors is necessary for the preparation of monodisperse colloidal QDs. In addition, it is difficult to control the reaction process using organometallic precursors with extremely high reactivity, such as $Zn(CH_3)_2$ and $Cd(CH_3)_2$. Metal oxides and fatty acid salts can be used as substitutes for these organometallic precursors because of their balanced reactivity with anionic precursors (e.g. selenium-octadecene solution) in the alternative synthetic pathway [10]. The reactivity of the precursors can also be further tuned by the addition of activators such as aliphatic amines [17–20]. In addition, the choice of precursors and the molar ratio between cationic and anionic precursors can also affect the fluorescence QY and crystalline phase of colloidal QDs [21]. Figure 3.4 illustrates X-ray diffraction (XRD) patterns of CdSe nanocrystals with a diameter of 5 nm grown using cadmium carboxylate precursors with or without aliphatic amines along with wurtzite crystal species.

3.2.3 Ligating and Non-ligating Solvents

Ligating and non-ligating solvents play different roles in the synthesis of colloidal QDs (Table 3.1). The ligand solvents or ligands (e.g. tertiary phosphines, fatty acids, alkyl amines, phosphonic acids, and trioctyl phosphine oxides) have different

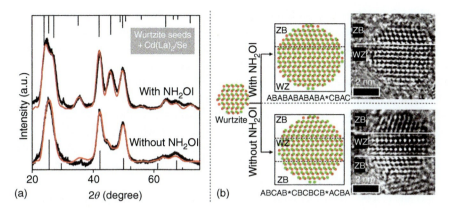

Figure 3.4 (a) XRD patterns of CdSe nanocrystals (5 nm in size) grown with wurtzite seeds using a cadmium carboxylate precursor with (or without) amine. Red lines are the computer simulation results. (b) Schematic diagram of the CdSe nanocrystals used for XRD simulations and the corresponding typical HR-TEM images. Source: Gao and Peng [21]/ American Chemical Society.

Table 3.1 Some ligating and non-ligating solvents used in the synthesis of colloidal quantum dots.

Solvent	Function
Oleylamine	Ligand
Trioctylphosphine	Ligand
1-Octadecene	Reaction medium
Oleic acid	Ligand
Tri-n-octylphosphine oxide	Ligand

bonding strengths to the cation and anion atoms on the surface of the QDs, thus affecting the growth rates, fluorescence QYs, and crystalline phases of the different crystalline surfaces of colloidal QDs. In contrast, non-coordinating solvents neither participate in the reaction nor act as ligands for monomers and nanocrystals. They can act as reaction media to help tune the reactivity of the monomer by varying the concentration of precursors and ligands during the reaction. In addition, the chemical reactivity of the monomer can be controlled by adjusting the ligand strength and concentration to provide the necessary balance between the nucleation process and the nanocrystal growth process, thus becoming a key factor in the synthesis of high-quality colloidal nanocrystals with narrow-size distribution. Long-chain alkanes and olefins are used as typical uncoordinated solvents because they are liquid at high temperatures and have good stability to withstand the reaction temperatures of semiconductor nanocrystal synthesis because they do not participate in the reaction. Non-coordinated organic solvents (e.g. 1-octadecene, tetradecane, n-eicosatetraane, and 1-eicosatene) have been used in the synthesis of colloidal QDs. However, 1-octadecene is most widely used due to its liquid state at room temperature and its ability to be used as a reaction medium at temperatures below 300 °C. In addition, because 1-octadecene is liquid at room temperature and has a nonpolar structure, it is easy to remove after synthesis. The 1-octadecene can also be removed from the reaction mixture using methanol and hexane. The nanocrystals will remain in the hexane layer, while excess unreacted monomers and ligands will remain in the methanol layer. Further methanol or acetone can be added to the hexane layer to precipitate the nanocrystals, which proved to be gentler on the nanocrystals and more effective than directly precipitating the nanocrystals from the mixture using an excess of acetone. In addition, tetradecane was found to be an excellent solvent at reaction temperatures below 200 °C. This is because tetradecane can be used as a high-purity reagent at operating temperatures and has a lower viscosity compared to 1-octadecene [22].

3.2.4 Mechanism of Nucleation and Growth of Colloidal Quantum Dots

A comprehensive understanding of the nucleation process and growth mechanism of QDs is essential to control the size, shape, and composition of QDs. As shown in Figure 3.5, according to the LaMer diagram established in 1950, the formation of QDs consists of three different stages: pre-nucleation (stage I), nucleation (stage II), and crystal growth (stage III) [23]. LaMer's pioneering study of the various stages of colloidal QD formation from homogeneous reaction solutions suggests that the nucleation and crystal growth processes must be temporarily separated to obtain monodisperse colloidal QDs. Typically, the "hot injection synthesis method" is used to achieve a rapid burst of nucleation and a temporal separation of the crystal growth phase. During the pre-nucleation phase, the precursors are converted to monomers when the monomer concentration exceeds a critical concentration

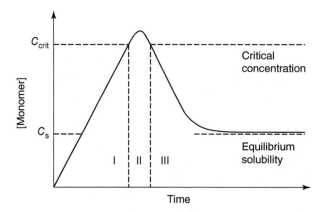

Figure 3.5 Three phases of pre-nucleation (phase I), nucleation (phase II), and crystal growth (phase III) in the LaMer model depicting the growth of monomer concentration with time. Monomer concentration increases in phase I, nucleation occurs in phase II, and nanocrystals grow in phase III until equilibrium is established. Source: Adapted content from Kim et al. [23].

of C_{crit}, which triggers the instantaneous formation of a large number of nuclei. Subsequently, as the monomer concentration decreases to C_{crit}, the crystal growth phase begins. Currently, the "hot injection synthesis method" is the most widely used method for the synthesis of monodisperse colloidal QDs. Another synthesis method is the "direct heating method," in which the reaction mixture is raised from room temperature to the reaction temperature. The pre-nucleation and nucleation periods of the "direct heating" method are much longer than those of the "hot injection synthesis" method. In 2006, Hyeon and coworkers discovered that $CdCl_2$-aliphatic amine complexes can be used as soft colloidal templates for the synthesis of two-dimensional fibrillated zincite nanosheets [24]. The cationic precursor with a bilayer template structure was prepared by mixing CdCl with octylamine, and then the anionic precursor octylammonium selenocarbamate was injected into the cationic precursor, which could form zinc blende CdSe nanosheets at a suitable reaction temperature. Subsequently, Buhro and coworkers found that two "magic nanoclusters" series could be synthesized using a soft colloidal template approach and that the "magic nanoclusters" could be grown as nuclei to form two-dimensional nanocrystals such as colloidal QD nanosheets and nanoribbons [25].

3.2.5 Size Distribution Focus and Size Distribution Scatter

The formation of colloidal monodisperse high-quality nanocrystals depends on the so-called "size distribution focusing" principle. Typically, controlled synthesis requires rapid injection of precursors, allowing rapid formation of a large number of nuclei, followed by a relatively slow and prolonged crystal growth process as shown in Figure 3.6. Initially, nuclei with a relatively narrow size distribution will be formed, and then a slow growth phase will provide sufficient time to regulate the growth of the nanocrystals. At a given monomer concentration, there is a critical size value at equilibrium. Also, when the size of the nanocrystals is smaller than the critical size value, the nanocrystals will have a negative growth rate, while larger nanocrystals will grow at a rate strongly dependent on their size. When the size is larger than the critical size value, smaller-sized nanocrystals will grow much faster than larger-sized nanocrystals due to their greater chemical reactivity for reaction-controlled or diffusion-controlled growth processes, allowing

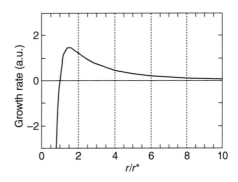

Figure 3.6 Relationship between crystal growth rate as a function of crystal size according to the Sugimoto model. Source: Peng et al. [19]/American Chemical Society.

the final nanocrystals to have a narrow size distribution range. As the monomer concentration decreases due to the depletion of crystal growth, the critical size value will become larger than the current average size value of the nanocrystals. As the smaller nanocrystals dissolve and then disappear, the size distribution of the nanocrystals will widen, but the larger nanocrystals will still grow, which is called "size distribution scattering" or "Ostwald ripening." However, the size distribution range can be refocused by additional injection of monomers at the growth temperature. Therefore, the "focus time" and "focus size" of the nanocrystals can be adjusted by changing the initial monomer concentration due to the time of depletion of the monomer concentration change.

3.2.6 Crystalline Species-Mediated Growth and Orientation of Nanocrystals Attachment Growth

Crystalline seed-mediated growth and directed attachment growth of nanocrystals are two very interesting ways of crystal growth. Stoykovich and coworkers successfully grew CdSe nanosheets, nanocubes, and nanorods by using 2–3 nm fibrous zincite CdSe nanocrystals as crystal seeds, as shown in Figure 3.7 [26]. Both "magic nanoclusters" and nanocrystals can be used as crystal seeds. Buhro and coworkers investigated the crystal growth process of converting $(CdSe)_{34}$ "magic nanoclusters" into two-dimensional CdSe nanosheets with fibrillated zincite crystal structures [27, 28].

The directional attachment growth of nanocrystals is a crystalline species-mediated growth in a broader sense. In different syntheses of two-dimensional and one-dimensional colloidal QDs, such as CdSe nanosheets, PbS nanosheets, CdSe nanorods, and ZnSe nanowires, it has been widely observed that the directional attachment growth mode occurs during the synthesis of nanosheets. Directed attachment growth occurs due to the high surface energy of certain crystalline surfaces and dipole–dipole interactions of nanocrystals along certain crystallographic orientations. Banfield and coworkers used molecular energy calculations to study the processes of directional attachment of nanoparticles and asymmetric crystal formation [29]. It was found that directional-specific remote interatomic interactions and a reduction in surface energy predict the development of crystal morphology and explain how directional adsorption produces crystals with lower symmetry than the initial material. Furthermore, the results suggest that Coulomb interactions instead of van der Waals interactions control the growth of ionic nanocrystals by directional attachment, as shown in Figure 3.8a. In 2008, Weller and coworkers found that PbS nanocrystals could form ultrathin PbS nanosheets by the directional attachment growth in two-dimensional planes driven by oleic acid ligands on the [001] crystal plane, as shown in Figure 3.8b [30]. Recently, it was reported by Peng and coworkers that CdSe nanocrystals with diameters between 1.7 and 2.2 nm could be used as crystal seeds to grow two-dimensional CdSe nanocrystals with zinc blende crystal structures supplemented with cadmium acetate to form single-site intermediates followed by directional attachment [31]. First, the crystal seeds are grown into single-dot intermediates by intraparticle ripening,

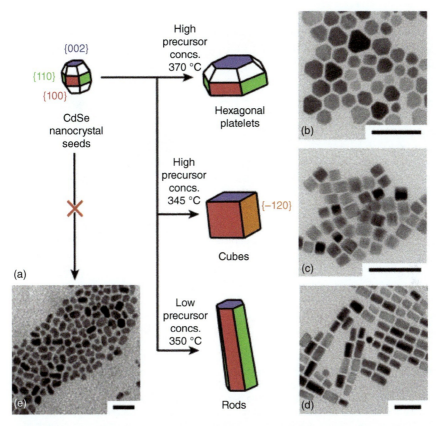

Figure 3.7 (a) Schematic diagram illustrating the relationship between CdSe nanocrystal seeds and the geometry and boundary crystal planes (colored) of the nanocrystals produced in the seed-mediated synthesis. Typical TEM images of CdSe nanocrystals with (b) hexagonal flakes, (c) cubes, and (d) rods. (e) TEM images obtained from a similar synthesis performed without the use of CdSe nanocrystal species, resulting in elongated and distorted particles. Scale bars correspond to 50 nm, respectively. Source: Rice et al. [26]/American Chemical Society.

and then these single-dot intermediates are transformed into two-dimensional crystal embryos by attachment. Next, two-dimensional nanocrystals are formed by directed attachment and in-particle ripening of two-dimensional crystal embryos. A schematic diagram of the synthesis process is shown in Figure 3.8c.

3.2.7 Synthesis Methods and Band Gap Regulation Engineering of Nuclear-Shell Quantum Dots

To improve the photoluminescence quantum yield (PLQY) efficiency and stability of the core QDs, the core surface of the QDs can be enhanced by coating it with a wide bandgap shell material. Due to the small size of the QDs, the surface-to-volume ratio of the QDs is high, which leads to defects on the core surface. Surface passivation requires an effective material to eliminate surface defects, thereby improving

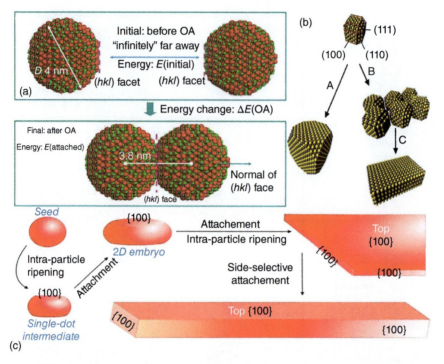

Figure 3.8 (a) This figure shows the directional attachment process, where two particles of 4 nm diameter are attached to the (*hkl*) surface. Source: Zhang and Banfield [29]/American Association. (b) Schematic diagram of large particle (A) and thin sheet formation (B, C) formed from small PbS quantum dots. Source: Schliehe et al. [30]/American Association for the Advancement of Science. (c) Schematic diagram of crystal species-mediated growth of sphalerite CdSe nanosheets. Source: Chen et al. [31]/American Chemical Society.

stability and PLQY. Various types of core/shell QDs exist, depending on the type of material, including organic/inorganic, organic/organic, inorganic/inorganic, and organic/inorganic materials. The shell material of core/shell QDs can be selected according to the desired application. Depending on the band arrangement between the valence and conduction bands (CBs) of the constituent materials, core/shell systems can be defined as type I, type II, or quasi-II as shown in Figure 3.9. In type I (e.g. $CuInS_2$/ZnS, $AgInS_2$/ZnS, InP/ZnS, CdSe/ZnS) core/shell systems, the conduction band (CB) of the core is lower in energy compared to the shell, while the valence band (VB) of the core is higher in energy compared to the shell. In all cases, holes and electrons are confined to the nucleus. To improve PLQY and chemical stability, shell coatings have been applied to the surface of the core. In type II, the VB edge or the CB edge of the shell material is located in the bandgap of the core. When the nanocrystals are excited, the resulting staggered band arrangement leads to a spatial separation of holes and electrons in different regions of the core/shell structure. Unlike type I QDs, one of the electrons or holes is confined to the core and the other to the shell. In the type I structure, CdS, and ZnS while in the type II structure, CdSe and ZnSe are used as shell materials for synthetic core/shell QDs.

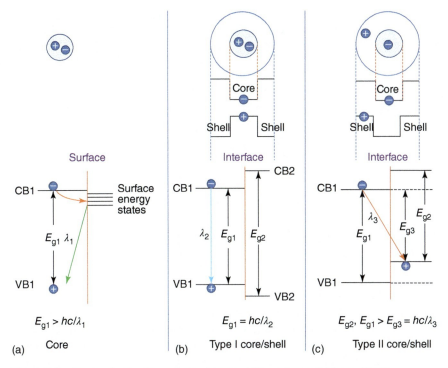

Figure 3.9 Electronic structure of core (a), type I (b), and type II (c) core/shell heterostructures.

3.2.7.1 Non-alloyed Core–Shell Quantum Dots

There are four main synthetic approaches to synthesizing high-quality core/shell QDs (Figure 3.10) [32]. The simplest type of synthesis is the so-called one-pot scheme, where all precursors of the core and shell are mixed prior to heating to form the target core/shell QDs. While it may be simpler and not produce core/shell QDs with the desired control for industrial production, this method has produced some cadmium-based core/shell QDs that still perform well for the fabrication of quantum dot light-emitting diodes (QDLEDs) with good performance. The other three methods aim to provide the necessary control over the epitaxial growth of the shell. SILAR (sequential ion layer adsorption and reaction) is by far the most applied of all four options and achieves controlled epitaxy by alternating the addition of cationic and anionic precursors. TC-SP (thermocyclically coupled single precursor) applies a single source of precursor, and the reaction temperature is programmed to avoid self-core aggregation of the precursor. Slow dropping has been studied more extensively and has a long research history; it avoids the self-nucleation of precursors by slowly adding a dilute solution of precursors. Each of these three methods has produced core/shell QDs with almost ideal and stable photoluminescence properties, and from an industrial point of view, scale-up of all three methods should be achievable. In fact, SILAR has already been adopted by industry to produce core/shell QD backlights for LCDs [33].

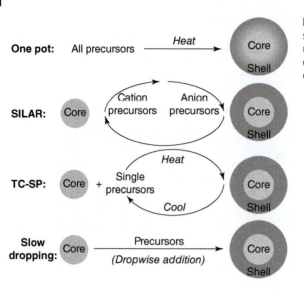

Figure 3.10 Four types of synthesis schemes widely used in high-quality core/shell QDs. Source: Shu et al. [32]/John Wiley & Sons.

3.2.7.2 Alloy Core–Shell Quantum Dots

It is well-known that the Auger effect accounts for efficiency drop in LEDs at high driving currents band engineering strategy. In order to minimize the Auger effect, core/shell QDs for optoelectronic applications require improved core/shell interface layers to form the alloy structure and the development of band engineering strategies for this purpose. Beginning by overcoating InAs QDs with a lattice-matched CdSe shell to grow InAs/CdSe core/shell QDs, an outer CdS shell was chosen to grow onto InAs/CdSe QDs to further confine the charge carriers, resulting in InAs/CdSe/CdS QDs, as shown in Figure 3.11 [34]. The synthesis was achieved in a two-step manner

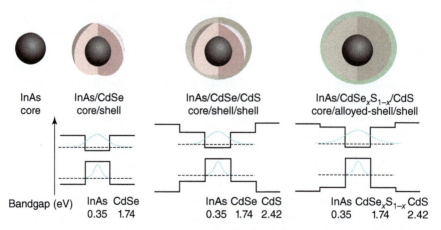

Figure 3.11 Band engineering strategies in InAs-based QDs. The top row shows the composition of heterostructures, and the lower panel, the band alignment. InAs/CdSe core/shell QDs have a quasi-type-II band alignment, while growth of an additional outer CdS shell leads to further confinement of carriers in InAs/CdSe/CdS core/shell/shell QDs, alloying of an intermediate layer leads to a spatial distribution of electron and hole in InAs/CdSe$_x$S$_{1-x}$/CdS core/alloyed-shell/shell QDs. Source: Sagar et al. [34]/American Chemical Society.

with an intentional sharp interface between CdSe and CdS. By preheating InAs QDs in a mixture of oleylamine and ODE at 270 °C while continually supplying Cd, Se, and S sources, a smooth alloyed interface of InAs/CdSe$_x$S$_{1-x}$/CdS QDs is achieved.

3.2.8 Surface Chemistry of Colloidal Quantum Dots

3.2.8.1 Covalent Bond Classification Method

The surface chemistry of QDs affects their electronic structure, fluorescence QY, and colloidal stability due to the interaction of surface inorganic or organic ligands coordinated to the surface atoms of QDs with surface atoms or solvents. Consistent with the covalent bond classification (CBC) approach proposed by Green, ligands can be bonded to metal centers through three basic types of interactions and are classified according to the nature and number of these interactions. The symbols Z, X, and L represent the three interaction types corresponding to 0-electron neutral, 1-electron, and 2-electron ligands, respectively, and are clearly distinguished according to the molecular orbital representation of the bond as shown in Figure 3.12. L-type ligands are bonds that interact with the metal center through qualitative covalent bonds, where both electrons are provided by the L-type ligand. These L-type ligands are Lewis bases with lone pairs of electrons, such as primary amines (R-NH$_2$) and trialkylphosphines (R$_3$P trialkylphosphines). An X-type ligand is a bond that interacts with a metal center through a normal 2-electron covalent bond consisting of a metal-given electron and an electron from the X-type ligand. Simple examples of X-type ligands are Cl and H. For Z-type ligands, they also interact with the metal center through a ligand covalent bond with the two electrons provided by the metal. Z-type ligands are Lewis acids with empty orbitals that can accept a pair of electrons from the metal.

Owen and coworkers explained the surface chemistry of colloidal QDs using the CBC method and gave the binding rules between nanocrystals and ligands as shown in Figure 3.13 for the CBC method [35]. They also investigated the stoichiometry and ligand exchange of metal-sulfide nanocrystals by NMR spectroscopy. The relative displacement ability was found to depend mainly on geometrical factors such as spatial effects, chelation, and soft/hard matching with cadmium ions. In addition, they found that most un-liganded "hanging" atoms do not form surface traps

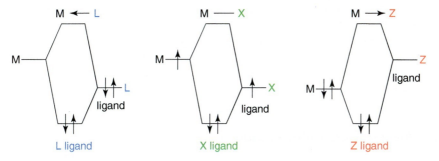

Figure 3.12 Three basic types of molecular orbitals for the interaction between metal centers and ligands.

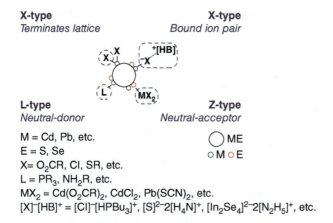

M = Cd, Pb, etc.
E = S, Se
X = O_2CR, Cl, SR, etc.
L = PR_3, NH_2R, etc.
MX_2 = $Cd(O_2CR)_2$, $CdCl_2$, $Pb(SCN)_2$, etc.
$[X]^-[HB]^+$ = $[Cl]^-[HPBu_3]^+$, $[S]^{2-}2[H_4N]^+$, $[In_2Se_4]^{2-}2[N_2H_5]^+$, etc.

Figure 3.13 Binding rules between nanocrystals and ligands according to the covalent bonding classification method. Source: Anderson et al. [35]/American Chemical Society.

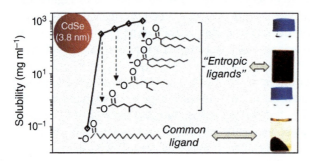

Figure 3.14 Solubility of CdSe nanocrystals encapsulated with different entropic ligands. Source: Yang et al. [36]/American Chemical Society.

and that X- and L-type ligands favor the electronic structure stability of colloidal QDs. However, Z-type ligands will induce intermediate energy gap states, which are located on the 4p lone pairs of the surface atoms of the 2-liganded sulfide.

3.2.8.2 Entropic Ligands

Solution processability of colloidal QDs is the key to solution-processable optoelectronic, biomedical labeling, and printable electronics applications. From the perspective of physical chemistry, Xiaogang Peng and coworkers proposed the concept of "entropic ligands" to represent a class of ligands that give colloidal QDs excellent solubility. Due to the bending and rotational entropy of C—C σ bonds in organic ligands, "entropic ligands" can maximize the intra-molecular entropic effect and thus increase the solubility of various colloidal QDs by 10^2–10^6 times as shown in Figure 3.14 where the solubility of CdSe nanocrystals coated with different entropic ligands is presented schematically [36].

3.3 Material Characterization

The main material characterization techniques used for QDs include ultraviolet–visible (UV–Vis) absorption spectroscopy, fluorescence spectroscopy, nuclear magnetic resonance (NMR) spectroscopy, Fourier transform infrared (FTIR) spectroscopy, transmission electron microscopy (TEM), X-ray photoelectron spectroscopy (XPS), powder X-ray diffraction (XRD), small-angle X-ray scattering (SAXS) and wide-angle X-ray scattering (WAXS), and X-ray absorption fine structure (XAFS) spectroscopy. These characterization methods can help researchers to investigate the optical properties, chemical composition, crystal structure, and morphology of QDs, as well as the interaction between QDs and their surface ligands, etc.

3.3.1 Ultraviolet–Visible (UV–Vis) Absorption and Fluorescence Spectra

UV–Vis absorption and fluorescence spectroscopy are commonly used techniques to study the optical properties of QDs. UV–Vis absorption spectroscopy is used to determine the absorption spectra of QDs, which provide information about the electronic structure of the QDs. In this technique, a beam of light is passed through a sample of QDs, and the amount of light absorbed by the sample is measured. The absorption spectra of QDs typically show a strong absorption band corresponding to the bandgap energy of the QD, which can be used to determine the size and composition of the QDs. The absorption spectra can also be used to study the effects of surface ligands and other surface modifications on the electronic structure of the QDs. Fluorescence spectroscopy is used to study the emission properties of QDs. In this technique, a sample of QDs is excited with a light source, and the emitted fluorescence is measured. The fluorescence spectra of QDs typically show a broad and tunable emission peak that corresponds to the bandgap energy of the QD. The fluorescence spectra can be used to study the effects of size, composition, surface ligands, and other surface modifications on the emission properties of the QDs. Additionally, fluorescence lifetime measurements can be used to study the radiative and non-radiative decay pathways of QDs, which can provide information about their photophysical properties.

UV–Vis absorption spectroscopy measures the leap from the ground state to the excited state of a substance, which is complementary to fluorescence spectroscopy, which records the transition from the excited state to the ground state, as shown in Figure 3.15. Peng and coworkers used two independent methods to determine the molar extinction coefficients of CdX (X = Te, Se, and S) nanocrystals based on the Lambert–Beer law and absorption spectra. They found that the molar extinction coefficients of colloidal QDs are in the strong quantum-limited size range and that the size of QDs is proportional to the square of the cubic function, as shown by the absorption spectra [37]. After determining the molar extinction coefficients of colloidal QDs, UV–Vis absorption spectra can be used to derive molar concentrations and provide quantitative analysis of relevant data. It is worth noting that the extinction coefficients of CdSe-based QDs are independent of various factors, such as the nature of the surface capping group, the refractive index of the solvent, the PLQY,

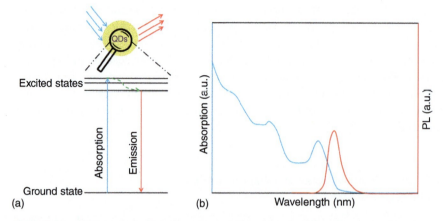

Figure 3.15 (a) Schematic diagram of absorption and emission transition of colloidal quantum dots; (b) UV–Vis absorption spectrum and fluorescence emission spectrum of colloidal quantum dots.

and the synthesis method used to produce the QDs. Similarly, studies on CdSe quantum rods [38], CuInS QDs [39], and Pb-based QDs [40–42] have also confirmed the relationship between the size of QDs and the molar extinction coefficient. Recently, Peng and coworkers developed a method to determine the extinction coefficient of each ZnX (X = S, Se) unit of zinc blende ZnX (X = S, Se) QDs [43]. They found that the magnitude of the extinction coefficient of Zn-ZnSe QDs terminating at the surface of Zn atoms follows a simple exponential relationship with the QD size, consistent with the quantum confinement effect. However, for Se-ZnSe QDs terminating at the surface of Se atoms, the relationship includes an additional quadratic term between QD size and the extinction coefficient. These findings provide a valuable insight into the optical properties and synthesis of zinc blende ZnX (X = S, Se) QDs. Interestingly, for ZnSe QDs of the same size, it has been found that the unit extinction coefficient of Se-ZnSe QDs is consistently larger than that of Zn-ZnSe QDs, eventually approaching the limiting value of ZnSe for the same size, as shown in Figure 3.16.

Fluorescence spectroscopy is a common method used to measure the light emitted by colloidal QDs. By analyzing the fluorescence spectrum of colloidal QDs, we can determine the wavelength and full width at half maximum (FWHM) of the emission peak, which provides valuable information about the optical properties of these nanocrystals. Interestingly, for two-dimensional QDs (also known as QD nanosheets or nanoribbons), the wavelength shift between the absorption and emission peaks is almost zero [44, 45]. From the shape of the fluorescence spectrum, we can also determine whether there is fluorescence emission due to surface defects of colloidal QDs. Time-resolved fluorescence spectroscopy can be used to study the scintillation behavior and fluorescence lifetime of individual colloidal QDs, which provides insights into the fundamental photophysical properties of QDs. The transient fluorescence spectra obtained from this technique can also be used to study the energy transfer and charge carrier dynamics in QDs, as shown in Figure 3.17 [46].

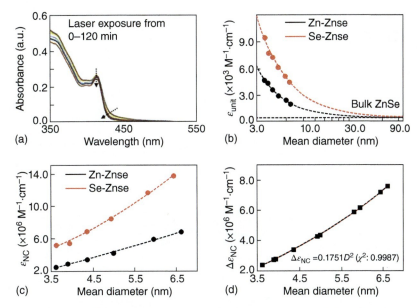

Figure 3.16 (a) Evolution of the UV–Vis absorption spectrum of Se-ZnSe colloidal quantum dots exposed to laser irradiation under ambient conditions. (b) Convergence of the unit extinction coefficient of Zn-/Se-ZnSe colloidal quantum dots. (c) The extinction coefficient per molar nanocrystal of ZnSe colloidal quantum dots in relation to their size. (d) The extinction coefficient of the surface state of ZnSe colloidal quantum dots versus their size. All dashed lines in the figure correspond to the fitting functions recorded in the text. Source: Lin et al. [43]/Springer Nature.

3.3.2 Nuclear Magnetic Resonance Spectroscopy

NMR spectroscopy is another technique that can be applied to QD analysis. NMR spectroscopy is a powerful tool for studying the electronic and molecular structure of materials. It is based on the measurement of the magnetic properties of atomic nuclei in a magnetic field. NMR spectroscopy can be used to study the electronic structure of QDs by providing information about the chemical bonding and the arrangement of atoms in the QD core and surface. The technique can also provide information about the size, shape, and distribution of QDs in a sample. Additionally, it can be used to investigate the interactions between QDs and other materials or molecules.

There are several types of NMR spectroscopy that can be used in QD analysis, including proton NMR, carbon-13 NMR, and phosphorus-31 NMR. These techniques can provide information about the chemical composition of QDs and their ligands. For example, proton NMR can be used to study the ligand exchange dynamics of QDs, while carbon-13 NMR can be used to study the interactions between QDs and organic molecules. Overall, NMR spectroscopy is a powerful tool that can provide valuable information for understanding the electronic and molecular properties of QDs.

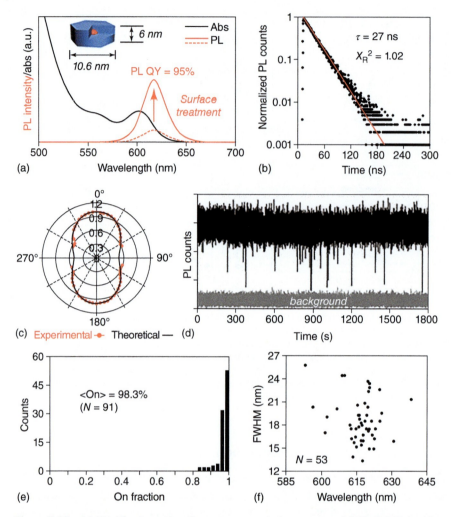

Figure 3.17 (a) UV–Vis absorption/fluorescence emission spectra of CdSe@CdS dot/sheet core–shell structured quantum dots in the overall case; (b) transient fluorescence spectra; (c) fluorescence intensity of quantum dot films at different angles with the fluorescence value of 0° set to 1; (d) representative fluorescence intensity traces; (e) individual CdSe@CdS dot/sheet core–shell structured quantum dots of "bright" state time distribution; (f) statistics of half-peak full-width values and fluorescence emission peak positions. Source: Wang et al. [46]/American Chemical Society.

Solution NMR spectroscopy has been used to observe and quantitatively analyze the binding kinetics of ligands on the surface of colloidal QDs [47–51]. For example, by measuring and analyzing NMR and fluorescence spectra, the chemical equilibrium constants for the adsorption and desorption processes of CdSe-amine nanocrystal ligands were found to be approximately 50–100, as shown in Figure 3.18 [52]. Owen and coworkers discovered that CdSe nanocrystals with surface-modifying aliphatic amines were able to react with oleic acid, *n*-octadecyl phosphonic acid, or carbon dioxide to form ion pairs of surface-bound *n*-alkyl

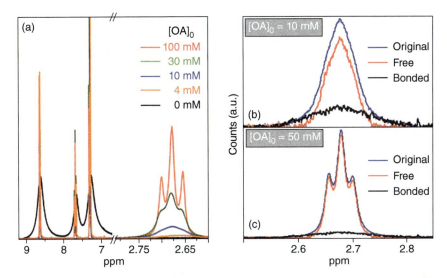

Figure 3.18 NMR spectra of pyridine-treated CdSe nanocrystals (optical density of 10) after equilibration at a given concentration of oleylamine. ^{1}H NMR spectra. (b) and (c) NMR spectra in the range of spectra related to the R hydrogen of oleylamine are fitted. Source: Ji et al. [52]/American Chemical Society.

ammonium oleate, phosphonate, and carbamate with a greater affinity than primary n-alkyl amines [53]. Furthermore, a comprehensive set of solid-state nuclear magnetic resonance (SSNMR) techniques were employed to accurately measure the atomic structures and ligand–ligand interactions at the surfaces of the nanocrystals, as illustrated in Figures 3.19 and 3.20 [54]. The structure of the core–shell interface in CdSe/CdS QDs can be analyzed and distinguished using dynamic nuclear polarization-enhanced PASS-PIETA NMR spectroscopy [55].

3.3.3 Fourier Transform Infrared Spectroscopy (FTIR)

FTIR spectroscopy can be used to analyze the surface chemistry and functionalization of QDs. In FTIR, infrared radiation is passed through the QD sample, and the resulting spectrum provides information on the functional groups and chemical bonds present on the surface of the QD. QD functionalization can be achieved through the attachment of various ligands or molecules to the QD surface, and FTIR can be used to analyze the chemical bonds between these surface ligands and the QD core. Also, it can be used to study the effect of surface ligands on the stability and optical properties of QDs.

For example, FTIR can be used to confirm the presence of ligands on the QD surface after ligand exchange reactions. FTIR can also be used to study the interaction between QDs and biomolecules, such as proteins or DNA, which can have important applications in biosensing and biomedical imaging.

Overall, FTIR is a useful technique for the analysis of QD surface chemistry and functionalization and can provide valuable information for the design and

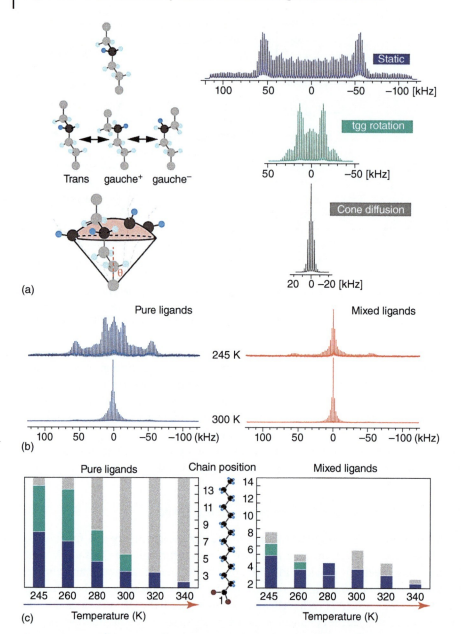

Figure 3.19 2H NMR lineshapes and chain flexibility. (a) The three distinct dynamic modes of methylene units and the corresponding 2H NMR patterns under 2 kHz magic-angle spinning. These dynamic modes could present in a hydrocarbon chain at different temperatures or at different positions, e.g. the middle segment or the free end. (b) 2H NMR patterns for nanocrystal-ligands complexes with pure ligands ($f_{He} = 0$) and nanocrystal-ligands complexes with mixed ligands ($f_{He} = 0.68$) with fully deuterated myristates at 245 and 300 K. (c) The histograms of methylene flexibility along the myristate ligand at variable temperatures, based on the deconvolutions of 2H patterns. The blue, green, and gray bars represent static deuterium, tgg rotation, and cone diffusion, respectively. Source: Pang et al. [54]/CC BY 4.0/Public Domain/Springer Nature.

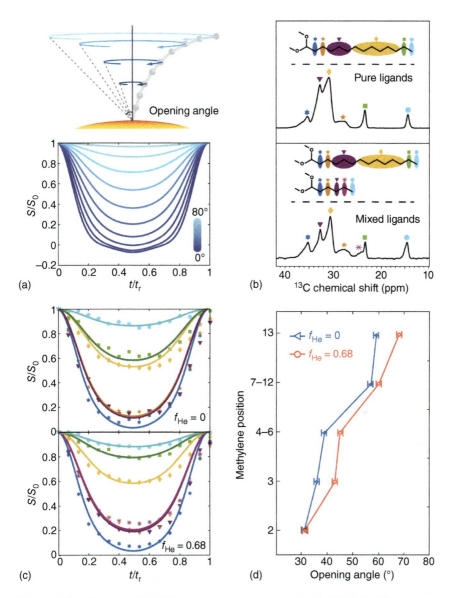

Figure 3.20 1H–13C DIPSHIFT results and opening angles. (a) The cone diffusion model of methylenes and the simulated DIPSHIFT curves for different opening angles. (b) 13C solid-state NMR spectra obtained at 300 K. The peaks correspond to the chain segments labeled with the same symbols. (c) The DIPSHIFT results (dots) and fitting curves (solid lines). The colors correspond to the chain segments labeled in (b). (d) The opening angles of each chain segment of myristate ligands assume the cone diffusion motion. The fitting standard deviation is smaller than 1°. Source: Pang et al. [54]/CC BY 4.0/Public Domain/Springer Nature.

optimization of QD-based materials for various applications. The anharmonic oscillator model describes the infrared vibrational modes in molecules, where the frequencies v of the fundamental bending and stretching modes are derived from the following equations:

$$v = 1303\sqrt{\frac{F}{\mu}}, \quad \mu - \frac{1}{\mu_1} + \frac{1}{\mu_2} \tag{3.1}$$

where F is the force constant; μ_1 and μ_2 are the masses of the atoms involved. The infrared (IR) spectrum is typically divided into five main spectral regions based on wavenumber, which are: below 1000 cm^{-1} (e.g. C–X: X = Cl, Br, I, and heavier atoms), 1000–1500 cm^{-1} (E—X single bonds: E = B, C, N, O), 1500–2000 cm^{-1} (E=X double bonds: E = X = C, N, O), 2000–2700 cm^{-1} (E≡X triple bond: E = X = C, N, O), and 2700–4000 cm^{-1} (E–H stretching: E = B, C, N, O). When molecules absorb infrared radiation, there is a net change in their dipole moment, leading to vibrational or rotational motions. These motions can be divided into two categories based on whether the bond length and bond angle change: bending (scissoring, rocking, wagging, and twisting) and stretching (symmetric and asymmetric). FTIR spectroscopy can be used to analyze the types of functional groups on the surface of nanocrystals based on their characteristic absorption peaks.

3.3.4 X-Ray Photoelectron Spectroscopy (XPS)

XPS, also known as electron spectroscopy for chemical analysis (ESCA), is a surface analysis technique that can be used to identify the chemical composition of a material. It involves shining a beam of X-rays on the surface of a sample, which causes the emission of electrons. The energy of these emitted electrons can be analyzed to determine the identity and concentration of the elements present on the surface of the sample.

For analyzing QDs, XPS can be used to study the chemical composition and electronic structure of the QD surface. This information can be important for understanding the stability, reactivity, and electronic properties of the QDs. XPS can also be used to study the ligand shell that typically surrounds QDs, providing information on the nature of the ligands and their bonding to the QD surface.

In addition to providing information on the surface chemistry of QDs, XPS can also be used to study the oxidation state and electronic structure of individual atoms within the QD. This can be achieved through angle-resolved XPS, which involves varying the angle of the X-ray beam to measure the photoelectron emission at different depths within the sample. By analyzing the energy and intensity of the photoelectron signal as a function of depth, it is possible to obtain information about the electronic structure of the QD interior.

Overall, XPS is a powerful technique for analyzing the surface chemistry and electronic properties of QDs, providing important insights into their behavior and potential applications. XPS is a surface-sensitive analytical technique that measures the number and kinetic energy of electrons emitted from a sample surface when it is irradiated with an X-ray beam. XPS is capable of analyzing the elemental composition of the top 0–10 nm of a sample's surface. Figure 3.21 displays the XPS spectra of core/shell QDs at different photoelectron kinetic energies and their corresponding QD sizes [56]. The XPS spectra of Cd 3d electrons and Se 3d electrons of zinc blende and wurtzite CdSe are shown in Figure 3.22 [57].

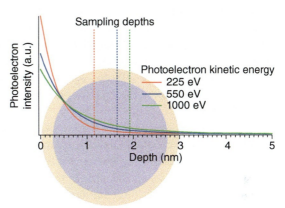

Figure 3.21 Schematic representation of the XPS signal at three photoelectron kinetic energies versus quantum dot size for core/shell structured quantum dots. Source: Clark and Flavell [56]/John Wiley & Sons.

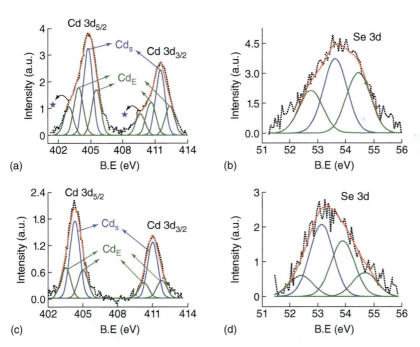

Figure 3.22 XPS counts from (a) Cd 3d electrons of zinc blend CdSe, (b) Se 3d electrons of zinc blende CdSe, (c) Cd 3d electrons of wurtzite CdSe, and (d) Se 3d electron of wurtzite CdSe. The experimental data as obtained are presented as black dotted traces and deconvoluted into Gaussian components. The peaks are characterized as (a) Cd bound to Se (blue traces) and (b) Cd bound to other elements (green traces; marked by asterisk represents Cd bound to oxygen). The red trace represents the Gaussian resultant fit. Source: Subila et al. [57]/American Chemical Society.

3.3.5 Transmission Electron Microscopy

TEM is a powerful tool for analyzing the structural and morphological properties of QDs. TEM allows for high-resolution imaging of the crystal structure and lattice fringes of individual QDs, as well as their size, shape, and distribution. TEM can also be used to study the interparticle spacing and orientation of QDs in ordered arrays, which is important for applications in optoelectronics and photonics.

One common technique for TEM analysis of QDs is high-resolution transmission electron microscopy (HRTEM), which can reveal the lattice structure of the QD with atomic resolution. Electron diffraction can also be used to determine the crystallographic orientation of the QD, and selected area electron diffraction (SAED) can provide information about the crystal structure and orientation of a large number of QDs in a sample.

TEM can also be used to study the interactions between QDs and other materials, such as metals and shell QDs. For example, TEM can be used to visualize the formation of core–shell QDs, where a shell material is deposited onto the surface of the QD to improve its optical and electronic properties. TEM can also be used to study the behavior of QDs under different conditions, such as exposure to different solvents or temperatures, and can provide important insights into their stability and performance.

TEM can be classified into two categories: low-resolution and high-resolution TEM. Other techniques, such as high-angle annular dark field scanning (HAADF-STEM), SAED, and energy dispersive X-ray spectroscopy (EDS), are also available for analyzing the morphology, crystal structure, and elemental distribution of samples.

3.3.6 Small-Angle X-Ray Scattering and Wide-Angle X-Ray Scattering

SAXS and WAXS are two techniques commonly used for the analysis of QDs. SAXS is used to study the size, shape, and distribution of nanoparticles, including QDs, in solution. By measuring the intensity of the scattered X-rays at small angles, information can be obtained about the size and shape of the particles. This technique is useful in the characterization of the polydispersity and size distribution of QD samples.

WAXS is used to determine the crystal structure of QDs. By measuring the intensity of the scattered X-rays at larger angles, information can be obtained about the interatomic distances and orientations in the crystal lattice. This technique is useful in the characterization of the crystalline structure and orientation of QD samples.

Both SAXS and WAXS can provide valuable information about the properties and structure of QDs and are often used in conjunction with other techniques such as XRD, TEM, and SEM to provide a more complete characterization of QD samples.

SAXS is a scattering phenomenon that occurs when X-rays are transmitted through a specimen at a small angle of 2–5° relative to the original beam. This characterization method can be used to analyze the structure of large crystalline substances and to measure the size, shape, and distribution of ultrafine cavities in particles or solids with particle sizes of tens of nanometers or less. For polymer materials, it can be used to measure the size and shape of polymer particles or voids, analyze the polymer phase structure of a blend, analyze long periods, molecular chain length, and measure the glass transition temperature.

In QD analysis, SAXS can be used to analyze their particle size and distribution. For instance, Pulcinelli and coworkers used SAXS to study the synthesis process of colloidal ZnO nanocrystals, including the nucleation and growth of ZnO QDs, the growth of dense ZnO QD agglomerates, the growth of fractal agglomerates, and the secondary nucleation and growth of fractal agglomerates [58]. They analyzed the precursor species by recording in situ and simultaneous time-resolved monitoring of UV–Vis absorption spectra, combined with SAXS and XAFS spectra of ZnO nanocrystals.

Chen and coworkers used small-angle and wide-angle X-ray scattering (SAXS and WAXS, respectively) experimental methods to fully elucidate the self-assembly of truncated tetrahedral QDs into superstructures with translational alignment and consistent orientation order, as shown in Figure 3.23 [59].

3.3.7 X-Ray Diffractometer

X-ray diffractometry can be used to analyze the crystal structure and phase of QD samples. The technique involves irradiating the sample with X-rays and measuring the diffraction pattern of the scattered X-rays, which provides information about the arrangement of atoms in the crystal lattice. XRD is a diffraction phenomenon that occurs when X-rays pass through a sample in the angular range of 2θ from 5° to 165°, producing an XRD pattern. It is one of the most powerful analytical methods used to analyze the crystal structure of materials based on Bragg's law, which relates the X-ray interference pattern scattered by the crystal to the crystal lattice spacing, d_{hkl}, through the equation $n\lambda = 2d_{hkl} \sin \theta$ to explain it [60, 61], where λ is the wavelength of the incident X-rays; n is a positive integer; and d_{hkl} is the crystal face spacing, which is shown schematically in Figure 3.24.

The size of the crystal can be estimated by Scherrer's formula $\tau = \frac{K\lambda}{\beta} \cos \theta$, where τ is the average size of the ordered or crystalline region, which is less than or equal to the grain size; λ is the X-ray wavelength; K is called the dimensionless shape factor or Scherrer's constant, which has a value close to 0.9; β is the line at full wavelength broadening after deducting the instrumental line broadening in radians (FWHM) at half of the maximum intensity (FWHM); and θ is the Bragg angle. For colloidal QDs with sizes less than 20 nm, the diffraction peak broadening is quite ordered due to the reduced crystal size. Nevertheless, the SAED technique is more accurate for estimating the crystal structure of particularly small QDs due to its ability to isolate the diffraction pattern of single crystals.

By analyzing the diffraction pattern, it is possible to determine the crystal structure, orientation, and size of QD samples. This information can be used to identify the QD material and the crystallographic phase, as well as to determine the size and shape of the QDs. Additionally, X-ray diffractometry can be used to study the effects of thermal and chemical treatments on QD samples and to determine the degree of crystallographic disorder or amorphous character of the QD material.

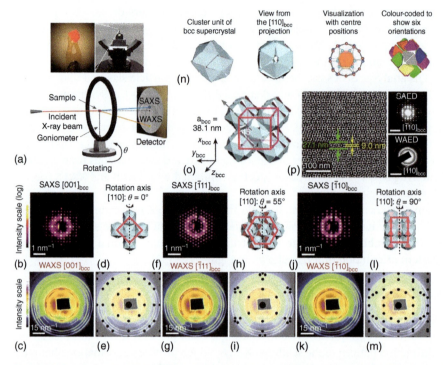

Figure 3.23 Characterization of 3D cluster-based bcc single supercrystals. (a) Photograph of a piece of bcc supercrystal loaded on a goniometer (top) and a schematic illustration of the synchrotron-based rotational X-ray scattering setup (bottom). (b–m) 2D images of SAXS (b, f, j) and WAXS (c, g, k) patterns at three representative crystallographic orientations: $[001]_{bcc}$ (b–e), $[\bar{1}11]_{bcc}$ (f–i) and $[\bar{1}10]_{bcc}$ (j–m). The simulated WAED patterns (e, i, m) were generated from computer models (d, h, l) at the three crystallographic orientations. (n) Computer-generated models of a cluster unit from the $[001]_{bcc}$ projection (left) and from the $[110]_{bcc}$ projection (middle left), a polyhedron ($3^{24}\,5^{12}\,6^2$) created by connecting TTQD center points (middle right), and a cluster model with six-crystal orientation domains classified by color (right). (o) Computer-generated model of a unit cell of the 3D cluster-based bcc supercrystals. (p) TEM image of a monolayer of the cluster unit viewed from the $[\bar{1}10]_{bcc}$ projection (left) and the corresponding SAED (top right; scale bar, 0.1 nm^{-1}) and WAED (bottom right; scale bar, 2 nm^{-1}) patterns. Source: Nagaoka et al. [59]/Springer Nature.

3.3.8 X-Ray Absorption Fine Structure Spectra

XAFS spectroscopy can be used to study the local atomic structure of materials, including QD films. XAFS measures the absorption of X-rays by a sample as a function of energy and provides information about the local atomic environment around a specific element in the sample.

XAFS is a powerful technique that is widely used in materials science research and can provide valuable information about the local atomic structure of a wide range of materials, including QD films. XAFS spectroscopy and extended X-ray absorption fine structure analysis (EXAFS) have been used to analyze surfactant-free ZnS

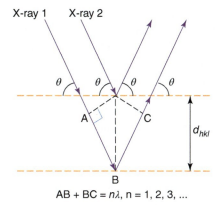

Figure 3.24 Schematic diagram of Bragg's law.

QDs, showing a reversible structural transformation during methanol desorption and water adsorption that significantly reduces surface and interior disorder [62]. Colloidal InAs QDs have also been analyzed using X-ray absorption near edge spectroscopy (XANES), providing evidence for surface reconstruction and relaxation [63].

3.3.9 Measurement of Fluorescence Quantum Yield

The measurement of QD fluorescence QY is a commonly used technique to evaluate the efficiency of QDs as light-emitting materials. The fluorescence QY is defined as the ratio of the number of photons emitted by a fluorophore to the number of photons absorbed by the same fluorophore. There are several methods to measure the QY of QDs, including the integrating sphere method, the comparative method, and the absolute method. The integrating sphere method is a widely used approach that involves measuring the total fluorescence intensity of a sample using a spectrometer equipped with an integrating sphere. The comparative method involves comparing the fluorescence of the sample to a reference sample of a known QY. The absolute method uses a spectrophotometer to measure the absorbance of the sample, and a calibrated fluorometer to measure the fluorescence intensity.

In general, the measurement of QD fluorescence QY requires careful selection and preparation of the sample, accurate instrumentation, and precise calibration of the experimental setup. The fluorescence QY of a molecule or material is defined as the ratio of the number of photons emitted to the number of photons absorbed and is an important characteristic of luminescent materials used to understand the molecular behavior and interactions of many key materials.

3.4 Conclusion and Outlook

The unique properties of colloidal QDs have attracted significant attention in recent decades, making the synthesis of zero-dimensional nanodots, one-dimensional nanorods or nanowires, and two-dimensional nanosheets or nanoribbons one of the hot research topics. The progressive advancements in colloidal synthesis, material characterization methods, and fundamental physical and chemical concepts have propelled QDs toward large-scale commercialization in various fields, such as display and lighting. This chapter aims to provide a comprehensive and fundamental understanding of colloidal QDs, covering two key aspects: colloidal synthesis and material characterization.

However, several challenges still need to be addressed for the future development of colloidal QDs, including:

(i) Enhancing the synthesis chemistry of cadmium-free and heavy metal-free colloidal QDs to achieve good stability and high fluorescence QYs across the entire visible spectrum of the fluorescence emission band, particularly in the blue light region;

(ii) Scaling up synthesis processes to meet the material demands of the optoelectronics market. This includes the selection of appropriate ligands and precursors, and adopting more environmentally friendly synthesis methods to reduce the cost of manufacturing highly monodisperse QDs while maintaining high quality. Implementation of continuous flow automated synthesis equipment, along with machine learning and artificial intelligence, could also significantly reduce production costs;

(iii) Addressing the issue of luminescence degradation that occurs when QDs are used as a light-emitting layer in light-emitting diodes;

(iv) Further improving the optoelectronic properties of colloidal QDs, such as enhancing their PLQY, increasing their carrier mobility, and reducing the number of trap states, in order to achieve even higher device performance and efficiency;

(v) Exploring new applications of colloidal QDs, such as in quantum computing, single photon emitters, and advanced sensing and imaging technologies, to expand the potential markets for these materials.

In terms of synthesis, there has been significant progress in developing methods for producing high-quality colloidal QDs with precisely controlled size, shape, and composition. This has enabled the preparation of QDs with uniform emission properties, narrow size distributions, and high PLQYs, which are essential for the development of efficient and stable LEDs. Furthermore, there have been advances in the synthesis of environmentally friendly and nontoxic colloidal QDs that can replace traditional toxic QDs in optoelectronic applications.

In terms of characterization, researchers have been developing advanced techniques to investigate the optical and electronic properties of colloidal QDs. For example, transient absorption spectroscopy and time-resolved photoluminescence spectroscopy have been used to study the dynamics of charge carrier relaxation and recombination in QDs, which is important for understanding the fundamental properties of QD materials and optimizing their use in LEDs.

The future of colloidal QDs is bright, with ongoing research and development pushing the boundaries of what is possible with these unique and versatile materials. With continued improvements in synthesis and characterization, as well as advancements in device architectures and applications, colloidal QDs are poised to play a significant role in many areas of science and technology in the years to come.

Looking toward the future, the development of new synthesis and characterization techniques will be essential for the continued progress in the field of colloidal QDLEDs. For example, the development of methods for producing QDs with improved stability and durability, as well as the ability to fabricate high-performance LEDs on a large scale, will be important for commercializing this technology. Additionally, the integration of QDs with other materials and structures to enhance their performance in LEDs is an exciting area of research, such as the use of QD-nanowire hybrid structures or plasmonic nanoparticles to improve the efficiency of QDLEDs.

References

1 Walter, P., Welcomme, E., Hallégot, P. et al. (2006). Early use of PbS nanotechnology for an ancient hair dyeing formula. *Nano Letters* 6 (10): 2215–2219.
2 Ekimov, A.I. and Onushchenko, A.A. (1981). Quantum dimensional effect in three-dimensional microcrystals of semiconductors. *Pis'ma v Zhurnal Eksperimental'noi i Teoreticheskoi Fiziki* 34: 363–366.
3 Efros, A.L. and Efros, A.L. (1982). Interband absorption of light in a semiconductor sphere. *Soviet Physics Semiconductors-USSR* 16: 772.
4 Rossetti, R., Ellison, J.L., Gibson, J.M., and Brus, L.E. (1984). Size effects in the excited electronic states of small colloidal CdS crystallites. *The Journal of Chemical Physics* 80 (9): 4464–4469.
5 Reed, M.A., Randall, J.N., Aggarwal, R.J. et al. (1988). Observation of discrete electronic states in a zero-dimensional semiconductor nanostructure. *Physical Review Letters* 60: 535.
6 Li, L.-s., Hu, J., Yang, W., and Alivisatos, A.P. (2001). Band gap variation of size- and shape-controlled colloidal CdSe quantum rods. *Nano Letters* 1 (7): 349–351.
7 Brus, L. (1986). Electronic wave functions in semiconductor clusters: experiment and theory. *The Journal of Physical Chemistry* 90 (12): 2555–2560.
8 Owen, J. and Brus, L. (2017). Chemical synthesis and luminescence applications of colloidal semiconductor quantum dots. *Journal of the American Chemical Society* 139 (32): 10939–10943.
9 Pu, Y., Cai, F., Wang, D. et al. (2018). Colloidal synthesis of semiconductor quantum dots toward large-scale production: a review. *Industrial and Engineering Chemistry Research* 57 (6): 1790–1802.
10 Qu, L., Peng, Z.A., and Peng, X. (2001). Alternative routes toward high quality CdSe nanocrystals. *Nano Letters* 1 (6): 333–337.
11 Ghosh, S. and Manna, L. (2018). The many "facets" of halide ions in the chemistry of colloidal inorganic nanocrystals. *Chemical Reviews* 118 (16): 7804–7864.
12 Xing, B., Li, W., Dou, H. et al. (2008). Systematic study of the properties of CdSe quantum dots synthesized in paraffin liquid with potential application in multiplexed bioassays. *The Journal of Physical Chemistry C* 112 (37): 14318–14323.

13 Pradhan, N. and Efrima, S. (2003). Single-precursor, one-pot versatile synthesis under near ambient conditions of tunable, single and dual band fluorescing metal sulfide nanoparticles. *Journal of the American Chemical Society* 125 (8): 2050–2051.

14 Cao, Y.C. and Wang, J. (2004). One-pot synthesis of high-quality zinc-blende CdS nanocrystals. *Journal of the American Chemical Society* 126 (44): 14336–14337.

15 Liu, T.-Y., Li, M., Ouyang, J. et al. (2009). Non-injection and low-temperature approach to colloidal photoluminescent PbS nanocrystals with narrow bandwidth. *The Journal of Physical Chemistry C* 113 (6): 2301–2308.

16 Du, Y., Xu, B., Fu, T. et al. (2010). Near-infrared photoluminescent Ag_2S quantum dots from a single source precursor. *Journal of the American Chemical Society* 132 (5): 1470–1471.

17 Sun, M. and Yang, X. (2009). Phosphine-free synthesis of high-quality CdSe nanocrystals in noncoordination solvents: "activating agent" and "nucleating agent" controlled nucleation and growth. *The Journal of Physical Chemistry C* 113 (20): 8701–8709.

18 Manna, L., Scher, E.C., and Alivisatos, A.P. (2000). Synthesis of soluble and processable rod-, arrow-, teardrop-, and tetrapod-shaped CdSe nanocrystals. *Journal of the American Chemical Society* 122 (51): 12700–12706.

19 Peng, X., Wickham, J., and Alivisatos, A.P. (1998). Kinetics of II–VI and III–V colloidal semiconductor nanocrystal growth: "focusing" of size distributions. *Journal of the American Chemical Society* 120 (21): 5343–5344.

20 Li, L.S., Pradhan, N., Wang, Y., and Peng, X. (2004). High quality ZnSe and ZnS nanocrystals formed by activating zinc carboxylate precursors. *Nano Letters* 4 (11): 2261–2264.

21 Gao, Y. and Peng, X. (2014). Crystal structure control of CdSe nanocrystals in growth and nucleation: dominating effects of surface versus interior structure. *Journal of the American Chemical Society* 136 (18): 6724–6732.

22 Li, Y., Pu, C., and Peng, X. (2017). Surface activation of colloidal indium phosphide nanocrystals. *Nano Research* 10 (3): 941–958.

23 Kim, H.H., Park, S., Yi, Y. et al. (2015). Inverted quantum dot light emitting diodes using polyethylenimine ethoxylated modified ZnO. *Scientific Reports* 5: 8968.

24 Joo, J., Son, J.S., Kwon, S.G. et al. (2006). Low-temperature solution-phase synthesis of quantum well structured CdSe nanoribbons. *Journal of the American Chemical Society* 128 (17): 5632–5633.

25 Liu, Y.-H., Wang, F., Wang, Y. et al. (2011). Lamellar assembly of cadmium selenide nanoclusters into quantum belts. *Journal of the American Chemical Society* 133 (42): 17005–17013.

26 Rice, K.P., Saunders, A.E., and Stoykovich, M.P. (2013). Seed-mediated growth of shape-controlled wurtzite CdSe nanocrystals: platelets, cubes, and rods. *Journal of the American Chemical Society* 135 (17): 6669–6676.

27 Wang, Y., Zhou, Y., Zhang, Y., and Buhro, W.E. (2015). Magic-size II–VI nanoclusters as synthons for flat colloidal nanocrystals. *Inorganic Chemistry* 54 (3): 1165–1177.

28 Zhou, Y., Jiang, R., Wang, Y. et al. (2019). Isolation of amine derivatives of (ZnSe)$_{34}$ and (CdTe)$_{34}$. Spectroscopic comparisons of the (II–VI)$_{13}$ and (II–VI)$_{34}$ magic-size nanoclusters. *Inorganic Chemistry* 58 (3): 1815–1825.

29 Zhang, H. and Banfield, J.F. (2012). Energy calculations predict nanoparticle attachment orientations and asymmetric crystal formation. *The Journal of Physical Chemistry Letters* 3 (19): 2882–2886.

30 Schliehe, C., Juarez, B.H., Pelletier, M. et al. (2010). Ultrathin PbS sheets by two-dimensional oriented attachment. *Science* 329 (5991): 550.

31 Chen, Y., Chen, D., Li, Z., and Peng, X. (2017). Symmetry-breaking for formation of rectangular CdSe two-dimensional nanocrystals in zinc-blende structure. *Journal of the American Chemical Society* 139 (29): 10009–10019.

32 Shu, Y., Lin, X., Qin, H. et al. (2020). Quantum dots for display applications. *Angewandte Chemie International Edition* 59 (50): 22312–22323.

33 Shu, Y., Lin, X., Qin, H. et al. (2020). Quantum dots for display applications. *Angewandte Chemie* 132 (50): 22496–22507.

34 Sagar, L.K., Bappi, G., Johnston, A. et al. (2020). Suppression of auger recombination by gradient alloying in InAs/CdSe/CdS QDs. *Chemistry of Materials* 32 (18): 7703–7709.

35 Anderson, N.C., Hendricks, M.P., Choi, J.J., and Owen, J.S. (2013). Ligand exchange and the stoichiometry of metal chalcogenide nanocrystals: spectroscopic observation of facile metal-carboxylate displacement and binding. *Journal of the American Chemical Society* 135 (49): 18536–18548.

36 Yang, Y., Qin, H., Jiang, M. et al. (2016). Entropic ligands for nanocrystals: from unexpected solution properties to outstanding processability. *Nano Letters* 16 (4): 2133–2138.

37 Yu, W.W., Qu, L., Guo, W., and Peng, X. (2003). Experimental determination of the extinction coefficient of CdTe, CdSe, and CdS nanocrystals. *Chemistry of Materials* 15 (14): 2854–2860.

38 Shaviv, E., Salant, A., and Banin, U. (2009). Size dependence of molar absorption coefficients of CdSe semiconductor quantum rods. *ChemPhysChem* 10 (7): 1028–1031.

39 Xia, C., Wu, W., Yu, T. et al. (2018). Size-dependent band-gap and molar absorption coefficients of colloidal CuInS$_2$ quantum dots. *ACS Nano* 12 (8): 8350–8361.

40 Dai, Q., Wang, Y., Li, X. et al. (2009). Size-dependent composition and molar extinction coefficient of PbSe semiconductor nanocrystals. *ACS Nano* 3 (6): 1518–1524.

41 Moreels, I., Lambert, K., Smeets, D. et al. (2009). Size-dependent optical properties of colloidal PbS quantum dots. *ACS Nano* 3 (10): 3023–3030.

42 Peters, J.L., de Wit, J., and Vanmaekelbergh, D. (2019). Sizing curve, absorption coefficient, surface chemistry, and aliphatic chain structure of PbTe nanocrystals. *Chemistry of Materials* 31 (5): 1672–1680.

43 Lin, S., Li, J., Pu, C. et al. (2020). Surface and intrinsic contributions to extinction properties of ZnSe quantum dots. *Nano Research* 13 (3): 824–831.

44 Ithurria, S., Tessier, M.D., Mahler, B. et al. (2011). Colloidal nanoplatelets with two-dimensional electronic structure. *Nature Materials* 10 (12): 936–941.

45 Bertrand, G.H.V., Polovitsyn, A., Christodoulou, S. et al. (2016). Shape control of zincblende CdSe nanoplatelets. *Chemical Communications* 52 (80): 11975–11978.

46 Wang, Y., Pu, C., Lei, H. et al. (2019). CdSe@CdS dot@platelet nanocrystals: controlled epitaxy, monoexponential decay of two-dimensional exciton, and nonblinking photoluminescence of single nanocrystal. *Journal of the American Chemical Society* 141 (44): 17617–17628.

47 Hens, Z., Moreels, I., and Martins, J.C. (2005). In situ ^{1}H NMR study on the trioctylphosphine oxide capping of colloidal InP nanocrystals. *ChemPhysChem* 6 (12): 2578–2584.

48 Owen, J.S., Park, J., Trudeau, P.-E., and Alivisatos, A.P. (2008). Reaction chemistry and ligand exchange at cadmium–selenide nanocrystal surfaces. *Journal of the American Chemical Society* 130 (37): 12279–12281.

49 Gomes, R., Hassinen, A., Szczygiel, A. et al. (2011). Binding of phosphonic acids to CdSe quantum dots: a solution NMR study. *The Journal of Physical Chemistry Letters* 2 (3): 145–152.

50 Hens, Z. and Martins, J.C. (2013). A solution NMR toolbox for characterizing the surface chemistry of colloidal nanocrystals. *Chemistry of Materials* 25 (8): 1211–1221.

51 Anderson, N.C. and Owen, J.S. (2013). Soluble, chloride-terminated CdSe nanocrystals: ligand exchange monitored by 1H and 31P NMR spectroscopy. *Chemistry of Materials* 25 (1): 69–76.

52 Ji, X., Copenhaver, D., Sichmeller, C., and Peng, X. (2008). Ligand bonding and dynamics on colloidal nanocrystals at room temperature: the case of alkylamines on CdSe nanocrystals. *Journal of the American Chemical Society* 130 (17): 5726–5735.

53 Chen, P.E., Anderson, N.C., Norman, Z.M., and Owen, J.S. (2017). Tight binding of carboxylate, phosphonate, and carbamate anions to stoichiometric CdSe nanocrystals. *Journal of the American Chemical Society* 139 (8): 3227–3236.

54 Pang, Z., Zhang, J., Cao, W. et al. (2019). Partitioning surface ligands on nanocrystals for maximal solubility. *Nature Communications* 10 (1): 2454.

55 Piveteau, L., Ong, T.-C., Walder, B.J. et al. (2018). Resolving the core and the surface of CdSe quantum dots and nanoplatelets using dynamic nuclear polarization enhanced PASS–PIETA NMR spectroscopy. *ACS Central Science* 4 (9): 1113–1125.

56 Clark, P.C.J. and Flavell, W.R. (2019). Surface and interface chemistry in colloidal quantum dots for solar applications studied by X-ray photoelectron spectroscopy. *The Chemical Record* 19 (7): 1233–1243.

57 Subila, K.B., Kishore Kumar, G., Shivaprasad, S.M., and George Thomas, K. (2013). Luminescence properties of CdSe quantum dots: role of crystal structure and surface composition. *The Journal of Physical Chemistry Letters* 4 (16): 2774–2779.

58 Caetano, B.L., Santilli, C.V., Meneau, F. et al. (2011). In situ and simultaneous UV–vis/SAXS and UV–vis/XAFS time-resolved monitoring of ZnO quantum dots formation and growth. *The Journal of Physical Chemistry C* 115 (11): 4404–4412.

59 Nagaoka, Y., Tan, R., Li, R. et al. (2018). Superstructures generated from truncated tetrahedral quantum dots. *Nature* 561 (7723): 378–382.

60 Patterson, A. (1939). The Scherrer formula for X-ray particle size determination. *Physical Review* 56 (10): 978.
61 Holzwarth, U. and Gibson, N. (2011). The Scherrer equation versus the 'Debye-Scherrer equation'. *Nature Nanotechnology* 6 (9): 534.
62 Tajoli, F., Dengo, N., Mognato, M. et al. (2020). Microfluidic crystallization of surfactant-free doped zinc sulfide nanoparticles for optical bioimaging applications. *ACS Applied Materials and Interfaces* 12 (39): 44074–44087.
63 Ishii, M., Ozasa, K., and Aoyagi, Y. (2003). Selective X-ray absorption spectroscopy of self-assembled atom in InAs quantum dot. *Microelectronic Engineering* 67–68: 955–962.

4

Red Quantum Dot Light-Emitting Diodes

4.1 Background

Red quantum dot light-emitting diodes (QDLEDs), which emit light in the 620–760 nm range, are an important component of QDLEDs that cover the visible spectrum, along with blue and green QDLEDs. Figure 4.1 displays the progress of red QDLEDs since their invention with the horizontal axis indicating the year and the vertical axis indicating the external quantum efficiency (EQE), and the center point of each colored circle representing the EQE of a device and the year of publication. As shown in the figure, the highest EQE for red QDLEDs has improved significantly from less than 0.2% in 1997 to 23.1% in 2018. The graph also illustrates that around 2015, the EQE of QDLEDs experienced a rapid increase due to innovations in device structure, such as the use of inorganic oxides (ZnO/SnO$_2$) as the electron transport layer. Consequently, QDLED device performance, such as efficiency and brightness, has increased rapidly.

In this chapter, a detailed overview of the fundamental concepts and recent advancements in red QDLED materials and devices will be provided. Two main aspects will be introduced: firstly, the commonly used light-emitting materials for red QDLEDs, such as CdSe and CdTe, and their physicochemical properties. Various strategies to optimize the material properties, such as surface ligand modifications, formation of core–shell structures, and other techniques, will be discussed. Secondly, the evolution and development of red QDLED device structures, including the use of inorganic oxides as electron transport layers, and other recent innovations will be highlighted. Finally, the challenges and opportunities that red QDLEDs are facing will be summarized.

Here are some key breakthrough technology developments in red QDLED during the years 2000 and 2023, along with some device performance data:

- 2000: The first red QDLEDs were reported using core–shell CdSe/ZnS QDs, achieving an EQE of 0.01% and a peak emission wavelength of around 620 nm;
- 2007: A red QDLED with an EQE of 0.23% was reported using a multi-layered QD structure, where the QD layers were separated by organic spacer layers. The peak emission wavelength was around 630 nm;

Colloidal Quantum Dot Light Emitting Diodes: Materials and Devices, First Edition. Hong Meng.
© 2024 WILEY-VCH GmbH. Published 2024 by WILEY-VCH GmbH.

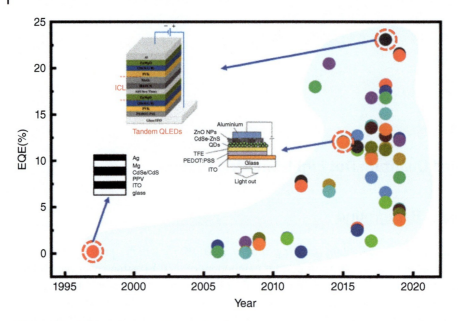

Figure 4.1 EQE development for red QDLEDs.

- 2010: A red QDLED with an EQE of 4.4% was reported using a CdSeTe/CdS QD structure with an optimized ZnO electron injection layer. The peak emission wavelength was around 635 nm;
- 2013: A red QDLED with an EQE of 9.4% was reported using a CdSe/CdZnS QD structure with an optimized electron injection layer. The peak emission wavelength was around 640 nm.
- 2016: A red QDLED with an EQE of 19.3% was reported using a CdSe/CdZnS QD structure and an optimized hole transport layer. The peak emission wavelength was around 650 nm.
- 2020: A red QDLED with an EQE of 23.1% was reported using a CdSe/CdS/ZnS QD structure and a hole-transporting host-guest system. The peak emission wavelength was around 650 nm.

4.2 Red Light Quantum Dot Materials

The core component of a QDLED is the quantum dot (QD) light-emitting material. QDs are typically referred to as nanosized crystals that are synthesized and processed in a solution [1]. These are colloidal QDs that are uniformly dispersed in a solution and have a layer of organic ligands covering their surfaces. The ligands are attached to the surface of the dots via ligand bonds.

The most commonly used QDs are semiconductor nanoparticles of group II–VII (CdSe, CdS, ZnSe, CdS, PbS, PbSe), group III–VI (InP, InAs), or group I–III–VII (CuInS$_2$, AgInS$_2$). By tuning the size, composition, and morphology of the QDs,

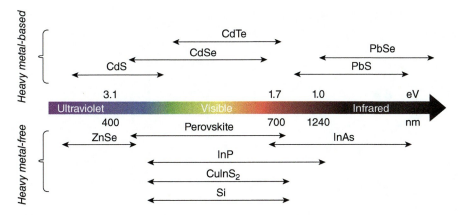

Figure 4.2 Fluorescence emission range of QDs. Source: Shang and Ning [4]/CC BY 4.0/Public domain/Oxford University Press.

a range of quantum dots with different bandgaps can be obtained, leading to light emission at different wavelengths [2, 3]. Quantum dots are available in a range of wavelengths, with the upper part of the spectrum consisting of semiconductor QDs containing heavy metals, and the lower part of the spectrum containing no heavy metals. Among them, the group of II–VI semiconductor QDs that emit in the red-light spectrum (640–760 nm) include CdSe, CdTe, perovskite, InAs, InP, Si, and $CuInS_2$. Among the many semiconductor QDs that can emit red light, group II–VI semiconductor QDs are more mature and have better performance, especially those containing cadmium elements (CdSe, CdTe), which are the most promising light-emitting materials that can enter the market. Thus, this section will focus on CdSe and CdTe semiconductor QDs. Figure 4.2 illustrates the wavelength range of these QDs.

4.2.1 Materials

CdTe bulk material has a sphalerite crystal structure with a forbidden bandgap of 1.49 eV and CdSe bulk material has a zinc blende crystal structure with a forbidden bandgap of 1.75 eV. CdSe also has a wurtzite crystal structure with a larger forbidden bandgap of 1.9 eV. The choice of crystal structure and size can significantly affect the optical and electronic properties of the QD material (Figure 4.3).

Researchers used single-component QDs as a luminescent layer to achieve electroluminescence. In 1994, Alivisatos et al. used a polymer-QD bilayer structure, (ITO/CdSe/PPv/Mg and ITO/PPv/CdSe/Mg), where CdSe QDs acted as both the light-emitting and electron-transporting layers, to achieve the first QDLED in history with a tunable light-emitting color from red to yellow and a device efficiency of only 0.22% [5]. The device efficiency was very low because the exciton was not well confined to the QD layer. This was due to the strong parasitic emission of the polymer on the electroluminescence spectrum.

Passivation of QDs involves the use of surface ligands to cover the surface of the QDs, which can help to reduce the density of surface defects and trap states. This

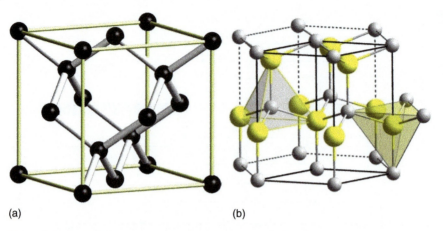

Figure 4.3 Crystal structures of zinc blende (a) and wurtzite (b).

passivation layer can also improve the stability and prevent agglomeration of the QDs. Commonly used ligands include organic molecules such as oleic acid, oleylamine, and trioctylphosphine oxide (TOPO) and inorganic ligands such as halides and chalcogenides. By modifying the surface of the QDs, the energy level of the surface states can be controlled and reduced, leading to improved radiative recombination efficiency and higher photoluminescence (PL) intensity. Additionally, the surface modification of QDs can also improve the compatibility of QDs with other materials and enable better charge transfer and transport in device structures [6, 7], which has been demonstrated in many studies.

4.2.2 Quantum Dot Structure Design and Optimization

Surface ligand modification is one approach to reduce or eliminate the surface defects of QDs. This involves replacing the original ligands with new ones that have better passivating abilities. The new ligands can effectively passivate the surface defects of QDs and thus improve their luminescence stability and quantum yield. For example, the commonly used original ligands for CdSe QDs are TOPO and hexadecylamine (HDA). These ligands can effectively protect the QDs during the synthesis process, but they cannot passivate the surface defects of QDs very well. To improve the luminescence efficiency and stability of CdSe QDs, new ligands such as mercaptopropionic acid (MPA), thioglycolic acid (TGA), and 3-mercaptopropyl trimethoxysilane (MPTMS) have been used to replace the original ligands. These new ligands have better passivating abilities and can effectively reduce the surface defects of QDs, resulting in improved luminescence stability and quantum yield [8–11]. Growth of a shell layer around the QDs can help to passivate the surface defects and improve the quantum yield and stability of the QDs. Core–shell QDs are typically made by first synthesizing the core of the QD, which is then coated with a shell material through various techniques such as successive ionic layer adsorption and reaction (SILAR), cation exchange, and thermal decomposition. The shell material can be chosen based on its ability to passivate the surface defects, and can also provide additional functionality such as chemical stability, improved charge transport, or enhanced quantum confinement [12–15]. Whether ligand modification on the surface of the QDs or covering them with wide bandgap materials, the aim is to eliminate defects in the QDs. This effectively confines electrons and holes within the QD and enables the electron wave function of the conduction band to be off-domain to the shell material, reducing the crossover of the hole and electron wave functions within the QD and improving the exciton radiation complex lifetime. By eliminating defects in the QDs, the exciton can be better confined within the QD, reducing non-radiative recombination pathways and increasing the radiative recombination rate. This leads to improved PL efficiency and stability. The use of surface ligand modification and core–shell structures are two common methods to achieve this goal.

4.2.3 Surface Ligands

Surface ligands on QDs play a crucial role in their stability, solubility, and optical properties. They act as a protective layer that passivates surface defects and prevents aggregation of the QDs. Additionally, the ligands provide a barrier between the QDs and the surrounding environment, preventing degradation due to exposure to air or moisture. The choice of ligands can also affect the electronic and optical properties of the QDs, such as their bandgap and luminescence. Therefore, ligands can be used to tune the properties of the QDs to meet specific application requirements as follows:

(1) Ligands are able to eliminate surface defects. Recently, Peng Xiaogang et al. pointed out that non-passivated S-sites and surface-adsorbed H_2S are two very deep hole-trapping centers for CdSe/CdS QDs, as shown in Figure 4.4a [16]. These hole traps can be eliminated by a two-step surface treatment, consisting of the removal of surface-adsorbed H_2S and surface-passivation of CdSe to eliminate unpassivated surface-S sites. This work demonstrates that suppressing surface traps by eliminating surface states is an effective way to improve the photoluminescence quantum yield (PLQY) of QDs, by reducing non-radiative quenching.
(2) The ligands play a crucial role in charge transfer and transport in QD devices. Studies have demonstrated that short-chain ligands are more favorable for efficient charge transfer and hole injection. The advancements in surface ligand chemistry have enabled simple ligand substitution reactions and the ability to design surface ligands for QDs in light-emitting diodes (LEDs) [20]. As shown in Figure 4.4b, Sun et al. used linker molecules of varying lengths (ranging from 3 to 8 –CH_2 groups) to adjust the distance between adjacent PbS QDs, in order to replace the long-chain ligands of the original oleate [17]. The precisely controlled distances between QDs optimized the balance between charge injection/transport and exciton radiation complexation, resulting in a significant improvement in the efficiency of the infrared LED. Shen et al. used 1-octanethiol instead of the pre-synthesized oleic acid ligand for blue QDs [21]. The authors concluded that the shorter 1-octanethiol ligand, which reduces

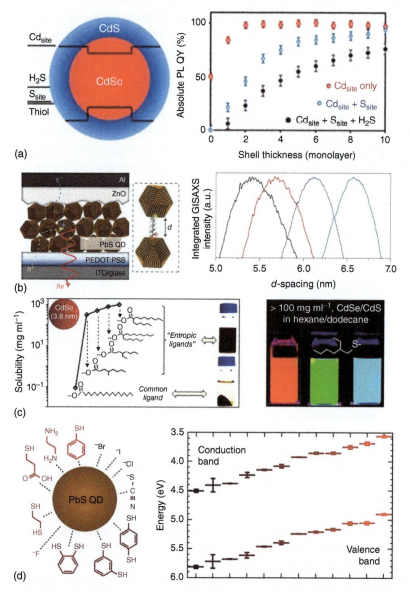

Figure 4.4 Quantum dot ligand engineering. (a) Schematic of surface-induced hole trapping in CdSe/CdS quantum dots (left) and its effect on PLQYs of CdSe/CdS quantum dots (right). Source: Pu and Peng [16]/American Chemical Society. (b) Schematic diagram of the device structure of NIR QDLEDs (left) and the inter-dot distance finely tailored in the QD film by varying the ligand length (right). Source: Sun et al. [17]/Springer Nature.
(c) Solubility at 303 K of 3.8 nm CdSe quantum dots coated with five different ligands (left) and a photograph of the CdSe/fluorescent ink. Source: Yang et al. [18]/American Chemical Society. (d) Schematic diagram of the various ligands of PbS quantum dots (left) and the corresponding energy level diagram of the PbS QD film (right). Source: Brown et al. [19]/American Chemical Society.

the point-to-point distance, improves hole injection into the QDs from the polymer hole transport layers and electron transport within the QD film. These advantages result in a lower turn-on voltage of 2.6 V and a maximum EQE of 12.2% for blue QDLEDs, which is a 70% increase in maximum EQE compared to devices using QDs with oleic acid pre-synthesized ligands.

There appears to be a trade-off between maintaining colloidal stability in the QD solution and optimizing the charge transport properties of the QD film. On one hand, shorter ligands can improve charge transport, leading to more efficient driving of QDLEDs [22]. On the other hand, in the conventional model, the spatial separation between the QDs provided by the surface ligands is responsible for the colloidal stability of the QD solution [23]. The surface ligands provide steric stabilization to the QDs in solution by creating a barrier between the particles and preventing them from aggregating or flocculating. The steric stabilization is achieved through repulsive forces between the ligands and between the ligands and solvent molecules, which counteract the attractive van der Waals forces between the QDs. So, in the conventional model, a balance needs to be struck between the ligand size, which affects charge transport properties, and the ligand coverage, which affects colloidal stability of the QD solution. Smaller ligands may reduce the solubility of QDs. In 2016, Peng and coworkers introduced the concept of entropic ligands, as shown in Figure 4.4c [18, 24]. The researchers used CdSe nanocrystals with n-alkanoic acid ligands as their experimental system and discovered the size- and temperature-dependent solubility of CdSe nanocrystals in organic solvents. The experimental results were quantitatively interpreted using a thermodynamic model based on the precipitation/dissolution phase transition. The solubility of QDs in molar volume was measured, and χ can be expressed by a simple Eq. (4.1).

$$\chi = e^{-\Delta^m H_{NC}/RT} e^{\Delta^m S_{NC}/R} \tag{4.1}$$

where e is the natural logarithm; $\Delta^m H_{NC}$ and $\Delta^m S_{NC}$ represent the partial molar mixing enthalpy and molar conformational entropy of QDs dissolved in the liquid, respectively; R is the ideal gas function, and T is the thermodynamic temperature of the ideal gas. The analysis revealed that the conformational entropy of the n-alkanoate chains released during dissolution, that is, the rotation and bending entropy associated with the C—C bond, exponentially increased the solubility of the QDs, while in the solids between adjacent particles strong chain interactions reduced solubility. This discovery leads to alkyl chains of entropic ligands with irregular branches that maximize intramolecular entropy and minimize the enthalpy that disrupts crystalline chain–chain interactions. The use of entropic ligands increases the solubility of CdSe nanocrystals to hundreds of mg ml^{-1}. The concept of entropic ligands proved to be effective when QDs were dissolved in various organic solvents. Therefore, the processability of colloidal nanocrystals in solution and the charge transport of the corresponding thin films can be improved simultaneously. The electrical measurement results of electronic devices show that the electrical conductivity of the CdSe nanocrystalline film coated with 2-ethyl ethyl sulfide is about 3 orders of magnitude higher than that of the CdSe nanocrystalline film coated with octadecyl sulfide. QLEDs were fabricated using CdSe/CdS core/shell QDs with different ligands, and the results showed that the use of ligand 2-ethyl ethyl sulfide increased the power efficiency of QLEDs by 30% due to the improved charge transport in the QD films.

The charge transport properties of QD films can also be tuned by using specially designed polymeric ligands with conducting and anchoring groups instead of synthetic insulating ligands. In addition to controlling the energy level structure of QD films, using specially designed polymeric ligands with conducting and anchoring groups can also improve the charge transport properties of the QD films, which is important for the performance of QDLEDs. However, the use of hybrid polymers requires careful consideration to avoid unwanted bursts or parasitic emissions that can reduce the overall efficiency of the devices.

(3) ligands affect the overall energy level structure of QD films.

Brown et al. demonstrated that ligand-induced surface dipoles are an effective strategy for controlling the absolute energy level of QD films, as shown in Figure 4.4d [19]. The authors showed that ligand exchange processing on PbS QDs membranes resulted in energy level transfers of up to 0.9 eV. The strength of the ligand-induced surface dipole can be modulated by the chemical binding group and dipole moment of the ligand. This finding allows for fine-band energy calibration of the PbS QDs used in solar cells, thereby improving the efficiency of the devices. This strategy was used in the preparation of QDLEDs by Yang et al. in combination with dimensional control of QDs to demonstrate the fine-tuning of bandgaps and band positions of PbS QDs films, allowing the QDs films to act as electron transport, emission, and hole transport layers in a single LED configuration [25].

4.2.4 Core–Shell Structure

Reiss et al. classified the core–shell QDs into three types, as shown in Figure 4.5, based on the relative positions of the forbidden bands and electron energy levels

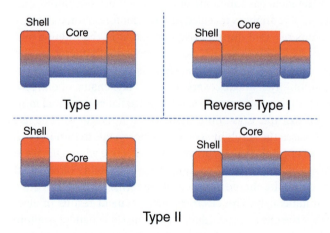

Figure 4.5 Classification of core–shell quantum dots (relationship between core–shell bandgaps). Source: Adapted from Reiss et al. [26].

of the semiconductor material [26]. These are type I, type II, and anti-type I structures.

In the type I structure, the forbidden band width of the shell material is greater than the forbidden bandwidth of the QD core. The shell is used to passivate the surface of the core to improve its optical properties. This QD shell physically separates the optically active core from its surrounding medium. As a result, the sensitivity of the optical properties to the local environment of the QD surface changes. For example, due to the presence of oxygen or water molecules they are reduced. For the core of the QDs, the core–shell block structure shows excellent stability in terms of photodegradation. At the same time, the growth of the shell layer reduces the number of dangling bonds on the surface that can act as trap states for charge carriers, thus reducing PLQY. The first reported use of this structure was for CdSe/ZnS core–shell QDs. The ZnS shell layer significantly improves the PLQY and the stability of the QDs against photobleaching. The growth of the shell layer is accompanied by a small redshift (5–10 nm) of the exciton peak at UV/Vis absorption spectra and PL wavelengths due to the partial leakage of excitons into the shell material.

The synthesis of highly luminescent CdSe@ZnS core–shell QDs with CdSe cores between 2.3 and 5.5 nm in diameter was first reported by B.O. Dabbousi et al. in 1997 [12]. The narrow PL (half-peak width less than 40 nm) of these synthesized dots spans most of the visible spectrum from blue to red with a quantum yield of 30–50% at room temperature. These materials have been characterized using a range of optical and structural techniques. Finally, it was determined that the ZnS cores were locally epitaxially grown on CdSe cores and that the structure of the ZnS shell layer affected the PL properties. In 2018, Wang et al. similarly used CdSe/ZnS QDs as the luminescent layer and introduced a double-layer PVK/TFB hole transport layer using HTL orthogonal solvents to achieve a QDLED with record performance in a full solution method inverted structure [27]. The result is a QDLED with a full solution inversion structure and record performance. The orthogonal solvent, 1,4-dioxane, prevents the HTL from attacking the QD layer and ensures a clean and intact interface between adjacent layers. The double PVK/TFB hole transport layer provides a progressive change in energy level to facilitate hole injection. As a result, the red QDLED has a peak current efficiency of 22.1 cd A^{-1} and a maximum EQE of 12.7%. The performance of their red QDLEDs is the highest among all solution-prepared inverted structure QDLEDs.

In type II structures, the conduction band or valence band of the shell material is located in the forbidden band of the QD nucleus, depending on the thickness of the shell material. In type II structures, the goal of the shell layer growth is to cause a significant redshift in the emission wavelength of the QDs. The bandgaps are staggered, resulting in QDs with an actual effective bandgap that is smaller than both the core and the shell. The interesting aspect of this structure is that it is possible to control the thickness of the shell layer and thus tune the emission color to be close to the desired spectral range, which is difficult to achieve with a single material, making this a way to make wide-band semiconductor QDs emit red light. Type II core–shell QDs have been developed for near-infrared emission, for example, using CdTe/CdSe or CdSe/ZnTe. Compared to Type I structures, the holes and electrons in

Type II structures are generally separated in the core and shell, so that the overlap of the wave functions of electrons and holes is lower, allowing for much longer PL decay times. Since one of the carriers (electron or hole) is located in the shell layer, the PLQY and photostability of type II core–shell QDs can be improved using overgrowth of a suitable material shell as in type I structures [9, 10, 28]. The electronic states of core/shell CdSe/CdS and CdSe/CdTe heterostructured QDs have been studied by Wang et al. through large-scale first principles calculations [29]. Based on the bias arrangement of their energy bands, CdSe/CdS is a Type I heterostructure, while CdSe/CdTe is a Type II heterostructure. They found only small differences between the two electronic states, but the hole wave function of the CdSe/CdS QD has been localized within the core, while the hole wave function of the CdSe/CdTe QD is localized within the shell layer. The hole state of the CdSe/CdTe quantum dot has a very different character from that of the CdSe and CdSe/CdS QDs.

In the inverse type I structure, the material with the narrower bandgap is overgrown onto the core with the wider bandgap. The charge carriers are at least partially delocalized in the shell layer, and the emission wavelength can be tuned by the shell layer thickness. In general, the bandgap is significantly redshifted with the shell layer thickness. Such structural core–shell QDs are CdS/HgS [30], CdS/CdSe [31], and ZnSe/CdSe [8]. The photostability and PLQY of these structures can be improved by continuing to grow a second shell layer of a larger bandgap semiconductor on top of the core/shell quantum dots.

As far as the QDs themselves are concerned, the core–shell structure (CdSe/CdS) QDs have high quantum efficiency (close to 100%), good stability, and can eliminate the Blinking phenomenon. The FRET phenomenon is easily induced after the QDs are made into a film, leading to a decrease in efficiency. It has been clearly shown that the FRET phenomenon has a strong size dependence. For core–shell QDs, the larger the size of the shell the less pronounced the FRET phenomenon. Such core–shell-structured QDs are well suited for use in red QDLEDs.

4.2.5 Alloy Core–Shell Structure

QDLED devices made from pure sphalerite phase CdSe/CdS core–shell QDs have been shown to have excellent performance in the red light part, but due to the similarity of the electronic structures of CdSe and CdS, this structure only has excellent performance in the long-wave part from orange to red; in addition to using CdS as a shell, ZnS is widely used due to its much different electronic structure from CdSe. ZnS materials have been extensively studied as a shell layer due to the fact that the fluorescence properties of QDs are regulated and their luminescence is usually determined by the size of CdSe.

However, as the lattice mismatch between CdSe and ZnS is approximately 12%, this causes a concentration of stress at the interface and the formation of intrinsically defective energy levels, resulting in lower quantum yields than CdSe/CdS QDs. In order to release the lattice stresses, the introduction of an alloy at the interface between the core/shell and the ability to reduce non-radiative intermittent complexes reduces the decay rate, thus increasing the EQE of the electroluminescence.

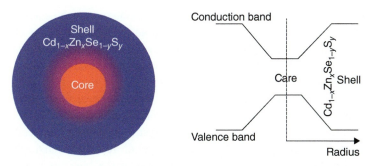

Figure 4.6 Schematic representation of the chemical composition and energy band structure of CdSe/Cd$_{1-x}$Zn$_x$Se$_{1-y}$S$_y$/ZnS alloy quantum dots. Source: Bae et al. [32]/American Chemical Society.

In this structure, the toxic Cd element is minimized and electrons and holes are confined within the core/alloy, providing more tunable colors; the alloy interface also provides a gradual potential barrier, thereby increasing electron/hole injection and improving the efficiency of the QDLED device. A typical core–shell alloy QD chemical composition and energy band structure are shown in Figure 4.6 [32].

For example, in 2018, Cao et al. used an energy design strategy to synthesize CdSe/Cd$_{1-x}$Zn$_x$Se/ZnSe QDs with a low bandgap ZnSe shell layer to achieve efficient hole injection [33]. The high PL efficiency was maintained by optimizing the composition gradient and thickness of the shells. These devices have a T$_{95}$ operating life of more than 2300 hours and an initial brightness of 1000 cd m^{-2}, which corresponds to a T$_{50}$ operating life of more than 2.2 million hours at 100 cd m^{-2}, fully meeting the industrial requirements for displays [34].

4.3 Red QDLED Devices

4.3.1 Red QDLED Device Architecture Development

Since the invention of colloidal QDLEDs in 1994, the luminance and EQE of the devices have been greatly improved, and the devices have undergone four structural developments and changes, as shown in Figure 4.7. Of course, the device structure of red QDLEDs has undergone the same development, and the following structure development will focus on the red QDLED device section [35].

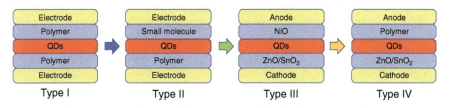

Figure 4.7 Evolution of the QDLED device structure (from Type I to Type IV).

Type I is the first QDLED device structure and utilizes a polymer as the carrier transport layer. This typical structure consists of a double layer or a mixture of both, containing CdSe single-core QDs and a polymer, sandwiched between two electrodes. However, this structure has a low EQE and a small maximum luminance due to the use of low-quantum-yield single-core CdSe QDs and the presence of significant intra-polymer parasitic electroluminescence. The most typical one was the first QDLED fabricated in 1994 [5]. The complexation of holes injected in the PPV layer and electrons injected into the CdSe nanocrystal multilayer produced the light emission. In the following year, B.O. Dabbousi et al. reported a QDLED consisting of a mixture of monodisperse CdSe QDs, polyvinyl carbazole (PVK), and oxadiazole derivatives (t-Bu PBD) as a light-emitting layer sandwiched between indium tin oxide (ITO) and Al electrodes [36]. The electroluminescence spectrum and the PL spectrum (FWHM of 40 nm) are almost identical at room temperature, and the color can be tuned by adjusting the size of the QDs to achieve red light. Electroluminescence measurements at 77K show that at low voltages only the QDs produce luminescence, while at high voltages both QDs and PVK produce electroluminescence. Also, variable temperature studies show that the lower the temperature tested, the more efficient the device.

In 2002, Coe et al. proposed a Type II device structure combining a single layer of QDs with a double layer of QDLEDs, combining the processing ease of organic materials with the efficient narrow-band luminescence of colloidal QDs [37]. This structure allows for the development of QDLEDs based on organic materials. This structure enables the addition of a single QD layer to the QDLEDS to separate the carrier transport process through the organic layer from the luminescence process, improving the EQE of the QDLEDS, with a luminescence efficiency (current efficiency of $1.6\,cd\,A^{-1}$ at $2000\,cd\,m^{-2}$) that is 25 times higher than the best previous QDLED [38]. The result is a 25-fold improvement. However, there is device leakage that can cause device instability when operating in air due to the use of organic layers.

Type III: In contrast to Structure Type II, Type III replaces the organic carrier transport layer with an inorganic carrier transport layer. This greatly improves the stability of the device in air and enables it to withstand higher current densities. Caruge et al. prepared an all-inorganic QDLED by sputtering using the inorganic carrier transport layers zinc tin oxide and nickel oxide as electron and hole transport layers, respectively [30]. The device was able to operate at current densities in excess of $3.5\,A\,cm^{-2}$ and a peak luminance of $1950\,cd\,m^{-2}$. This is a 100-fold improvement over previously reported structures. However, the efficiency of the device is not high enough due to QD destruction during sputtering of the oxide layer, carrier injection imbalance, and QD fluorescence quenching when the QD are surrounded by conducting metal oxides.

Type IV: QDLEDs with mixed organic and inorganic carrier transport layers, which is the most commonly used structure today. This structure generally uses an N-type inorganic metal oxide semiconductor as the electron transport layer and a P-type organic semiconductor as the hole transport layer. The hybrid structure of QDLEDs has high EQE and high brightness at the same time. Among them, Qian et al. reported a red hybrid structure QDLED with an EQE of 1.7% and a maximum luminance of 31 000 cd m^{-2}, respectively [31]. The mechanism of operation of Type IV QDLEDs relies more on charge injection than energy transfer. In 2019, Dongho Kim et al. developed a QDLED with InP/ZnSe/ZnS QDs as the light-emitting layer and a device structure of ITO/PEDOT:PSS/TFB/QDs/ZnMgO/Al [39]. The device can achieve an EQE of up to 21.4% and a maximum luminance of 100 000 cd m^{-2}. In addition, the device has an excellent lifetime of up to one million hours and is ready for commercialization. These performances are far superior to previously reported Cd-free QDLEDs and comparable to the most advanced Cd-based QDLEDs. The results of the study will therefore contribute to the production of Cd-free QDLEDs for next-generation displays.

A summary of the device structure and optoelectronic performance of a typical red QDLED is shown in Table 4.1. Currently, the red QDLED with the highest EQE of 23.1% is using a tandem device structure (discussed later), with each cell using a fourth device structure. There are many other efforts with EQE above 10%, most of which use a fourth device structure, further demonstrating the superiority of this device structure.

4.3.2 Common Device Structures

As described above, the fourth device structure is currently the most commonly used in QDLEDs. It should be noted that the fourth device structure comprises three different configurations: conventional, inverted, and tandem, as shown in Figure 4.8.

Conventional structures: Conventional QDLED structures typically comprise patterned ITO as the anode, an organic PEDOT:PSS layer as the hole injection layer, a polymer Poly-TPD or TFB layer as the hole transport layer, QDs as the light-emitting layer, ZnO nanocrystals as the electron transport layer, and aluminum electrodes as counter electrodes [21, 61, 69–75]. In 2009, Choi et al. utilized a similar device structure, ITO/PEDOT:PSS/TFB/QDs/ZnO/Al, as shown in Figure 4.8a, by cross-linking the colloidal QD layers and incorporating sol–gel TiO$_2$ layers for electron transport to significantly reduce the charge injection barrier in red QDLEDs [66]. This resulted in a device brightness of 12 380 cd m^{-2} and a driving voltage of only 1.9 V, with a power efficiency of 2.4 lm W^{-1}. Ten years later, Du et al. reported a red QDLED based on a CdSe/ZnSe core/shell

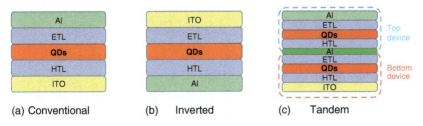

Figure 4.8 QDLED devices in (a) conventional structure; (b) inverted structure; and (c) series structure.

Table 4.1 Summary of typical QDLED device structures and their performance.

Device structure	QDs	EL max (nm)	Turn-on voltage (V)	Peak EQE (%)	Reference
ITO/PEDOT:PSS/PVK/QDs/ZnMgO/Al/HATCN/MoO$_3$/PVK/QDs/ZnMgO/Al	CdZnSe/ZnS	622	5.6	23.1	[40]
ITO/PEDOT:PSS/TFB/QDs/ZnO/Al	CdSe/ZnSe	—	—	21.6	[41]
ITO/PEDOT:PSS/TFB/QDs/ZnMgO/Al	InP/ZnSe/ZnS	630	1.8–2.0	21.4	[39]
ITO/PEDOT:PSS/Poly-TPD/PVK/QDs/PMMA/ZnO/Ag	CdSe/CdS	640	1.7	20.5	[42]
ITO/PEDOT:PSS/TFB/QDs/ZnO/Al	CdSe/Cd$_{1-x}$Zn$_x$Se/ZnSe	~625	1.7	15.1	[34]
ITO/PEDOT:PSS/Poly-TPD/PVK/QDs/PMMA/Zn$_{0.9}$Mg$_{0.1}$O/Al	CdSe/CdZnS	624	—	15.1	[43]
ITO/QDs/spiro-2NPB/LT-N125/LG-101/Al	CdSe/CdS	615	1.5	18	[44]
ITO/PEDOT:PSS/TFB/QDs/ZnO/Al	Cd$_{0.1}$Zn$_{0.9}$S/CdSe/CdS	629	2.4	12.44	[45]
ITO/ZnO/QDs/PVK/PMAH/Al	CdSe/ZnS	618	3.3	11.2	[27]
ITO/PEDOT:PSS/Poly-TPD/QDs/ZnMgO/Ag	InP/ZnSe/ZnS	630	1.8	12.2	[46]
ITO/PEDOT:PSS/TFB/F$_4$TCNQ/QDs/ZnO/Al	CdS/CdSe/ZnSe	612	1.55	12.05	[47]
ITO/PEDOT:PSS/TFB/QDs/ZnO/Al	Cd$_{1-x}$Zn$_x$Se$_{1-y}$S$_y$	631	1.7	12	[48]
ITO/PEDOT:PSS/TFB/QDs/ZnO/Al	Cd$_{1-x}$Zn$_x$S/ZnS	625	1.5	12	[49]
ITO/PEDOT/TFB/QDs/TiO$_2$/Al	CdSe/CdS/ZnS	615	—	—	[50]
ITO/ZnMgO/QDs/TCTA/TAPC/HATCN/Al	InP/ZnSe/ZnS	—	—	10.2	[51]
ITO/ZnO:CsN$_3$/TAPC/HAT-CN/MoO$_3$/Al	CdSe/ZnS	628	3.0	13.4	[52]
ITO/ZnO/QDs/CBP/MoO$_3$/Al	CdSe/Zn$_{0.5}$Cd$_{0.5}$S	625	2	7.4	[53]
ITO/ZnO/QDs/CBP/HAT-CN/Al	InP/ZnSe/ZnS	607	2.0	6.6	[33]
ITO/ZnO/QDs/CBP/MoO$_3$/Al	CdSe/CdS/ZnS	637	1.8	7.3	[54]

Device structure	QDs	EL peak (nm)	FWHM (nm)	EQE (%)	Ref.
ITO/MoO$_3$/NiO/QDs/ZnO/Al	CdSe/CdS/ZnS	630	2.8	5.51	[55]
ITO/ZnO/PEIE/QDs/CBP/MoO$_3$/Al	CuInS$_2$/ZnS	650	2.2	4.8	[56]
ITO/ZnO/QDs/CBP/MoO$_3$/Al	CuInS$_2$/ZnS	650	2	3.6	[56]
ITO/ZnO/QDs/PVK/PEDOT:PSS/Al	CdSe/CdS/ZnS	~624	2	2.72	[57]
ITO/2-TPD/TCTA-VB/QDs/TPBi/CsF/Al	CdSe/CdS	611	6	1.1	[58]
ITO/2-TPD/TCTA-VB/QDs/TPBi/CsF/Al	CdSe/CdS/CdZS/ZnS	623	6	1.6	[58]
ITO/PEDOT/Poly-TPD/QDs/Alq$_3$/Ca/Al	CdSe/CdS/ZnS	619	3–4	—	[59]
ITO/PPV/CdSe/Mg	CdSe	580–620	4	—	[5]
ITO/PEDOT:PSS/Poly-TPD/QDs/ZnO/Al	CdSe/ZnS	600	1.7	1.7	[31]
ITO/PEDOT/TFB/QDs/TiO$_2$/Al	CdSe/CdS/ZnS	618	<2	1.6	[50]
ITO/PEDOT:PSS/poly-TPD/QDs/ZnO/Ag	CdSe/ZnS	~630	2	1.34	[60]
ITO/PEDOT:PSS/SpiroTPD/QDs/TPBi/Mg:Ag/Ag	ZnCdSe	650	4	1.0	[61]
ITO/PEDOT:PSS/CBP/QDs/TAZ/Alq$_3$/Mg:Ag/Ag	CdSe/ZnS	615	5	1.2	[62]
ITO/PPV/QDs/Mg/Ag	CdSe/CdS	613	—	0.2	[38]
ITO/NiO/QDs/Alq$_3$/Ag/Mg/Ag	CdSe/ZnS	625	6–7	0.18	[63]
ITO/α-NPD:F4-TCNQ/CBP/QDs/BCP/Alq$_3$/LiF/Al	CdSe/ZnS	~620	—	0.15	[64]
ITO/PEDOT:PSS/Poly-TPD/QDs/Alq$_3$/Ca/Al	(CuInS$_2$-ZnS)/ZnS	606	14	—	[65]
ITO/PEDOT:PSS/TFB/QDs/TiO$_2$/Al	CdSe/CdS/ZnS	615	1.9	—	[66]
ITO/SiO$_2$/QDs/Au	SWNT/PS-b-PPP:CdSe/ZnS	640	—	—	[67]
ITO/PEDOT:PSS/Poly-TPD/QDs/Alq$_3$/Ca/Al	CuInS$_2$-ZnS/ZnSe/ZnS	623	4.4	—	[68]

structure with the same device structure, ITO/PEDOT:PSS/TFB/QDs/ZnO/Al [41]. The use of Se in the entire core/shell region and an alloy bridging layer in the core/shell interface led to a maximum EQE of 21.6%, corresponding to a luminance of 13 300 cd m^{-2} and a maximum luminance of 356 000 cd m^{-2}. These studies demonstrate that, in a conventional QDLED device structure, the proper design of the light-emitting layer is critical for achieving high performance.

Inverted structure: relative to the standard device structure, patterned ITO is used as the electron injection electrode, the other materials are slightly changed accordingly, and the final electrode Al is used as the anode [44, 54, 57, 76, 77]. Lee et al. reported the use of solution-treated ZnO nanoparticles as the electron injection/transport layer [54], shown in Figure 4.8b, and achieved high brightness and high efficiency of inverted structure QDLEDs by optimizing the energy level of the organic hole transport layer. It succeeded in obtaining high brightness red QDLEDs with a maximum brightness of 23 040 cd m^{-2} and an EQE of 7.3%, respectively. It is also noteworthy that the devices exhibit extremely low turn-on voltage and long operating lifetime, which is mainly due to the direct recombination of excitons within the QDs via the inverted device structure. A new ITO/ZnO/PEIE/QDs/CBP/MoO$_3$/Al inverted device structure was reported by Ji et al. in 2020 to improve the efficiency of InP QD-based QDLEDs [78]. By introducing a thin layer of electron transport material, the buildup of holes at the interface between the transport layer and the QDs is greatly reduced, and the burst effect of holes on QD emission is suppressed. By embedding QDs at the electron-controlled interface of ZnO/TPBi, the EQE (current efficiency) is improved from 3.83% (5.17 cd/A) to 6.32% (8.54 cd/A) compared to the PN junction emitter at ZnO/CBP in the conventional device structure. The analysis shows that the internal quantum efficiency of the InP QD-based device is close to 100% (PLQY of 32%). This work provides an optional device structure for realizing highly efficient QDLED devices.

The tandem structure: The tandem structure refers to the connection of devices with conventional or inverted structures in series to form a QDLED, as illustrated in Figure 4.8c. One example of highly efficient tandem QDLEDs was developed by Sun et al. using interconnecting layers of the ZnMgO/Al/HATCN/MoO$_3$ structure, depicted in Figure 4.9 [40]. The interconnect layer (ICL) developed in this study demonstrates high transparency, efficient charge generation and injection, and excellent resistance to solvent damage during deposition of the upper functional layer. Red tandem QDLEDs incorporating these ICLs exhibit outstanding current efficiency and EQE, achieving values of 41.5 cd A^{-1} and 23.1%, respectively. The high efficiency is maintained over a wide range of luminosities from 10^2 to 10^4 cd m^{-2}. Notably, even at a high luminance of 20 000 cd m^{-2}, the EQE of the red QDLEDs remains at 99% of the maximum value.

4.4 Conclusion and Outlook

In this section, we have explored the energy level structures and optoelectronic properties of typical cadmium-containing red-light QD materials such as CdSe and CdTe,

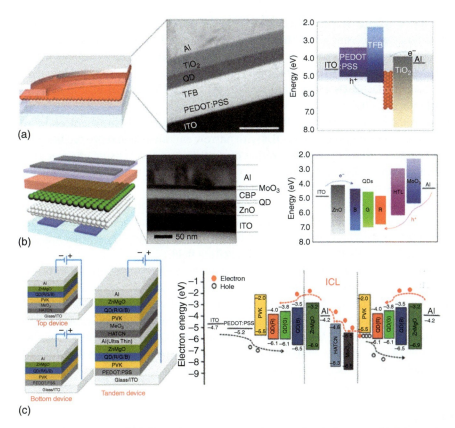

Figure 4.9 Common QDLED device structures and energy level diagrams. (a) Schematic diagram and electron microscopy of device cross-section and corresponding energy band diagram for conventional QDLED device structure, inverted QDLED device structure. Source: Cho et al. [66]/Springer Nature; (b) schematic diagram and electron microscopy of device cross-section and corresponding energy band diagram, inverted QDLED device structure. Source: Kwak et al. [54]/American Chemical Society; (c) schematic diagram of the device structure of tandem QDLEDs and energy band diagram of tandem QDLEDs. Source: Zhang et al. [40]/American chemical society.

and discussed various strategies to enhance their stability and performance. We have also focused on the development process of red QDLED device structures and identified some challenges that still need to be addressed, such as the use of heavy metals in some of the more efficient red QD materials, complex and costly synthesis procedures, and charge injection imbalances that can shorten the lifetime of QDLED devices.

The research on red colloidal QDs for LEDs has also made significant progress in recent years. The efficiency and stability of red QDLEDs have been greatly improved, making them increasingly competitive with traditional QDLED technology.

One of the key challenges in developing red QDLEDs is the low efficiency of red QDs, which is caused by the low PLQYs of the red QDs. Several research groups have focused on improving the PLQYs of red QDs by optimizing the size and composition of the QDs, as well as the surface passivation of the QDs.

Another challenge is the limited color gamut of red QDLEDs, which is due to the limited range of emission wavelengths of red QDs. To address this issue, researchers have explored various strategies to broaden the emission spectrum of red QDs, such as using core/shell structures or alloyed QDs. In addition, the development of multi-color QDLEDs by combining red, green, and blue QDLEDs is also an important research direction.

Despite these challenges, significant progress has been made in recent years in developing stable, highly efficient QDs with narrow emissivity and reproducible synthesis protocols. The future optimization of synthesis protocols using cheaper and more environmentally friendly chemicals will help to further reduce manufacturing costs and drive mass adoption of QDLEDs.

Overall, colloidal QDs possess unique optoelectronic properties and have a promising future in industrial applications. As research advances and the technology becomes increasingly mature, the demand for high-quality QDs will continue to grow, fueling further innovation in this exciting field.

In terms of future research direction, the development of high-performance red QDs with high PLQYs, long lifetime, and narrow emission linewidths is still an important challenge. In addition, the development of efficient and stable device architectures, as well as the exploration of new materials and device structures, is also crucial for advancing the commercialization of red QDLEDs. Furthermore, the integration of red QDLEDs with other technologies, such as micro-LEDs and flexible electronics, will also open up new application possibilities.

References

1 Peng, X. (2003). Mechanisms for the shape-control and shape-evolution of colloidal semiconductor nanocrystals. *Advanced Materials* 15: 459–463.
2 Guyot-Sionnest, P. (2008). Colloidal quantum dots. *Comptes Rendus Physique* 9 (8): 777–787.
3 Kumar, S. and Nann, T. (2006). Shape control of II–VI semiconductor nanomaterials. *Small* 2 (3): 316–329.
4 Shang, Y.Q. and Ning, Z.J. (2017). Colloidal quantum-dots surface and device structure engineering for high-performance light-emitting diodes. *National Science Review* 4 (2): 170–183.
5 Colvin, V.L., Schlamp, M.C., and Alivisatos, A.P. (1994). Light-emitting diodes made from cadmium selenide nanocrystals and a semiconducting polymer. *Nature* 370 (6488): 354–357.
6 Klimov, V.I., McBranch, D.W., Leatherdale, C.A., and Bawendi, M.G. (1999). Electron and hole relaxation pathways in semiconductor quantum dots. *Physical Review B* 60 (19): 13740–13749.
7 Fu, H.X. and Zunger, A. (1997). InP quantum dots: Electronic structure, surface effects, and the redshifted emission. *Physical Review B* 56 (3): 1496–1508.
8 Qu, L.H. and Peng, X.G. (2002). Control of photoluminescence properties of CdSe nanocrystals in growth. *Journal of the American Chemical Society* 124 (9): 2049–2055.

9 Pradhan, N., Reifsnyder, D., Xie, R.G. et al. (2007). Surface ligand dynamics in growth of nanocrystals. *Journal of the American Chemical Society* 129 (30): 9500–9509.

10 Hines, M.A. and Guyot-Sionnest, P. (1998). Bright UV-blue luminescent colloidal ZnSe nanocrystals. *Journal of Physical Chemistry B* 102 (19): 3655–3657.

11 Hines, D.A. and Kamat, P.V. (2014). Recent advances in quantum dot surface chemistry. *ACS Applied Materials and Interfaces* 6 (5): 3041–3057.

12 Dabbousi, B.O., RodriguezViejo, J., Mikulec, F.V. et al. (1997). (CdSe)ZnS core-shell quantum dots: synthesis and characterization of a size series of highly luminescent nanocrystallites. *Journal of Physical Chemistry B* 101 (46): 9463–9475.

13 Cumberland, S.L., Hanif, K.M., Javier, A. et al. (2002). Inorganic clusters as single-source precursors for preparation of CdSe, ZnSe, and CdSe/ZnS nanomaterials. *Chemistry of Materials* 14 (4): 1576–1584.

14 Blackman, B., Battaglia, D., and Peng, X.G. (2008). Bright and water-soluble near IR-emitting CdSe/CdTe/ZnSe Type-II/Type-I nanocrystals, tuning the efficiency and stability by growth. *Chemistry of Materials* 20 (15): 4847–4853.

15 Abdellah, M., Zidek, K., Zheng, K.B. et al. (2013). Balancing electron transfer and surface passivation in gradient CdSe/ZnS core-shell quantum dots attached to ZnO. *Journal of Physical Chemistry Letters* 4 (11): 1760–1765.

16 Pu, C.D. and Peng, X.G. (2016). To battle surface traps on CdSe/CdS core/shell nanocrystals: shell isolation versus surface treatment. *Journal of the American Chemical Society* 138 (26): 8134–8142.

17 Sun, L.F., Choi, J.J., Stachnik, D. et al. (2012). Bright infrared quantum-dot light-emitting diodes through inter-dot spacing control. *Nature Nanotechnology* 7 (6): 369–373.

18 Yang, Y., Qin, H.Y., Jiang, M.W. et al. (2016). Entropic ligands for nanocrystals: from unexpected solution properties to outstanding processability. *Nano Letters* 16 (4): 2133–2138.

19 Brown, P.R., Kim, D., Lunt, R.R. et al. (2014). Energy level modification in lead sulfide quantum dot thin films through ligand exchange. *ACS Nano* 8 (6): 5863–5872.

20 Owen, J. (2015). The coordination chemistry of nanocrystal surfaces. *Science* 347 (6222): 615–616.

21 Shen, H.B., Cao, W.R., Shewmon, N.T. et al. (2015). High-efficiency, low turn-on voltage blue-violet quantum-dot-based light-emitting diodes. *Nano Letters* 15 (2): 1211–1216.

22 Liu, Y., Gibbs, M., Puthussery, J. et al. (2010). Dependence of carrier mobility on nanocrystal size and ligand length in PbSe nanocrystal solids. *Nano Letters* 10 (5): 1960–1969.

23 Maslen, V.W. (1987). On the role of inner-shell ionization in the scattering of fast electrons by crystals. *Philosophical Magazine B: Physics of Condensed Matter: Statistical Mechanics, Electronic, Optical and Magnetic Properties* 55 (4): 491–496.

24 Yang, Y., Qin, H.Y., and Peng, X.G. (2016). Intramolecular entropy and size-dependent solution properties of nanocrystal-ligands complexes. *Nano Letters* 16 (4): 2127–2132.

25 Yang, Z.Y., Voznyy, O., Liu, M.X. et al. (2015). All-quantum-dot infrared light-emitting diodes. *ACS Nano* 9 (12): 12327–12333.
26 Reiss, P., Protiere, M., and Li, L. (2009). Core/Shell semiconductor nanocrystals. *Small* 5 (2): 154–168.
27 Liu, Y., Jiang, C.B., Song, C. et al. (2018). Highly efficient all-solution processed inverted quantum dots based light emitting diodes. *ACS Nano* 12 (2): 1564–1570.
28 Ji, X.H., Copenhaver, D., Sichmeller, C., and Peng, X.G. (2008). Ligand bonding and dynamics on colloidal nanocrystals at room temperature: the case of alkylamines on CdSe nanocrystals. *Journal of the American Chemical Society* 130 (17): 5726–5735.
29 Li, J. and Wang, L.-W. (2004). First principle study of core/shell structure quantum dots. *Applied Physics Letters* 84 (18): 3648–3650.
30 Caruge, J.M., Halpert, J.E., Wood, V. et al. (2008). Colloidal quantum-dot light-emitting diodes with metal-oxide charge transport layers. *Nature Photonics* 2 (4): 247–250.
31 Qian, L., Zheng, Y., Xue, J.G., and Holloway, P.H. (2011). Stable and efficient quantum-dot light-emitting diodes based on solution-processed multilayer structures. *Nature Photonics* 5 (9): 543–548.
32 Bae, W.K., Char, K., Hur, H., and Lee, S. (2008). Single-step synthesis of quantum dots with chemical composition gradients. *Chemistry of Materials* 20 (2): 531–539.
33 Cao, F., Wang, S., Wang, F. et al. (2018). A layer-by-layer growth strategy for large-size InP/ZnSe/ZnS core–shell quantum dots enabling high-efficiency light-emitting diodes. *Chemistry of Materials* 30 (21): 8002–8007.
34 Cao, W., Xiang, C., Yang, Y. et al. (2018). Highly stable QLEDs with improved hole injection via quantum dot structure tailoring. *Nature Communications* 9 (1): 2608.
35 Shirasaki, Y., Supran, G.J., Bawendi, M.G., and Bulovic, V. (2013). Emergence of colloidal quantum-dot light-emitting technologies. *Nature Photonics* 7 (1): 13–23.
36 Dabbousi, B.O., Bawendi, M.G., Onitsuka, O., and Rubner, M.F. (1995). Electroluminescence from Cdse quantum-dot polymer composites. *Applied Physics Letters* 66 (11): 1316–1318.
37 Coe, S., Woo, W.K., Bawendi, M., and Bulovic, V. (2002). Electroluminescence from single monolayers of nanocrystals in molecular organic devices. *Nature* 420 (6917): 800–803.
38 Schlamp, M.C., Peng, X.G., and Alivisatos, A.P. (1997). Improved efficiencies in light emitting diodes made with CdSe(CdS) core/shell type nanocrystals and a semiconducting polymer. *Journal of Applied Physics* 82 (11): 5837–5842.
39 Won, Y.-H., Cho, O., Kim, T. et al. (2019). Highly efficient and stable InP/ZnSe/ZnS quantum dot light-emitting diodes. *Nature* 575 (7784): 634–638.
40 Zhang, H., Chen, S.M., and Sun, X.W. (2018). Efficient red/green/blue tandem quantum-dot light-emitting diodes with external quantum efficiency exceeding 21%. *ACS Nano* 12 (1): 697–704.

41 Shen, H.B., Gao, Q., Zhang, Y.B. et al. (2019). Visible quantum dot light-emitting diodes with simultaneous high brightness and efficiency. *Nature Photonics* 13 (3): 192–197.

42 Dai, X.L., Zhang, Z.X., Jin, Y.Z. et al. (2014). Solution-processed, high-performance light-emitting diodes based on quantum dots. *Nature* 515 (7525): 96–99.

43 Zhang, Z., Ye, Y., Pu, C. et al. (2018). High-performance, solution-processed, and insulating-layer-free light-emitting diodes based on colloidal quantum dots. *Advanced Materials* 30 (28): 1801387.

44 Mashford, B.S., Stevenson, M., Popovic, Z. et al. (2013). High-efficiency quantum-dot light-emitting devices with enhanced charge injection. *Nature Photonics* 7 (5): 407–412.

45 Bai, J., Chang, C., Wei, J. et al. (2019). High efficient light-emitting diodes with improved the balance of electron and hole transfer via optimizing quantum dot structure. *Optical Materials Express* 9 (7): 3089–3097.

46 Li, Y., Hou, X., Dai, X. et al. (2019). Stoichiometry-controlled InP-based quantum dots: synthesis, photoluminescence, and electroluminescence. *Journal of the American Chemical Society* 141 (16): 6448–6452.

47 Nam, S., Oh, N., Zhai, Y., and Shim, M. (2015). High efficiency and optical anisotropy in double-heterojunction nanorod light-emitting diodes. *ACS Nano* 9 (1): 878–885.

48 Manders, J.R., Qian, L., Titov, A. et al. (2015). High efficiency and ultra-wide color gamut quantum dot LEDs for next generation displays. *Journal of the Society for Information Display* 23 (11): 523–528.

49 Yang, Y., Zheng, Y., Cao, W. et al. (2015). High-efficiency light-emitting devices based on quantum dots with tailored nanostructures. *Nature Photonics* 9 (4): 259–266.

50 Kim, T.-H., Cho, K.-S., Lee, E.K. et al. (2011). Full-colour quantum dot displays fabricated by transfer printing. *Nature Photonics* 5 (3): 176–182.

51 Lee, C.Y., Naik Mude, N., Lampande, R. et al. (2019). Efficient cadmium-free inverted red quantum dot light-emitting diodes. *ACS Applied Materials and Interfaces* 11 (40): 36917–36924.

52 Pan, J., Wei, C., Wang, L. et al. (2018). Boosting the efficiency of inverted quantum dot light-emitting diodes by balancing charge densities and suppressing exciton quenching through band alignment. *Nanoscale* 10 (2): 592–602.

53 Lim, J., Jeong, B.G., Park, M. et al. (2014). Influence of shell thickness on the performance of light-emitting devices based on CdSe/Zn$_{1-x}$Cd$_x$S core/shell heterostructured quantum dots. *Advanced Materials* 26 (47): 8034–8040.

54 Kwak, J., Bae, W.K., Lee, D. et al. (2012). Bright and efficient full-color colloidal quantum dot light-emitting diodes using an inverted device structure. *Nano Letters* 12 (5): 2362–2366.

55 Yang, X., Zhang, Z.-H., Ding, T. et al. (2018). High-efficiency all-inorganic full-colour quantum dot light-emitting diodes. *Nano Energy* 46: 229–233.

56 Yuan, Q., Guan, X., Xue, X. et al. (2019). Efficient CuInS$_2$/ZnS quantum dots light-emitting diodes in deep red region using PEIE modified ZnO. *Electron Transport Layer* 13 (5): 1800575.

57 Zhang, H., Li, H., Sun, X., and Chen, S. (2016). Inverted quantum-dot light-emitting diodes fabricated by all-solution processing. *ACS Applied Materials and Interfaces* 8 (8): 5493–5498.

58 Jing, P., Zheng, J., Zeng, Q. et al. (2009). Shell-dependent electroluminescence from colloidal CdSe quantum dots in multilayer light-emitting diodes. *Journal of Applied Physics* 105 (4): 044313.

59 Sun, Q., Wang, Y.A., Li, L.S. et al. (2007). Bright, multicoloured light-emitting diodes based on quantum dots. *Nature Photonics* 1 (12): 717–722.

60 Liu, Y., Li, F., Xu, Z. et al. (2017). Efficient all-solution processed quantum dot light emitting diodes based on inkjet printing technique. *ACS Applied Materials and Interfaces* 9 (30): 25506–25512.

61 Anikeeva, P.O., Halpert, J.E., Bawendi, M.G., and Bulović, V. (2009). Quantum dot light-emitting devices with electroluminescence tunable over the entire visible spectrum. *Nano Letters* 9 (7): 2532–2536.

62 Kim, L., Anikeeva, P.O., Coe-Sullivan, S.A. et al. (2008). Contact printing of quantum dot light-emitting devices. *Nano Letters* 8 (12): 4513–4517.

63 Zhao, J., Bardecker, J.A., Munro, A.M. et al. (2006). Efficient CdSe/CdS quantum dot light-emitting diodes using a thermally polymerized hole transport layer. *Nano Letters* 6 (3): 463–467.

64 Rizzo, A., Mazzeo, M., Palumbo, M. et al. (2008). Hybrid light-emitting diodes from microcontact-printing double-transfer of colloidal semiconductor CdSe/ZnS quantum dots onto organic layers. *Advanced Materials* 20 (10): 1886–1891.

65 Chen, B., Zhong, H., Zhang, W. et al. (2012). Highly emissive and color-tunable $CuInS_2$-based colloidal semiconductor nanocrystals: off-stoichiometry effects and improved electroluminescence performance. *Advanced Functional Materials* 22 (10): 2081–2088.

66 Cho, K.S., Lee, E.K., Joo, W.J. et al. (2009). High-performance crosslinked colloidal quantum-dot light-emitting diodes. *Nature Photonics* 3 (6): 341–345.

67 Cho, S.H., Sung, J., Hwang, I. et al. (2012). High performance AC electroluminescence from colloidal quantum dot hybrids. *Advanced Materials* 24 (33): 4540–4546.

68 Tan, Z., Zhang, Y., Xie, C. et al. (2011). Near-band-edge electroluminescence from heavy-metal-free colloidal quantum dots. *Advanced Materials* 23 (31): 3553–3558.

69 Lee, K.H., Lee, J.H., Kang, H.D. et al. (2014). Over 40 cd/A efficient green quantum dot electroluminescent device comprising uniquely large-sized quantum dots. *ACS Nano* 8 (5): 4893–4901.

70 Bae, W.K., Kwak, J., Lim, J. et al. (2010). Multicolored light-emitting diodes based on all-quantum-dot multilayer films using layer-by-layer assembly method. *Nano Letters* 10 (7): 2368–2373.

71 Pal, B.N., Ghosh, Y., Brovelli, S. et al. (2012). 'Giant' CdSe/CdS core/shell nanocrystal quantum dots as efficient electroluminescent materials: strong influence of shell thickness on light-emitting diode performance. *Nano Letters* 12 (1): 331–336.

72 Shen, H.B., Wang, S., Wang, H.Z. et al. (2013). Highly efficient blue-green quantum dot light-emitting diodes using stable low-cadmium quaternary-alloy ZnCdSSe/ZnS core/shell nanocrystals. *ACS Applied Materials and Interfaces* 5 (10): 4260–4265.

73 Shen, H.B., Bai, X.W., Wang, A. et al. (2014). High-efficient deep-blue light-emitting diodes by using high quality $Zn_xCd_{1-x}S$/ZnS core/shell quantum dots. *Advanced Functional Materials* 24 (16): 2367–2373.

74 Cao, F., Wang, H., Shen, P. et al. (2017). High-efficiency and stable quantum dot light-emitting diodes enabled by a solution-processed metal-doped nickel oxide hole injection interfacial layer. *Advanced Functional Materials* 27 (42): 1704278.

75 Liu, G.H., Zhou, X., and Chen, S.M. (2016). Very bright and efficient microcavity top-emitting quantum dot light-emitting diodes with Ag electrodes. *ACS Applied Materials and Interfaces* 8 (26): 16768–16775.

76 Castan, A., Kim, H.M., and Jang, J. (2014). All-solution-processed inverted quantum-dot light-emitting diodes. *ACS Applied Materials and Interfaces* 6 (4): 2508–2515.

77 Fu, Y., Kim, D., Moon, H. et al. (2017). Hexamethyldisilazane-mediated, full-solution-processed inverted quantum dot-light-emitting diodes. *Journal of Materials Chemistry C* 5 (3): 522–526.

78 Wang, Y.C., Chen, Z.J., Wang, T. et al. (2020). Efficient structure for InP/ZnS-based electroluminescence device by embedding the emitters in the electron-dominating interface. *Journal of Physical Chemistry Letters* 11 (5): 1835–1839.

5

Green Quantum Dot LED Materials and Devices

5.1 Background

In the past decades, quantum dots (QDs) have attracted a lot of attention from researchers due to their excellent optoelectronic properties, wide range of luminescence in the visible wavelength band, narrow luminescence spectrum (FWHM), photoluminescence quantum yield (PLQY) (close to 100%), and high efficiency. Therefore, the use of QDs as a light-emitting layer for the preparation of quantum dot light-emitting diodes (QDLEDs) has significant advantages [1–3]. Compared to organic light-emitting diodes (OLEDs), QDLEDs perform better in terms of color saturation and preparation cost, thanks to the excellent optoelectronic properties of QDs, their high stability, and the fact that they can be prepared by the solution method at low cost. As a result, QDLEDs have great potential to become the next generation of wide-color gamut display devices for applications [4]. Much work has therefore been carried out to improve red [5–7], Green [8–11], and blue [11–13] trichromatic QDLEDs. To date, the external quantum efficiency (EQE) of red, green, and blue QDLEDs has exceeded 21%, with green QDLEDs in particular achieving an EQE of 27.6% [4, 11]. At the same time, the human eye is very sensitive to green light, so the development of high-efficiency and high-stability green QDLEDs is of great interest.

The colloidal QDs commonly used in green QDLEDs are CdSe/ZnS [14, 15], $Zn_{1-x}Cd_xSe/ZnS$ [16], ZnCdS [17], ZnCdSSe/ZnS [18], $Cd_{1-x}Zn_xSe_{1-y}S_y$ [19], etc. Because of their large specific surface area, QDs are already compromised by their surroundings, so QDs are usually applied in QDLEDs in a core/shell structure. Therefore, wrapping the QD with a shell containing organic chains or inorganic components to maintain its stability and reduce surface defect states is necessary to improve the performance of QDs. Also, the thickness of the shell has an impact on the performance of the device, to some extent, it can suppress intermittent combination and promote charge injection and transport [19, 20]. As shown in Figure 5.1, QDs can be classified into discrete core/shell structure, transition core/shell structure, and alloyed core/shell structure according to the core/shell structure, and the performance of QDs can be improved by adjusting the energy band structure. There are various methods to synthesize QDs with different structures, including thermal injection [22], colloidal chemistry [23], and

Colloidal Quantum Dot Light Emitting Diodes: Materials and Devices, First Edition. Hong Meng.
© 2024 WILEY-VCH GmbH. Published 2024 by WILEY-VCH GmbH.

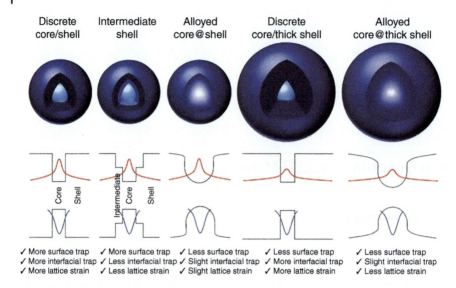

Figure 5.1 Schematic representation of the morphology, energy band structure, electron and hole wave functions, and their properties of different core/shell structures of QDs. Source: Moon et al. [21]/John Wiley & Sons.

molecular beam epitaxy (MBE) [15]. In 2009, the Deng group reported a mild, green, phosphine-free, and low-cost method to synthesize alloyed ZnCdSSe QDs for the first time [24]. They took advantage of the difference in the relative chemical reactivity of Zn, Cd, S, and Se by injecting Zn/Cd precursors into S/Se precursors in paraffin solutions. The resulting QDs were prepared with a fluorescence quantum yield of 40–65%. The QDs emit green light when they are composed of more CdSe than ZnS because the energy band of CdSe is narrower than that of ZnS. The Hsu group synthesized $Cd_{1-x}Zn_xSe$ QDs by a one-step thermal injection method [22]. They successfully synthesized QDs that emit different wavelengths of light by adjusting the reaction time. They then chemically etched $Cd_{1-x}Zn_xSe$ QDs by dissolving benzoyl peroxide (BPO) in a toluene–methanol mixture as an etching solution. The etched $Cd_{1-x}Zn_xSe$ QDs are coated with a thin ZnS shell, forming a Type-I core/shell structure, and the combination region of excitons is confined in the $Cd_{1-x}Zn_xSe$ core. The resulting QDLEDs assembled with QDs exhibit highly saturated electroluminescent properties. The Araki group reported CdSe/ZnSe QDs by the MBE method and assembled the devices [15]. They investigated the effect of the number of layers of CdSe and the thickness of the ZnSe shell on the performance of the QDLEDs. The devices assembled from QDs consisting of five layers of CdSe cores and 10 nm thick ZnSe showed the strongest green light emission at 530 nm due to the suitable crystallinity of CdSe at the right thickness shell.

The application of colloidal QDs to light-emitting devices was first reported by the Colvin group in 1994. They used CdSe QDs as the light-emitting layer and *p*-phenylene vinylene (PPV) as the hole transport material (HTM), and assembled organic/inorganic hybrid structures for the devices. They achieved the transition from red to yellow light by adjusting the size of the CdSe QDs. Although the device has an EQE of less than 0.01%, it is driven by a voltage of only 4 V. More interestingly, the light is emitted by the CdSe QDs at low voltages; at high voltages, it appears as green light emitted by the PPV. Next, researchers began to look at more types of QDs and optimize their synthesis. In 2011, the Qian group reported ZnO nanoparticles as an electron transport layer (ETL) to improve stability and power efficiency (PE) during the process [25]. They investigated the effect of ZnO thickness on device performance and found that the driving voltage of the device increased with increasing ZnO thickness, but the EQE of the device reached a maximum of 1.4% with a ZnO thickness of 35 nm. Meanwhile, they assembled devices using CdSe/ZnS QDs as the light-emitting layer and emitted green light at 540 nm. The driving voltage, maximum brightness, PE, and EQE of the devices were 1.8 V, 68 000 cd m^{-2}, 8.2 lm W^{-1}, and 1.8%, respectively. The Shen group used a phosphine-free method to synthesize $Zn_{1-x}Cd_xSe/ZnSe/ZnSe_xS_{1-x}/ZnS$ core/multi-shell QDs and assembled devices to investigate the effect of shell thickness on the quantum yield and stability of the QDs [26]. Based on the absence of a significant shift in the peak of the photoluminescence (PL) spectra, it can be inferred that the outer wrapped shell prevents the outward carrier migration and leakage. At the same time, the increasing thickness of the shell layer causes a gradual change in the lattice parameters, which facilitates the passivation of the defect state on the surface of the QD and increases its fluorescence quantum yield (PLQY). In 2012, the Kwak group reported the construction of QDLED devices with inverted structure consisting of ITO/ZnO/CdSe@ZnS QDs/CBP/MoO$_3$/Al for the first time [27]. They found that ZnO as an ETL had little effect on the exciton combination between adjacent QDs, and the devices achieved a maximum luminance of 218 800 cd m^{-2} and a current efficiency (CE) of 19.2 cd A^{-1}. Since ZnO exhibits efficient electron injection and transport, the optoelectronic performance of QDLEDs depends on the HTM. Poly(9-vinylcarbazole) (PVK) were mixed with small molecules as a HTM to improve hole migration [28]. It was found that PVK blended with 4,4′,4″-tris(N-carbazole)-triphenylamine (TCTA) at 20 wt% gave the best performance with an EQE of 27% and a maximum brightness of 40 900 cd m^{-2}. In the same year, the Jun group reported the size and thickness of the multilayer shells to improve the performance of green QDLED devices through a multi-shell passivation strategy [20]. These alloyed core/multi-shell structure QDs have PL peaks at 520 nm and a PLQY close to 100%. They assembled green QDLED devices with CdSe//ZnS/CdSZnS QDs as the light-emitting layer, and the electroluminescence (EL) spectrum peaks were at 530 nm with a FWHM of 35 nm. The device has a turn-on voltage of 3.3 V, and the International Commission on Illumination (CIE) 1931 (x, y) coordinates are (0.209, 0.742). The Lee group proposed the first one-step synthesis method to prepare QDs with gradients in shell chemistry in 2014 [29]. QDs with ZnS double shells hinder non-radiative energy transfer and oscillatory combinations between neighboring QDs. In 2015, the Manders group synthesized $Cd_{1-x}Zn_xSe_{1-y}S_y$ QDs to achieve an EQE of more than 20% for green QDLEDs for the first time [10]. They assembled bottom emission devices with a structure of ITO/PEDOT/PSS/TFB/QDs/ZnO/Al. By tuning the assembly environment, QDs synthesis method, and effective charge injection, the green QDLEDs exhibit a EQE of 21%, CE of 82 cd A^{-1}, and PE of 79.8 lm W^{-1} with a turn-on voltage of only 3.5 V. In addition, the QDLEDs have a lifetime T_{50} of over 100 000 hours after simple packaging. In 2016, the Peng group achieved vacuum-free preparation of QDLEDs using eutectic gallium–indium (EGaIn) instead of conventional metal cathodes, showing excellent device performance [30]. Meanwhile, the device fabrication process does not require organic solvents and thus can avoid high-temperature thermal annealing treatment. In 2017, the Zhang group constructed a tandem QDLED through an interconnect layer (ICL) to greatly improve the optoelectronic performance of the device, which achieved an EQE of 27.6% [11]. In 2018, the Vasan group reported the best-performing all-inorganic green QDLED using alloyed CdSe/ZnS QDs as the emitting layer [31]. The CE, maximum luminance, and EQE of this device reached 144 cd A^{-1}, 116 000 cd m^{-2}, and 11.4%, respectively. Meanwhile, the Su group used forward aging – involving an interfacial reaction between the Al cathode and ZnMgO – to promote electron injection and prevent exciton quenching [4]. In 2019, the Li group used a shell layer adjustment strategy to optimize the device structure [32], which achieved an EQE of 23.9%, which was the highest value for green QDLEDs prepared by the all-solution method. In addition, the Yang group achieved both high efficiency and long lifetime of green QDLED devices through shell layer engineering [32]. The device structure is an inverted structure, presenting a CE of 96.42 cd A^{-1}, a maximum luminance of 70 650 cd m^{-2}, and a T_{50} lifetime of 4943.6 hours [9]. Since then, many scientists have remained committed to improving the efficiency and operating time of green QDLEDs for use in next-generation display devices, while exploring the factors and mechanisms that lead to the roll-off of device efficiency. Table 5.1 summarizes the works on green QDLEDs.

Figure 5.2 illustrates the development of the EQE of green QDLEDs. In 2009, the Anikeeva group chose organic materials as charge transport materials, and the EQE of the devices was 2.6% [35]. They synthesized tightly stacked monolayers of ZnSe/CdSe/ZnS core-bilayer shell QDs clad by hexylphosphonic acid and TOPO. Meanwhile, Alq$_3$ was replaced by 2,2′,2″-(1,3,5-triphenyl)-tris(L-phenyl-1-*H*-benzimidazole) (TPBi) as ETL to improve the exciton energy transfer from TPBi to QDs. In 2011, ZnO became the preferred ETL in later QDLED devices, although the EQE of the devices in this literature was only 1.8% [25]. In 2012, the Kwak group reported green QDLEDs with an inverted structure, which is a good match for thin-film transistors (TFTs) [27]. Thanks to the direct combination of QDs in the inverted structure, this green QDLED has an EQE of 5.8% and a turn-on voltage of only 2.4 V, which is consistent with the bandgap energy of the QD. In 2014, the EQE of the green QDLED was increased to 12.6% by shell engineering [29]. They compared two QDs: CdSe@ZnS QDs and CdSe@ZnS/ZnS QDs and found that green light QDLEDs based on CdSe@ZnS/ZnS QDs performed better because the outermost shell in CdSe@ZnS/ZnS QDs effectively reduced the Förster resonance energy transfer (FRET) process between neighboring QDs. In 2015, the EQE of green QDLEDs assembled from chemical composition gradient QDs successfully exceeded 20% [10]. In 2018, the Fu group balanced the carrier injection and transport through interfacial engineering, and the EQE of this green light QDLED

Table 5.1 Summary of works on green QDLEDs.

Device structure	QDs	EL peak (nm)	Turn-on voltage (V)	Peak EQE (%)
ITO/PEDOT:PSS/PVK/QDs/ZnMgO/Al/HATCN/MoO$_3$/PVK/QDs/ZnMgO/Al	CdZnSeS/ZnS	534	6.1	27.6
ITO/ZnO/QDs/PEIE/Poly-TPD/PMA/Al	CdSeZnS/ZnS/ZnS	524	—	25.04
ITO/ZnO/QDs/PEIE/Poly-TPD/PMA/Al	CdSeZnS/ZnS	532	—	5.67
ITO/ZnO/QDs/PEIE/Poly-TPD/PMA/Al	CdSeZnS	540	—	2.15
ITO/ZnO/QDs/PEIE/Poly-TPD/MoO$_x$/Al	CdSe@ZnS/ZnS	522	—	24.8
ITO/ZnO/QDs/PEIE/Poly-TPD/MoO$_x$/Al	CdSe@ZnS/ZnS	522	—	21.7
ITO/ZnO/QDs/PEIE/Poly-TPD/MoO$_x$/Al	CdSe@ZnS/ZnS	522	—	18.8
ITO/ZnO/QDs/PEIE/Poly-TPD/MoO$_x$/Al	CdSe@ZnS/ZnS	522	—	16.1
ITO/PEDOT:PSS/TFB/QDs/ZnO/Al	ZnCdSe/ZnSe/ZnSeS/ZnS	530	2.3	23.9
ITO/ZnMgO/QDs/PVK/PEDOT:PSS/ZnMO/QDs/TCTA/NPB/HATCN/EGaIn	CdZnSeS/ZnS	538	6.3	23.68
ITO/PEDOT:PSS/TFB/QDs/ZnO/Al	CdSe/ZnSe	—	—	22.9
ITO/PEDOT:PSS/TFB/QDs/ZnO NPs/Al	Zn$_x$Cd$_{1-x}$S$_y$Se$_{1-y}$	525.8	2.1	21
ITO/PEDOT:PSS/TFB/QDs/ZnO/Al	Cd$_{1-x}$Zn$_x$Se$_{1-y}$S$_y$	526	2.3	18.3
ITO/PEDOT:PSS/Poly-TPD/PVK/QDs/Zn$_{0.9}$Mg$_{0.1}$O/Al	CdZnSeS/ZnS	526	—	18.1
ITO/PEDOT:PSS/Poly-TPD/PVK/QDs/ZnO/Al	CdZnSeS/ZnS	526	—	11.1
ITO/PEDOT:PSS/TFB/QDs/ZnO NP/Al	Zn$_{1-x}$Cd$_x$Se/ZnS	532	2.2	16.5

(continued)

Table 5.1 (Continued)

Device structure	QDs	EL peak (nm)	Turn-on voltage (V)	Peak EQE (%)
ITO/ZnMgO/QDs/PVK/PEDOT:PSS/ZnMO/QDs/TCTA/NPB/HATCN/Al	CdZnSeS/ZnS	538	6.1	16.76
ITO/ZnO NPs/QDs/PEIE/poly-TPD/MoO$_x$/Al	CdSe@ZnS/ZnS	528	3.1–3.2	15.6
ITO/PEDOT:PSS/TFB/thick-shell QDs/ZnO/Al	Zn$_{1-x}$Cd$_x$Se/ZnS	517	2.35	15.4
ITO/PEDOT:PSS/TFB/QDs/ZnO/Ag	CdSe/ZnS	536	2.8	15.45
ITO/ZnMgO/QDs/PVK/PEDOT:PSS/ZnMO/QDs/PVK/PEDOT:PSS/Al	CdZnSeS/ZnS	538	7.3	13.65
ITO/PEDOT:PSS/TFB/QDs/ZnO/Al	Cd$_{1-x}$Zn$_x$S/ZnS	537	2.0	14.5
ITO/PEDOT:PSS/PVK/QDs/ZnO/EGaIn	CdZnSeS/ZnS	522	4.0	12.85
ITO/PEDOT:PSS/PVK/QDs/ZnO/Al	CdSe@ZnS/ZnS	516	—	12.6
ITO/NiO/PVK/QDs/ZnO/Al	CdSe/ZnS	538	3.5	10.5
ITO/ZnO/QDs:PVK/CBP/MoO$_3$/Al	—	528	—	10
ITO/ZnO/QDs/CBP/MoO$_3$/Al	—	528	—	7.5
ITO/ZnO:CsN$_3$/TAPC/HATCN/MoO$_3$/Al	CdSe/ZnS	532	2.6	9.1
ITO/s-NiO$_x$/Al$_2$O$_3$/QDs/ZnO/Al	ZnCdSSe/ZnS	527	—	8.1
ITO/PEDOT:PSS/TFB/QDs/ZnO/Al	Cd$_{1-x}$Zn$_x$S/ZnS	534	2.2	7.5
ITO/PEDOT/TFB/QDs/TiO$_2$/Al	CdSeS/ZnS	530	—	—
ITO/PEDOT/Poly-TPD/QDs/Alq$_3$/Ca/Al	CdSe/ZnS	525	3–4	—
ITO/MoO$_3$/NiO/QDs/ZnO/Al	CdSe/ZnS	508	3.0	6.52
ITO/s-MoO$_x$/NiO$_x$/Al$_2$O$_3$/QDs/ZnO/Al	CdSe/ZnS	534	4.7	5.5
ITO/s-MoO$_x$/NiO$_x$/QDs/ZnO	CdSe/ZnS	534	4.3	4.3
ITO/NiO$_x$/QDs/ZnO/Al	CdSe/ZnS	532	3.9	1.7

Device structure	QDs	EL peak (nm)	EQE (%)	CE (cd/A)
ITO/MoO₃/NiO/QDs/ZnO/Al	CdSe/ZnS	516	3.1	5.48
ITO/ZnO/QDs/PVK/PVK/TFB/PMAH/Al	CdSe/ZnS	528	2.7	5.29
ITO/ZnO/QDs/CBP/MoO₃/Al	CdSe/ZnS/ZnS	515	3.5	6.35
ITO/ZnO/QDs/CBP/MoO₃/Al	CdSe/ZnS/ZnS	515	2.5	4.63
ITO/ZnO/QDs/PVK/PMAH/Al	CdSe/ZnS	528	5.1	3.83
ITO/ZnO/QDs/CBP/MoO₃/Al	CdSe@ZnS	520	2.4	5.8
ITO/PEDOT:PSS/Poly-TPD/QDs/ZnO/Al	CdSe/ZnS	540	1.8	1.8
ITO/PEDOT:PSS/poly-TPD/QDs/TPBI/LiF/Al	CdSe@ZnS	528	3.5	1.4
ITO/PEDOT:PSS/Spiro TPD/QDs/TPBi/Mg:Ag/Ag	ZnSe/CdSe/ZnS	545	5	2.6
ITO/CBP/QDs/TAZ/Alq₃/Mg:Ag/Ag	$Cd_xZn_{1-x}Se/Cd_yZn_{1-y}S$	527	—	0.5
ITO/PEDOT:PSS/CBP/QDs/TAZ/Alq₃/Mg:Ag/Ag	ZnSe/CdSe/ZnS	~525	4	0.5
ITO/PEDOT:PSS/TFB/QDs/TiO₂/Al	CdSeS/ZnS	530	—	—
ITO/SiO₂/QDs/Au	SWNT/PS-b-PPP:CdSe/ZnS	540	—	—
ITO/PEDOT:PSS/Poly-TPD/QDs/Alq₃/Ca/Al	CdSe/ZnS	525	—	—
ITO/PEDOT:PSS/Poly-TPD/QDs/ZnO/Al	ZnCdSeS	~530	2.6	—
ITO/PEDOT:PSS/PVK/Poly-TPD/TPD/TCTA/CBP/QDs/ZnO/Al	CdSeZnS	~525	5.5	27
ITO/PEDOT:PSS/poly-TPD/QDs/ZnO:Au/Al	CdSe/ZnS	532	3.0	—
ITO/PEDOT:PSS/PF8Cz/QDs/Zn₀.₈₅Mg₀.₁₅O/Al	CdSe/CdZnSe/ZnS	537	2.05	28.7

reached 22.4% [33]. They introduced PVK to remove excess electrons and ethoxylated polyethyleneimine (PEIE) to reduce the hole injection potential. Devices with sandwich structures (PVK/QD/PEIE) show high luminance of 72 814 cd m^{-2} and CE of 89.8 cd A^{-1}. In 2019, inverted green QDLEDs based on QDs with double-shell structures exhibit a high EQE of 25.04%, showing great potential in next-generation displays [9]. CdSeZnS/ZnS/ZnS QDs with high stability and PLQY of 85% can be synthesized by precisely controlling the precursors of the synthesized ZnS shell. In addition, the dual ZnS shell facilitates the suppression of intermittent combination, FRET processes, and efficiency roll-off at high current densities. Devices based on these QDs show an EL peak at 524 nm with a FWHM of 21 nm and a high CE of 96.42 cd A^{-1}. In 2022, a green QDLED achieved a record-breaking peak EQE efficiency of 28.7% and demonstrated excellent stability, with an extrapolated T_{95} lifetime of 580 000 hours [34]. This was ascribed by using a shallower LUMO level of HTM PF8Cz with simultaneous low electron affinity and reduced energetic disorder. Although green QDLEDs in conventional or inverted structures are now achieving high EQE, more innovative work is needed to achieve their high efficiency and long lifetime. The lifetime of QDLEDs for commercialization requires a $T > 10\,000$ hours at 100 cd m^{-2}. The causes of efficiency roll-off and rapid lifetime decay can be attributed to exciton quenching caused by trap states or high fields [33]. The causes are non-radiative intermittent combinations caused by bandgap mismatch between QDs and charge transport layers (CTLs) [8]. The rapid decay in lifetime can be attributed to exciton quenching caused by trapped states or high fields, non-radiative oscillation combination caused by bandgap mismatch between QDs and CTLs, FRET processes caused by close packing of QDs, and unbalanced charge injection and transport [21]. In addition, the surrounding environment (e.g. light or humidity) as well as the synthesis and device assembly processes have an impact on device performance. To address these issues, strategies such as shell engineering, interface engineering, and device structure optimization have been adopted [36]. Considering the low cost and environmental protection, new cadmium-free QDs have been proposed, such as InP QDs [37, 38], ZnSe QDs [39, 40], CuInS QDs [41], and AgInZnS QDs [42].

Here are some key breakthroughs in green QDLED technology development from 2000 to 2023:

1) 2004: Development of green QDLEDs with high efficiency and narrow spectral linewidths.
    - CIE value: (0.27, 0.65)
    - Emission wavelength: 540 nm
    - FWHM: 24 nm
    - Efficiency: 0.21%
    - Lifetime: 1000 hours
2) 2009: Demonstration of green QDLEDs with a peak EQE of 10.6%.
    - CIE value: (0.27, 0.63)
    - Emission wavelength: 532 nm
    - FWHM: 24 nm

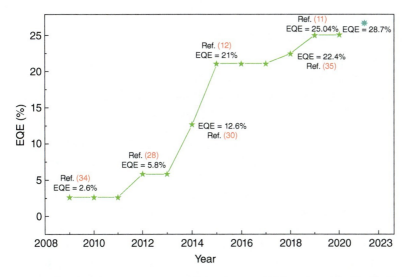

Figure 5.2 Development process of EQE of green QLED devices [11, 12, 28, 30, 33, 34].

- Efficiency: 10.6%
- Lifetime: Not reported

3) 2012: Development of green QDLEDs with a high EQE and improved device stability.
   - CIE value: (0.28, 0.59)
   - Emission wavelength: 542 nm
   - FWHM: 24 nm
   - Efficiency: 12.8%
   - Lifetime: >10 000 hours

4) 2016: Demonstration of highly efficient and stable green QDLEDs based on a novel electron injection layer.
   - CIE value: (0.25, 0.63)
   - Emission wavelength: 540 nm
   - FWHM: 24 nm
   - Efficiency: 17.3%
   - Lifetime: 20 000 hours at 1000 cd m^{-2}

5) 2020: Development of green QDLEDs with a high EQE.
   - CIE value: (0.27, 0.61)
   - Emission wavelength: 538 nm
   - FWHM: 20 nm
   - Efficiency: 21.4%
   - Lifetime: Not reported

6) 2022: Development of green QDLEDs with a record-high EQE of 28.7% and a narrow FWHM.
   - CIE value: Not reported
   - Emission wavelength: 537 nm
   - FWHM: 26 nm

- Efficiency: 28.7%
- Lifetime: T_{95} at $100\,\text{cd}\,\text{m}^{-2}$ is 580 000 hours

Overall, the development of green QDLEDs has progressed significantly over the years, with improvements in efficiency, stability, and spectral linewidth. The most recent breakthroughs have achieved record-high efficiencies and narrowed FWHMs, making green QDLEDs even more promising for future display and lighting applications.

In this section, we will discuss the development of cadmium-based green QDLEDs in terms of both materials and device structures. First, we will discuss the development of cadmium-based core/shell-structured QDs, including their synthesis methods. Then, we will discuss the development of green QDLEDs in terms of the conventional, inverted, and tandem structures of the devices. In addition, we will discuss the factors affecting the performance of green QDLEDs and propose some strategies to improve the optoelectronic performance of the devices. Finally, we will discuss the factors that limit the practical application of green QDLEDs, such as lifetime and efficiency roll-off. We hope that this chapter will provide some constructive ideas and directions for the future development of green QDLEDs.

5.2 Commonly Used Luminescent Layer Materials in Green QDLEDs

QDs have become a hot research topic due to their unique quantum-confined effect and excellent optoelectronic properties. However, the surface defects of QDs limit their practical applications because QDs have poor fluorescence stability when exposed to light and humidity. Therefore, it is often used to cover the core QDs with an inorganic shell layer to maintain their stability and optoelectronic properties. As shown in Figure 5.3, the core/shell structure of QDs can be classified into type I, reverse type I, and type II based on the relative energy levels of the core and shell layer materials [43]. For type I structure QDs, the energy gap of the nuclear material is narrower than that of the shell material. The role of the shell layer is mainly to confine the excitons in the core after passivating the defects on the surface of the core. Reverse type I structure QDs, which have a wider energy gap in the nuclear material than in the shell material, and the excitons are pooled onto the shell material. In this case, different color emissions can be achieved by adjusting the thickness of the shell material. In the case of type II QDs, the bottom of the conduction band or the top of the valence band of the nuclear material lies between the energy gaps of the shell material.

5.2.1 Discrete Core/Shell Quantum Dots

Initially, CdSe was usually chosen as the core material and CdS or ZnSe as the shell layer material for green QDLEDs [35, 43]. In 2002, the Matsumura group synthesized CdSe/ZnSe QDs by MBE [44]. The QDs emitted green light at 520 nm and had

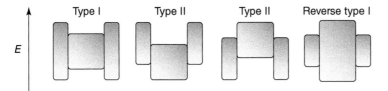

Figure 5.3 Schematic diagram of the energy level distribution of QDs systems with different core/shell structures. Source: Ji et al. [43]/The Royal Society of Chemistry.

a lifetime of more than 10 hours at 77 K. However, CdS did not match the lattice parameters of the core material, which led to a reduction in device performance. Next, researchers found that ZnS can also be used as a shell layer material, which can effectively increase the exciton combination rate and suppress Auger combination process. The Kim group assembled green QDLEDs with CdSe/ZnS quantum dots as the light-emitting layer and Au nanoparticle-doped ZnO as the ETL with localized surface plasmon treatments (LSPs) [45]. Compared to the QD films without Au nanoparticles, the QD films with Au nanoparticle grains showed a 4.12-fold increase in PL intensity at 535 nm, as the excitons in the QDs formed a strong resonant coupling with the LSPs of the Au nanoparticles in ZnO. However, CdS and ZnSe cannot confine the electrons and holes within the CdSe core QD because the bandgaps of CdS and ZnSe are not sufficient to provide the necessary potential barriers.

5.2.2 Alloyed Core/Shell Quantum Dots

Unfortunately, there is still a mismatch in lattice parameters between ZnS and QDs, producing lattice strain and low PLQY. QDs with a gradient distribution in the chemical composition of the shell were created in an attempt to resolve the mismatch in lattice parameters. Three methods for synthesizing alloyed QDs were summarized by Kim's group, including annealing of QDs with discrete core/shell structures, replacement of cations in the precursor, and adjustment of the anion/cation ratio in the precursor [46]. They successfully synthesized $Zn_xCd_{1-x}Se$ alloyed QDs by adjusting the Zn/Cd ratio in the precursor, and the EL peak was located at 534 nm. Bae et al. modified the CdSe@ZnS QDs with 1-octanethiol ligands, which still had high PLQY even after purification [47]. The assembled device structure was ITO/PEDOT:PSS/poly-TPD/QDs/TPBI/LiF/Al with a turn-on voltage of 3.5 V and an EQE of 1.4%. The CdSe/ZnSe core/shell QDs were alloyed at high temperatures by Panda's group [48]. The photostability and PLQYs of the QDs were better than those of the discrete CdSe/ZnSe QDs because the alloyed dots effectively alleviated the lattice strain between the discrete core/shell QDs.

5.2.3 Core/Multilayer Shell Quantum Dots

The core/shell/shell structure is also beneficial for limiting the electron and hole wave functions while enhancing the PLQY of QDs. The Jun group reported an alloyed core/multi-shell QD, CdSe//ZnS/CdSZnS QDs, and discussed the effect of multi-shell thickness on device performance [20]. They found that a thicker inorganic shell layer favors the strong photostability of the QDs and yields a higher PLQY. In addition, Yang's group reported CdSeZnS/ZnS/ZnS QDs with a double ZnS shell structure [9]. They demonstrated that the outermost ZnS shell layer is important for suppressing the FRET process and protecting the longer PL lifetime of the QD films with core/shell/shell structure. The device has the longest lifetime of all solution-prepared inverted QDLEDs at 4943.6 hours with a CIE color coordinate of (0.155, 0.787) and shell layers [49]. The PLQY of the CdSe/ZnSe/ZnS core/shell/shell structure QDs increased from 70% to 85% because ZnSe increased the quality of the ZnS shell layer. The researchers concluded that the intermediate ZnSe shell layer facilitates the crystallization of the ZnS shell and controls the particle shape, thus ensuring the quality of the shell layer and improving the performance of the QDs.

5.3 Development of Device Structures for Green QDLEDs

The device structure of QDLEDs is typically a "sandwich" structure, similar to that of OLEDs, consisting of an electrode, a hole injection layer (HIL), a hole transport layer (HTL), an emission layer (EML) for QDs, and an ETL. The electrodes consist of an anode (e.g. metallic Al or Ag) and a cathode (e.g. ITO).

QDLEDs can be divided into all-inorganic device structures and organic/inorganic hybrid device structures. Inorganic materials have higher stability and charge mobility than the prevailing organic materials, reducing the turn-on voltage of the device to a certain extent [50]. In all-inorganic QDLEDs, MoO_x is commonly used as a hole injection material, and NiO_x is a typical representative of HTMs. ZnO is the most commonly used electron transport material. For example, the Ji group prepared all-inorganic green QDLEDs of ITO/NiO/Al_2O_3/ZnCdSSe/ZnS QDs/ZnO/Al by an ultrasonic spraying process [51]. The devices emit green light at 530 nm with a maximum brightness of 20 000 cd m^{-2} and a CE of 20.5 cd A^{-1}. However, the performance of the all-inorganic QDLEDs still lags behind that of the organic/inorganic hybrid QDLEDs, which is related to the imbalanced carrier injection caused by the large energy barrier between the inorganic material and the QDs due to the strong interactions between the metal oxide and QDs. In addition, fully inorganic QDLEDs prepared by the pure solution method are susceptible to solvent erosion, which may cause current leakage. There are two main strategies to address this: firstly, to prepare high-quality HTLs with low defect states; secondly, to form a buffer layer by mixing HTL/ETL with metal or metal oxide to suppress exciton quenching, spatially separating the luminescent layer from the HTL/ETL fraction [52]. As shown in Figure 5.4a, in 2018, the Ji group inserted an ultrathin Al_2O_3 passivation layer between NiO and QDs and assembled the device structure as ITO/NiO/Al_2O_3/QDs/ZnO/Al [43]. The spectral peaks of the EL spectra at different voltages largely coincide, indicating that the effect of the external electric field is negligible, and the emission peak is located at 527.3 nm, as shown in Figure 5.4b. The formation of NiOOH on the s-NiO surface influenced by Al_2O_3 improves the

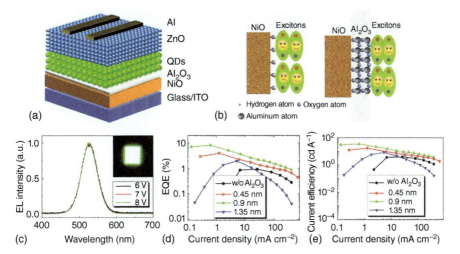

Figure 5.4 (a) Schematic diagram of the device structure; (b) schematic diagram of the passivation mechanism; (c) EL spectra of the device at different voltages; (d) characteristic curve of EQE–current density (J) of the device; (e) CE–J curve of the device. Source: Ji et al. [43]/Royal Society of Chemistry.

device's performance. The device achieves an EQE of 8.1%, an enhancement of over 800% over the original NiO-based QDLED. When assembling the device with 0.9 nm-thick Al_2O_3, the CE performance was 34.1 cd A^{-1} because exciton quenching was suppressed and electron leakage currents were reduced. In 2020, the Lin group used Al_2O_3 to reduce the strong interactions between QDs and metal oxides [53]. In addition, they used MoO_x as the HIL and assembled the device structure as ITO/MoO_x/NiO_x/Al_2O_3/QDs/ZnO/Al. The device has a CE of 20.4 cd A^{-1} and an EQE of 5.5%. The improved device performance is attributed to two factors: (i) Al_2O_3 suppresses exciton quenching and (ii) MoO_x promotes hole transport and attenuates current leakage.

In addition to the materials used in the functional layer, green QDLEDs can also be classified according to the device structure as conventional structures [47], inverted structures [54], and tandem structures [55]. The difference between conventional and inverted QDLEDs is the difference between the hole and the electron structure. The difference between conventional QDLEDs and inverted QDLEDs lies in the different electrodes where the holes and electrons are generated. In a conventional QDLED, holes are generated from the anode, e.g. ITO, and electrons are injected through the cathode, whereas the situation is reversed in an inverted QDLED. Furthermore, the advantages of QDLEDs with inverted structures over QDLEDs with conventional structures are their direct use in integrated circuits for n-type TFTs (e.g. oxide-based TFTs) and their low cost. The Zhang group synthesized colloidal CdSe/CdS core/shell nanosheets (NPLs) as a light-emitting layer with the device structure ITO/PEDOT:PSS/TFB/NPLs/ZnO/Al. The QDs emit saturated green light with a FWHM of 14 nm and a luminance of up to 33 000 cd m^{-2}. In 2012, the Lee research group first reported the green QLED with an inverted structure

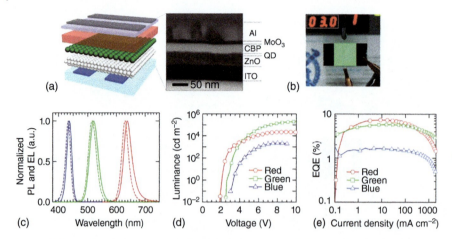

Figure 5.5 (a) Schematic diagram of the device structure (left) and cross-sectional TEM image (right) of an inverted QDLED. (b) Photograph of a green QDLED at an applied voltage of 3.0 V. (c) Quantum dot normalized PL spectrum (dashed line) and EL spectrum of the QDLED (solid line). (d, e) Characteristic curves of the luminance L–voltage V and EQE–J of the inverted QDLED, respectively. Source: Kwak et al. [27]/American Chemical Society.

to solve the problem that the organic HTL is easily eroded and hole injection is affected during the solution preparation process [27]. They constructed an inverted device: ITO/ZnO/CdSe@ZnS QDs/CBP/MoO$_3$/Al, as shown in Figure 5.5. The device emits green light at 520 nm with a turn-on voltage of only 2.4 V, a maximum brightness of 218 800 cd m^{-2}, and EQE of 5.8%. They attribute the high performance exhibited by this inverted structure of QDLED to the direct exciton combination occurring between the QDs and the auxiliary role of ZnO as an ETL. Although QDLEDs have made great strides in performance through the efforts of many researchers, they are still unable to outperform OLEDs. QDLEDs with tandem structures, which have been used in OLEDs with satisfactory results, may be an effective strategy to achieve both high efficiency and long lifetime. In 2017, the Zhang group reported an ICL-PEDOT:PSS/ZnMgO into the device to connect two electroluminescent cells, which can efficiently inject and transport carriers [55]. As shown in Figure 5.6, they used the solution method to construct a new ICL based on a P(positive)–N(negative) heterojunction to address the interfacial corrosion phenomenon in tandem QDLEDs. As shown in Figure 5.6b, each functional layer is clearly visible in the TEM cross-sectional image, indicating that the solvent used in the solution method preparation process did not erode the functional layers. In addition, they used a hybrid deposition technique where chlorobenzene (CB) was used to rinse the PVK and QD layers at the interface and the hydrophilic reagent isopropyl alcohol (IPA) was coated into the PEDOT:PSS to maintain the high quality of the functional layers. The performance of the tandem device using EGaIn as the anode electrode was dramatically improved compared to the single device with a turn-on voltage of 7.3 V and EQE, CE, and PE of 23.68%, 101 cd A^{-1} and 20.93 lm W^{-1}, respectively, as shown in Figure 5.6c–e. Later, the same group reported a tandem QDLED with a conventional structure that is suitable for both n-type and p-type

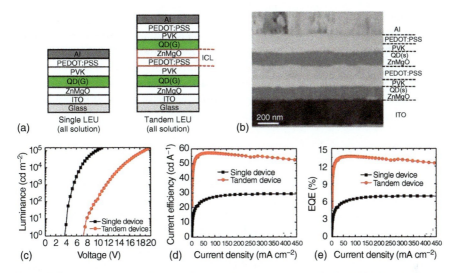

Figure 5.6 (a) Schematic diagram of the device structure of a single QDLED and a tandem-structured QDLED. (b) Cross-sectional TEM image of an inverted tandem structure QDLED. (c) L–V, (d) CE–J, and (e) EQE–J characteristic curves of a tandem structure QDLED. Source: Zhang et al. [55]/John Wiley & Sons.

TFTs [11]. They constructed a new ICL-ZnMgO/Al/HATCN/MoO$_3$ and assembled a tandem QDLED with a conventional structure. The QDLED device has a turn-on voltage of 6.1 V and exhibits extremely high CE and EQE of 121.5 cd A^{-1} and 27.6%, respectively.

5.4 Factors Affecting the Performance of Green QDLEDs

Factors affecting the photovoltaic performance of green QDLEDs (e.g. efficiency and lifetime) can be grouped into two main categories: materials and device structure [21, 36, 56, 57]. The causes of performance degradation of QDLEDs include poor material stability, charge transport imbalance, and the occurrence of exciton decay processes. Exciton decay can occur through (i) decreasing radiative combination rates, (ii) intermittent combination, (iii) non-radiative combination due to defective states, (iv) Förster resonance energy transfer (FRET) processes, (v) energy transfer to adjacent QDs or CTLs, etc. From material perspective, it is mainly the properties of the QDs themselves that affect device performance, such as the synthesis chemistry, including the size, structure, and composition of the QDs, and the surface chemistry of the QDs. Optimization of the shell layer thickness and structure is an effective way to suppress non-radiative combination and FRET processes and to improve the PLQY of QDs. In addition, the ligands encapsulated on the QDs are related to the stability of the solution-state QDs, the passivation of the trap state on the surface of the QDs, and the charge transport properties of the QD films. From device structure perspective, CTLs play a key role in the performance of QDLEDs. The stability of CTLs is also important for the operating time of QDLED devices.

Excitons are susceptible to quenching at the interface between the QD layer and the CTLs. At the same time, most QDLED devices are prepared using the solution method, so the solvents used in the preparation process may corrode the functional layer and thus impair the performance of the QDLEDs. In general, it is necessary to take measures to adjust the CTL energy band distribution to match well with the QD emitting layer. Strategies to improve device performance in terms of materials and device structure are discussed in more detail below.

5.4.1 QD Ligand Effect

Surface defects formed on the cores or shells of QDs during synthesis have a significant impact on the optoelectronic properties of QDLEDs, such as the efficiency and stability of the devices. Usually, QDs are coated with long-chain insulating organic ligands (e.g. oleylamine, oleic acid [OA], etc.) after synthesis, which is detrimental to carrier transport. Therefore, it is significant to select suitable ligands that not only improve the stability of QDs but also facilitate carrier transport [8, 23, 47, 54, 58–60]. For example, the Li group replaced the OA ligand with tris(mercaptomethyl)nonane (TMMN) to synthesize $Zn_{1-x}Cd_xSe/Zn$ core/shell QDs to improve device performance, as shown in Figure 5.7a [16]. A significant increase in hole current is achieved after ligand exchange to balance with electron transport, as shown in Figure 5.7b. The assembled device structure is ITO/PEDOT:PSS/TFB/QDs/ZnO/Al with EQE, PE, and CE of 16.5%, 57.6 lm W^{-1}, and 70.1 cd A^{-1}, respectively.

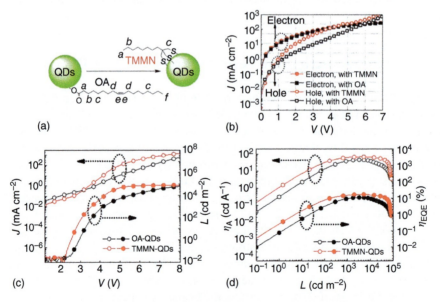

Figure 5.7 (a) Schematic diagram of ligand exchange during synthesis. (b) J–V characteristic curves of electron- and hole-only devices based on OA or TMMN ligands for 40 nm thick quantum dots. (c) J–V and L–V characteristic curves of QDLEDs based on OA or TMMN ligands. (d) CE–L and EQE–L characteristic curves of QDLEDs based on OA or TMMN ligands. Source: Li et al. [16]/John Wiley & Sons.

Surprisingly, the lifetime of the device exceeds 480 000 hours and it has a low turn-on voltage of 2.2 V. The Delville group proposes a new method for the synthesis of CdSe/ZnS thick-shell QDs using tri-*n*-octylphosphine (TOP) as ligand and solvent [58]. The TOP can effectively remove excess ions from the surface of the QDs and passivate the surface defects. The QDLED exhibits saturated green emission at 532 nm with a FWHM of 26 nm. In 2016, Kang et al. first proposed surface modification of CdSe/ZnS QDs with the bromine anion (Br^-) in cetyltrimethylammonium bromide (CTAB) instead of the OA ligand, as shown in Figure 5.8 [59]. The short ligand Br^- shortens the distance between adjacent QDs and accelerates the exciton energy transfer rate. The EL light intensity at 540 nm is higher for Br-QDLED than OA-QDLED, as shown in Figure 5.8b, because there are fewer defect states on the CdSe/ZnS QDs at the Br^- block. Compared to OA-QDLED, the maximum

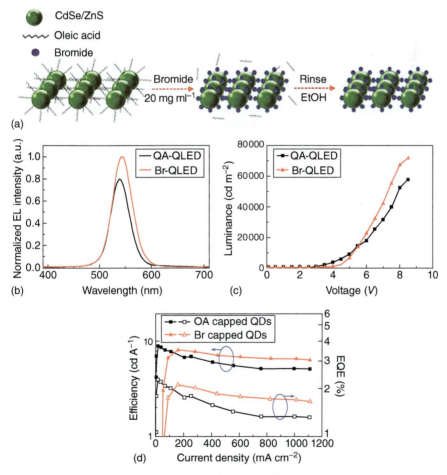

Figure 5.8 (a) Schematic representation of the surface ligand exchange from OA to Br^- after CTAB treatment. (b) Normalized EL spectra of OA- or Br- based QDLEDs. (c) L–V characteristic curves of OA- or Br-based QDLEDs. (d) CE–J and EQE–J characteristic curves of OA- or Br-based QDLEDs. Source: Kang et al. [59]/Springer Nature /CC BY 4.0.

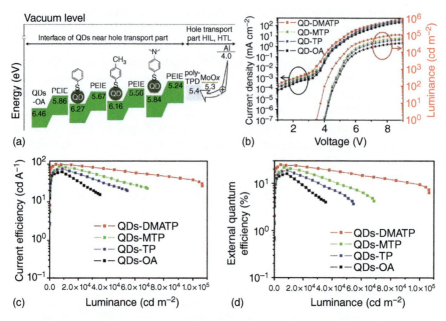

Figure 5.9 (a) Schematic diagram of the VBM and the energy level distribution at the quantum dot-hole transport layer interface based on measured quantum dots in inverted green QDLEDs. (b) J–V–L characteristic curves of QDLEDs. (c) CE–L characteristic curves of QDLEDs. (d) EQE–L characteristic curves of QDLEDs. Source: Moon and Chae [8]/John Wiley & Sons.

brightness of Br-QDLED is higher at 71 000 cd m^{-2}, with an EQE of 1.65%, which is associated with carrier injection balance, exciton combination enhancement, and reduced emission quenching, as shown in Figure 5.8c,d. The Br-QDLED has a slightly higher turn-on voltage of 3.0 V than that of the OA-QDLED. In addition, the Sargent group replaced the long organic ligands with chloride reagents (SOCl$_2$) to coat the CdSeZnS QDs [54]. The electrically conductive chloride ligands balanced the carrier transport and reduced the chance of non-radiative Auger combinations. Cl-QDLEDs have a low turn-on voltage of 2.5 V and a maximum brightness of 460 000 cd m^{-2}. As shown in Figure 5.9 the Moon group reported three short ligands-TP, 4-MTP, and 4DMATP thiophenol derivatives were selected to replace the OA ligands to coat the CdSe@ZnS/ZnS QDs [8]. The ligand exchange results in a significant reduction of the mismatch between the valence band maxima (VBM) and highest occupied molecular orbital (HOMO) energy levels while weakening the non-radiative oscillatory combination and facilitating carrier transport. Among these QDLEDs, the QDLED with the chosen 4-DMATP ligand exhibits the best optoelectronic performance. The maximum luminance, CE, and EQE of the device reached 106 400 cd m^{-2}, 98.2 cd A^{-1}, and 24.8%, respectively. In addition, the DMATP-QDLED has the lowest turn-on voltage of 3.5 V, which is due to the reduced energy barrier between the poly-TPD and QD layers, which accelerates the charge transfer.

5.4.2 QD Core/Shell Structure

The fundamental nature of QDs is the main factor affecting the performance of QDLED devices. The shell layer thickness of QDs with core/shell structure is related to EL brightness, PE, EQE, and charge injection capability of the device [9, 19, 20, 61]. It has been demonstrated that increasing the shell layer thickness is an effective means to suppress the Auger combination and FRET processes, regulate the energy band distribution of the core/shell QDs, and promote charge injection. For example, in 2013, the Shen group studied the effect of shell layer thickness on the performance of devices with the structure of ITO/PEDOT:PSS/TFB/QDs/ZnO/Al [17]. As shown in Figure 5.10, they set the shell layer thickness to 1.8, 2.1, 3.7, and 4.8 nm. Figure 5.10a summarizes some of the QD-related parameters. When the shell thickness is 2.1 nm, the QDLED has the lowest current density due to

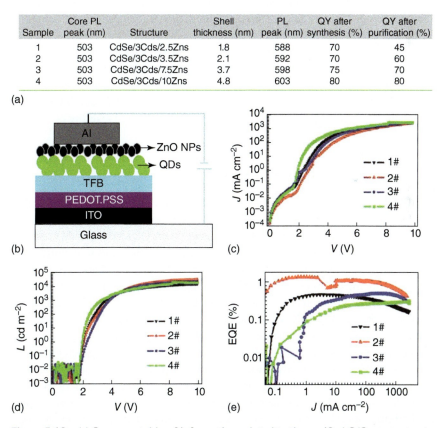

Sample	Core PL peak (nm)	Structure	Shell thickness (nm)	PL peak (nm)	QY after synthesis (%)	QY after purification (%)
1	503	CdSe/3Cds/2.5Zns	1.8	588	70	45
2	503	CdSe/3Cds/3.5Zns	2.1	592	70	60
3	503	CdSe/3Cds/7.5Zns	3.7	598	75	70
4	503	CdSe/3Cds/10Zns	4.8	603	80	80

Figure 5.10 (a) Summary table of information related to the purified CdSe core, structure, total shell layer thickness, PL position of the core/shell quantum dots, and QY. (b) Schematic diagram of the device structure. (c) Hop J–V characteristic curves of QDLEDs with different shell layer thicknesses. (d) L–V characteristic curves of QDLEDs with different shell thicknesses. (e) EQE–J characteristic curves of QDLEDs with different shell thicknesses. Source: Shen et al. [61]/American Chemical Society.

the carrier injection balance, as shown in Figure 5.10b, and the EQE also reaches a maximum value of 1.39%. Therefore, 2.1 nm is the optimum case thickness. In addition, the Shen group synthesized thick-shell structured $Zn_{1-x}Cd_xSe/ZnS$ core/shell QDs and found that their QY, luminescence efficiency, CE, and maximum brightness all increased with increasing shell thickness. The best-performing device had an EL peak at 517 nm, a FWHM of 45 nm, a turn-on voltage of only 2.35 V, and EQE and maximum luminance of 15.4% and 12 100 cd m^{-2}. However, too thick a shell may create resistance, resulting in a higher turn-on voltage required to achieve the same luminance. Another strategy for shell engineering is to use alloyed QDs with a gradient distribution of shell chemical composition. The middle shell acts as a connection point to alleviate the lattice mismatch between the core and the shell, improving the PL stability of the QDs. In 2015, Yang's group synthesized a shell layer ($Cd_{1-x}Zn_xSe_{1-y}S_y$) with a gradient distribution of chemical composition and coated the core layer material to obtain highly efficient and long-lived green QDLEDs [19]. The unique core/shell structure reduces the strain caused by lattice mismatch between the core and the shell, facilitating charge injection and transport and increasing the PLQY of the QDs. The EQE of the green QDLED is 14.5%, the turn-on voltage is as low as 2.0 V and the lifetime is up to 90 000 hours. In addition, the thermal stability of the QDs is also related to their shell layer structure [21]. If the shell layer of QDs is thermally unstable, the surface of QDs is prone to defects, which result in small energy bands and thus affect the effective combination of excitons. Therefore, it is also important to improve the thermal stability of QDs in terms of their chemical structure.

5.4.3 Optimization of the Device Structure

In addition to the properties of the QDs themselves, the degree to which the energy levels of each functional layer in the device are matched is an important factor affecting the optoelectronic performance of green QDLEDs. For this reason, considerable work has been done to balance charge injection and transport, such as by adjusting the energy level distribution of the HTL or ETL adjacent to the QD layer or introducing polymers as buffer layers [25, 28, 30, 33, 43, 45, 50, 51, 62–67]. There are two main approaches to accelerate charge transport: on the one hand, some reports deepen the HOMO of the HTL to lower the hole injection barrier; on the other hand, others report achieving carrier transport balance by slowing down the electron injection rate. The most commonly used ETL in green QDLEDs is ZnO, so many improvements have been based on this [25, 45, 62]. The Pan group doped ZnO nanoparticles (NPs) with certain concentrations of cesium azide (CsN_3) to slow down electron transport (see Figure 5.11) [62]. ZnO doped with different concentrations of CsN_3 exhibited different film morphology and roughness, with the quality of the film having a significant effect on current leakage. The best device performance was achieved when CsN_3 was doped at a concentration of 4 vol%. After doping, the conduction band minimum (CBM) of ZnO:CsN_3 is lower than that of ZnO, which widens the bandgap and thus increases the electron transport barrier. The device structure is ITO/ZnO:CsN_3/CdSe/ZnS

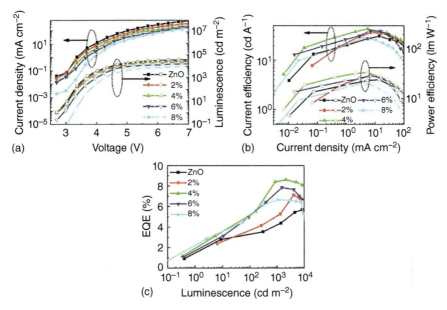

Figure 5.11 (a) J–V and L–V characteristic curves of QDLEDs with different CsN$_3$ doping concentrations in ZnO. (b) CE–J and PE–J characteristic curves of QDLEDs in ZnO with different CsN$_3$ doping concentrations. (c) EQE–L characteristic curves of QDLEDs with different CsN$_3$ doping concentrations in ZnO. Source: Pan et al. [62]/The Royal Society of Chemistry.

QDs/TAPC/HATCN/MoO$_3$/Al with a turn-on voltage of only 2.6 V, while exhibiting high CE, PE, and EQE of 43.1 cd A^{-1}, 33.6 lm W^{-1}, and 9.1%, respectively. Typically, the VBM of QDs (−6.0 V) is deeper than the HOMO energy level (−5.0 V) of organic HTL, which can successfully suppress hole transport [28]. Researchers tend to use small molecules to modify the HTL and adjust its bandgap. The Xu group introduced N,N′-bis(3-methylphenyl)-N,N′-bis(phenyl)benzidine (TPD) small molecules into PVK as a bilayer HTL, as shown in Figure 5.12 [65]. The addition of TPD effectively reduces the charge transfer resistance and improves the exciton compliance, as shown in Figure 5.12a. Compared to PVK-based QDLEDs, the EQE of QDLEDs with a double-layer HTL increases from 4.65% to 8.62%, the maximum luminance increases from 4114 to 56 157 cd m^{-2}, the CE increases from 5.96 to 23.05 cd A^{-1}, and the turn-on voltage decreases from 3.3 to 2.4 V. In inverted QDLEDs, where the solution method is commonly used to assemble the device. The solvent of the HTL may erode the lower EML layer and degrade the device's performance. Therefore, Chae's group inserted an ethoxylated polyethyleneimine (PEIE) as a buffer layer between the EML and the HTL [67]. As shown in Figure 5.13a,b, PEIE effectively enhances the VBM of the QDs and promotes hole migration because the addition of PEIE reduces the surface roughness of the QDs. They chose PEIE (15.5 nm) assembled devices with the structure of ITO/ZnO/CdSe@ZnS/ZnSQDs/PEIE/Poly-TPD/MoO$_x$/Al. The devices with PEIE introduced have lower current density than those without PEIE, which

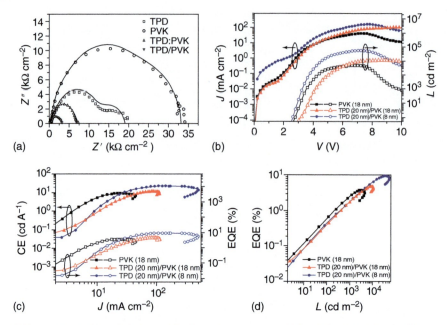

Figure 5.12 (a) Nyquist plots of the impedance spectra of QDLEDs with different HTLs at 3 V and their fitted curves. (b) J–V–L characteristic curves of QDLEDs based on pure PVK, TPD (20 nm)/PVK (18 nm) based dual HTL, and TPD (20 nm)/PVK (8 nm) based dual HTL. (c) CE–J and EQE–J characteristic curves of QDLEDs. (d) EQE–L characteristic curves of QDLEDs. Source: Li et al. [65]/American Chemical Society.

proves that PEIE can prevent the QDs layer from solvent corrosion during HTL deposition, as shown in Figure 5.13c. The device has a turn-on voltage of 3.3 V and CE, PE, EQE, and maximum luminance of 65.3 cd A^{-1}, 29.3 lm W^{-1}, 15.6%, and 110 205 cd m^{-2}, respectively. In addition to modifying the material of the functional layer, the preparation method of the device will also affect its performance. Vacuum evaporation is the most common method for preparing QDLEDs, but it is costly. While other functional layers can be prepared functionally using the solution method to reduce the cost, the metal cathodes still need to be prepared by vacuum evaporation [14, 33, 51, 68, 69]. In 2016, the Peng group reported EGaIn as a printable liquid metal cathode, which was prepared by a vacuum-free solution method [30]. They found that EGaIn could reduce electron injection and balance carrier transport. This green QDLED exhibited excellent optoelectronic performance with an EQE of 12.85% and a luminance of 41 160 cd m^{-2}, with a device turn-on voltage of 4.0 V.

5.4.4 Other Strategies to Improve Device Performance

As the operating time of the device increases, its related properties (such as EQE, CE, PE, and brightness) will decrease to a certain extent. In response, Acharya et al. first exploited the positive aging effect by using a special packaging technique to

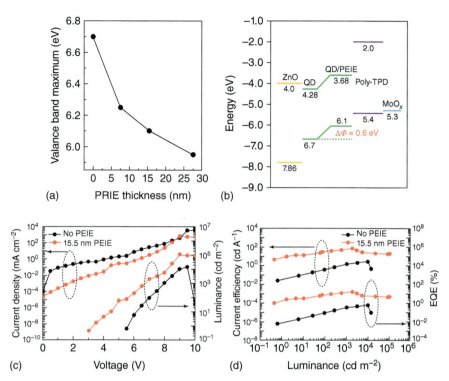

Figure 5.13 (a) Plot of VBM energy levels of quantum dot layers versus the variation of PEIE layer thickness. (b) Energy band distribution of multilayer QDLEDs with inverted structure. The VBM of quantum dots rises by 0.6 V when a 15.5 nm thick PEIE is added. (c) $J-L-V$ characteristic curves of QDLEDs without/with the addition of a 15.5 nm thick PEIE layer. (d) CE–EQE–L characteristic curve of QDLEDs without/with the addition of a 15.5 nm thick PEIE layer. Source: Kim et al. [67]/American Chemical Society.

improve the efficiency of the device after a long period of operation [70]. As shown in Figure 5.14a, the device structure was ITO/PEDOT:PSS/TFB/QDs/ZnO/Al, treated with acidic resin, and placed in a glove box for a few days. It was found that after positive aging, current leakage could be effectively suppressed and the brightness of the green QDLED was substantially enhanced, as shown in Figure 5.14c. In addition, the AlO_x interlayer formed by the reaction between Al and oxygen in ZnMgO reduces the electron injection barrier and thus inhibits the quenching of excitons by the metal electrode. The green QDLED has an EQE of up to 21%, a turn-on voltage as low as 2.1 V, and CE and PE of 82 cd A^{-1} and 80 lm W^{-1}, respectively. In 2018, the Chen group further explored the mechanism of positive aging to explain the exciton quenching phenomenon from a completely new perspective [4]. In addition, they concluded that post-annealing treatment is an effective method to promote the interfacial reaction between Al and ZnMgO [71]. The metallization of AlZnMgO reduces the interfacial contact resistance and balances the carrier transport. The annealed green QDLED exhibits a high EQE of 15.75%, a low drive voltage of 3.7 V, and a high luminance of 576 211 cd m^{-2}.

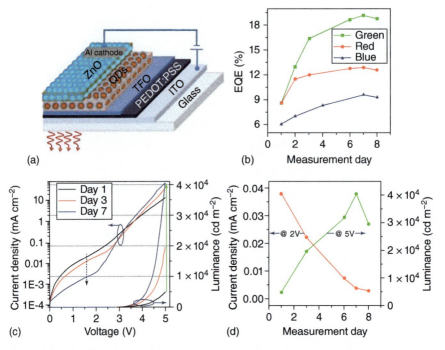

Figure 5.14 (a) Schematic diagram of the device structure. (b) Aging trend of red, green, and blue QDLEDs at room temperature after resin encapsulation. (c) Change in brightness and current density of green QDLEDs relative to applied voltage after aging at room temperature for one day, three days, or seven days. (d) Leakage current density of green QDLEDs measured at 2 V and luminance measured at 5 V over time. Source: Acharya et al. [70]/The Royal Society of Chemistry.

5.5 Summary and Outlook

In general, AM-QDLEDs offer many advantages over other lighting and display devices (such as LCDs and OLEDs), such as low cost, solution processability, high color purity, and high efficiency, which have attracted a great deal of interest from academia and industry. In this section, we discuss the development of green QDLEDs from the perspective of materials and device structures. On the one hand, we introduce in detail the types of QDs commonly used in green QLEDs, the core–shell structure, and the evolution of device structures; on the other hand, we summarize the strategies to improve the efficiency of green QLEDs, including ligand engineering, shell engineering, and interface engineering. The main causes of device efficiency roll-off are non-radiative combinations (including FRET process and AR process) and field effect-induced exciton quenching. QDs with high PLQY do not necessarily exhibit high efficiency because QDs with heterogeneous structures (e.g. core/shell structure and chemical composition) are prone to non-radiative recombination. Therefore, QDs with double shells or shells with a gradient distribution of chemical composition have been proposed to solve this problem. In addition, charge balance is also an important factor affecting device performance. There are two ways to balance charge injection and transport. The first is to mix small molecules or polymers (e.g. poly-TPD, TCTA) with HTL materials to accelerate the hole mobility and reduce the hole injection potential; the other is to introduce an insulating polymer (e.g. PEIE) between the QD layer and the ETL to slow down the electron transport.

However, the lifetime and stability of green QDLEDs are still factors that limit the further practical application of QDLEDs. Due to the large specific surface area of QDs, they are sensitive to light and humidity in the surrounding environment. In addition, the carrier transport balance caused by energy level mismatch and carrier migration also affects the lifetime and stability of the device. Joule heating and excitonic non-radiative processes generated during device operation can damage HTL materials. Based on the results of current research, there are three main approaches to achieve both long device lifetime and high performance. First, the properties of QDs should be improved. QDs with thick shells or shells with gradient distribution of chemical composition can solve the above problems, because thick shells can effectively increase the fluorescence quantum yield of QDs while suppressing non-radiative recombination of defect states; alloyed shells can relieve the core and shell materials lattice mismatch. Second, low hole mobility and high hole injection barrier are important factors limiting the efficiency of green QLEDs. Mixing the HTL with small molecules or developing new HTL materials are effective ways to modify the HTL. Third, reducing electron mobility is also feasible to balance charge transport, such as by inserting insulating polymers (such as PMMA and PVK) as buffer layers to improve device performance. Although researchers have done a lot of work, such as optimizing the synthesis of QDs and the CTL, and promoting the balance of charge injection and transport, in order to obtain high stability and long life of green QLEDs, the mechanism of their efficiency roll-off is still controversial. Whether the high voltage reduces the radiation rate or the Auger recombination process at high current density causes the efficiency roll-off is uncertain. In short, if green QLED is to be applied to real life, there is a long way to go, and researchers need to work hard to carry out more related innovations.

References

1 Shen, P., Cao, F., Wang, H. et al. (2019). Solution-processed double-junction quantum-dot light-emitting diodes with an EQE of over 40%. *ACS Applied Materials and Interfaces* 11 (1): 1065–1070.
2 Su, Q., Zhang, H., Xia, F. et al. (2018). Tandem red quantum-dot light-emitting diodes with external quantum efficiency over 34%. *SID Symposium Digest of Technical Papers* 49 (1): 977–980.
3 Dai, X., Zhang, Z., Jin, Y. et al. (2014). Solution-processed, high-performance light-emitting diodes based on quantum dots. *Nature* 515 (7525): 96–99.
4 Su, Q., Sun, Y., Zhang, H., and Chen, S. (2018). Origin of positive aging in quantum-dot light-emitting diodes. *Advanced Science (Weinheim)* 5 (10): 1800549.

5 Liang, F., Liu, Y., Hu, Y. et al. (2017). Polymer as an additive in the emitting layer for high-performance quantum dot light-emitting diodes. *ACS Applied Materials and Interfaces* 9 (23): 20239–20246.

6 Kim, H.-M., bin Mohd Yusoff, A.R., Youn, J.-H., and Jang, J. (2013). Inverted quantum-dot light emitting diodes with cesium carbonate doped aluminium-zinc-oxide as the cathode buffer layer for high brightness. *Journal of Materials Chemistry C* 1 (25): 3924–3930.

7 Sun, Y., Jiang, Y., Peng, H. et al. (2017). Efficient quantum dot light-emitting diodes with a $Zn_{0.85}Mg_{0.15}O$ interfacial modification layer. *Nanoscale* 9 (26): 8962–8969.

8 Moon, H. and Chae, H. (2020). Efficiency enhancement of all-solution-processed inverted-structure green quantum dot light-emitting diodes via partial ligand exchange with thiophenol derivatives having negative dipole moment. *Advanced Optical Materials* 8 (1): 1901314.

9 Yang, Z., Wu, Q., Lin, G. et al. (2019). All-solution processed inverted green quantum dot light-emitting diodes with concurrent high efficiency and long lifetime. *Materials Horizons* 6 (10): 2009–2015.

10 Manders, J.R., Qian, L., Titov, A. et al. (2015). High efficiency and ultra-wide color gamut quantum dot LEDs for next generation displays. *Journal of the Society for Information Display* 23 (11): 523–528.

11 Zhang, H., Chen, S., and Sun, X.W. (2018). Efficient red/green/blue tandem quantum-dot light-emitting diodes with external quantum efficiency exceeding 21%. *ACS Nano* 12 (1): 697–704.

12 Wang, L., Lin, J., Hu, Y. et al. (2017). Blue quantum dot light-emitting diodes with high electroluminescent efficiency. *ACS Applied Materials and Interfaces* 9 (44): 38755–38760.

13 Wang, O., Wang, L., Li, Z. et al. (2018). High-efficiency, deep blue $ZnCdS/Cd_xZn_{1-x}S/ZnS$ quantum-dot-light-emitting devices with an EQE exceeding 18%. *Nanoscale* 10 (12): 5650–5657.

14 Zou, Y., Ban, M., Cui, W. et al. (2017). A general solvent selection strategy for solution processed quantum dots targeting high performance light-emitting diode. *Advanced Functional Materials* 27 (1): 1603325.

15 Araki, Y., Ohkuno, K., Furukawa, T., and Saraie, J. (2007). Green light emitting diodes with CdSe quantum dots. *Journal of Crystal Growth* 301–302: 809–811.

16 Li, Z., Hu, Y., Shen, H. et al. (2017). Efficient and long-life green light-emitting diodes comprising tridentate thiol capped quantum dots. *Laser and Photonics Reviews* 11 (1): 1600227.

17 Shen, H., Lin, Q., Cao, W. et al. (2017). Efficient and long-lifetime full-color light-emitting diodes using high luminescence quantum yield thick-shell quantum dots. *Nanoscale* 9 (36): 13583–13591.

18 Choi, S., Moon, J., Cho, H. et al. (2019). Partially pyridine-functionalized quantum dots for efficient red, green, and blue light-emitting diodes. *Journal of Materials Chemistry C* 7 (12): 3429–3435.

19 Yang, Y., Zheng, Y., Cao, W. et al. (2015). High-efficiency light-emitting devices based on quantum dots with tailored nanostructures. *Nature Photonics* 9 (4): 259–266.

20 Jun, S. and Jang, E. (2013). Bright and stable alloy core/multishell quantum dots. *Angewandte Chemie International Edition in English* 52 (2): 679–682.

21 Moon, H., Lee, C., Lee, W. et al. (2019). Stability of quantum dots, quantum dot films, and quantum dot light-emitting diodes for display applications. *Advanced Materials* 31 (34): e1804294.

22 Pu, Y.-C. and Hsu, Y.-J. (2014). Multicolored $Cd_{1-x}Zn_xSe$ quantum dots with type-I core/shell structure: single-step synthesis and their use as light emitting diodes. *Nanoscale* 6 (7): 3881–3888.

23 Nguyen, H.T., Nguyen, N.D., and Lee, S. (2013). Application of solution-processed metal oxide layers as charge transport layers for CdSe/ZnS quantum-dot LEDs. *Nanotechnology* 24 (11): 115201.

24 Deng, Z., Yan, H., and Liu, Y. (2009). Band gap engineering of quaternary-alloyed ZnCdSSe quantum dots via a facile phosphine-free colloidal method. *Journal of the American Chemical Society* 131 (49): 17744–17745.

25 Qian, L., Zheng, Y., Xue, J., and Holloway, P.H. (2011). Stable and efficient quantum-dot light-emitting diodes based on solution-processed multilayer structures. *Nature Photonics* 5 (9): 543–548.

26 Shen, H., Wang, H., Zhou, C. et al. (2011). Large scale synthesis of stable tricolor $Zn_{1-x}Cd_xSe$ core/multishell nanocrystals via a facile phosphine-free colloidal method. *Dalton Transactions* 40 (36): 9180–9188.

27 Kwak, J., Bae, W.K., Lee, D. et al. (2012). Bright and efficient full-color colloidal quantum dot light-emitting diodes using an inverted device structure. *Nano Letters* 12 (5): 2362–2366.

28 Lee, K.-H., Lee, J.-H., Song, W.-S. et al. (2013). Highly efficient, color-pure, color-stable blue quantum dot light-emitting devices. *ACS Nano* 7 (8): 7295–7302.

29 Lee, K.H., Lee, J.H., Kang, H.D. et al. (2014). Over 40 cd/A efficient green quantum dot electroluminescent device comprising uniquely large-sized quantum dots. *ACS Nano* 8 (5): 4893–4901.

30 Peng, H., Jiang, Y., and Chen, S. (2016). Efficient vacuum-free-processed quantum dot light-emitting diodes with printable liquid metal cathodes. *Nanoscale* 8 (41): 17765–17773.

31 Vasan, R., Salman, H., and Manasreh, M.O. (2018). Solution processed high efficiency quantum dot light emitting diode with inorganic charge transport layers. *IEEE Electron Device Letters* 39 (4): 536–539.

32 Li, X., Lin, Q., Song, J. et al. (2020). Quantum-dot light-emitting diodes for outdoor displays with high stability at high brightness. *Advanced Optical Materials* 8 (2): 1901145.

33 Fu, Y., Jiang, W., Kim, D. et al. (2018). Highly efficient and fully solution-processed inverted light-emitting diodes with charge control interlayers. *ACS Applied Materials and Interfaces* 10 (20): 17295–17300.

34 Deng, Y., Peng, F., Lu, Y. et al. (2022). Solution-processed green and blue quantum-dot light-emitting diodes with eliminated charge leakage. *Nature Photonics* 16 (7): 505–511.

35 Anikeeva, P.O., Halpert, J.E., Bawendi, M.G., and Bulovic, V. (2009). Quantum dot light-emitting devices with electroluminescence tunable over the entire visible spectrum. *Nano Letters* 9 (7): 2532–2536.

36 Yang, Z., Gao, M., Wu, W. et al. (2019). Recent advances in quantum dot-based light-emitting devices: challenges and possible solutions. *Materials Today* 24: 69–93.

37 Zhang, H., Hu, N., Zeng, Z. et al. (2019). High-efficiency green InP quantum dot-based electroluminescent device comprising thick-shell quantum dots. *Advanced Optical Materials* 7 (7): 1801602.

38 Lim, J., Park, M., Bae, W.K. et al. (2013). Highly efficient cadmium-free quantum dot light-emitting diodes enabled by the direct formation of excitons within InP@ZnSeS quantum dots. *ACS Nano* 7 (10): 9019–9026.

39 Wang, A., Shen, H., Zang, S. et al. (2015). Bright, efficient, and color-stable violet ZnSe-based quantum dot light-emitting diodes. *Nanoscale* 7 (7): 2951–2959.

40 Xiang, C., Koo, W., Chen, S. et al. (2012). Solution processed multilayer cadmium-free blue/violet emitting quantum dots light emitting diodes. *Applied Physics Letters* 101 (5): 053303.

41 Chen, B., Zhong, H., Wang, M. et al. (2013). Integration of $CuInS_2$-based nanocrystals for high efficiency and high colour rendering white light-emitting diodes. *Nanoscale* 5 (8): 3514–3519.

42 Choi, D.B., Kim, S., Yoon, H.C. et al. (2017). Color-tunable Ag-In-Zn-S quantum-dot light-emitting devices realizing green, yellow and amber emissions. *Journal of Materials Chemistry C* 5 (4): 953–959.

43 Ji, W., Shen, H., Zhang, H. et al. (2018). Over 800% efficiency enhancement of all-inorganic quantum-dot light emitting diodes with an ultrathin alumina passivating layer. *Nanoscale* 10 (23): 11103–11109.

44 Kobayashi, M., Nakamura, S., Kitamura, K. et al. (1999). Luminescence properties of CdS quantum dots on ZnSe. *Journal of Vacuum Science and Technology, B: Microelectronics and Nanometer Structures–Processing, Measurement, and Phenomena* 17 (5): 2005–2008.

45 Kim, N.-Y., Hong, S.-H., Kang, J.-W. et al. (2015). Localized surface plasmon-enhanced green quantum dot light-emitting diodes using gold nanoparticles. *RSC Advances* 5 (25): 19624–19629.

46 Kim, J.-U., Lee, J.-J., Jang, H.S. et al. (2011). Widely tunable emissions of colloidal $Zn_xCd_{1-x}Se$ alloy quantum dots using a constant Zn/Cd precursor ratio. *Journal of Nanoscience and Nanotechnology* 11 (1): 725–729.

47 Bae, W.K., Kwak, J., Park, J.W. et al. (2009). Highly efficient green-light-emitting diodes based on CdSe@ZnS quantum dots with a chemical-composition gradient. *Advanced Materials* 21 (17): 1690–1694.

48 Panda, S.K., Hickey, S.G., Waurisch, C., and Eychmüller, A. (2011). Gradated alloyed CdZnSe nanocrystals with high luminescence quantum yields and stability for optoelectronic and biological applications. *Journal of Materials Chemistry* 21 (31): 11550–11555.

49 Talapin, D.V., Mekis, I., Götzinger, S. et al. (2004). CdSe/CdS/ZnS and CdSe/ZnSe/ZnS core–shell–shell nanocrystals. *The Journal of Physical Chemistry B* 108 (49): 18826–18831.

50 Kim, H.M., Kim, J., and Jang, J. (2018). Quantum-dot light-emitting diodes with a perfluorinated ionomer-doped copper-nickel oxide hole transporting layer. *Nanoscale* 10 (15): 7281–7290.

51 Cao, F., Wang, H., Shen, P. et al. (2017). High-efficiency and stable quantum dot light-emitting diodes enabled by a solution-processed metal-doped nickel oxide hole injection interfacial layer. *Advanced Functional Materials* 27 (42): 1704278.

52 Cho, S.H., Sung, J., Hwang, I. et al. (2012). High performance AC electroluminescence from colloidal quantum dot hybrids. *Advanced Materials* 24 (33): 4540–4546.

53 Xu, Q., Li, X., Lin, Q. et al. (2020). Improved efficiency of all-inorganic quantum-dot light-emitting diodes via interface engineering. *Frontiers in Chemistry* 8: 265.

54 Li, X., Zhao, Y.-B., Fan, F. et al. (2018). Bright colloidal quantum dot light-emitting diodes enabled by efficient chlorination. *Nature Photonics* 12 (3): 159–164.

55 Zhang, H., Sun, X., and Chen, S. (2017). Over 100 cd A^{-1} efficient quantum dot light-emitting diodes with inverted tandem structure. *Advanced Functional Materials* 27 (21): 1700610.

56 Dai, X., Deng, Y., Peng, X., and Jin, Y. (2017). Quantum-dot light-emitting diodes for large-area displays: towards the dawn of commercialization. *Advanced Materials* 29 (14): 1607022.

57 Davidson-Hall, T. and Aziz, H. (2020). Perspective: toward highly stable electroluminescent quantum dot light-emitting devices in the visible range. *Applied Physics Letters* 116 (1): 34090.

58 Hao, J., Liu, H., Miao, J. et al. (2019). A facile route to synthesize CdSe/ZnS thick-shell quantum dots with precisely controlled green emission properties: towards QDs based LED applications. *Scientific Reports* 9 (1): 12048.

59 Kang, B.H., Lee, J.S., Lee, S.W. et al. (2016). Efficient exciton generation in atomic passivated CdSe/ZnS quantum dots light-emitting devices. *Scientific Reports* 6: 34659.

60 Shen, H., Zhou, C., Xu, S. et al. (2011). Phosphine-free synthesis of $Zn_{1-x}CdxSe/ZnSe/ZnSe_xS_{1-x}$/ZnS core/multishell structures with bright and stable blue–green photoluminescence. *Journal of Materials Chemistry* 21 (16): 6046–6053.

61 Shen, H., Lin, Q., Wang, H. et al. (2013). Efficient and bright colloidal quantum dot light-emitting diodes via controlling the shell thickness of quantum dots. *ACS Applied Materials and Interfaces* 5 (22): 12011–12016.

62 Pan, J., Wei, C., Wang, L. et al. (2018). Boosting the efficiency of inverted quantum dot light-emitting diodes by balancing charge densities and suppressing exciton quenching through band alignment. *Nanoscale* 10 (2): 592–602.

63 Liu, Y., Jiang, C., Song, C. et al. (2018). Highly efficient all-solution processed inverted quantum dots based light emitting diodes. *ACS Nano* 12 (2): 1564–1570.

64 Huang, Q., Pan, J., Zhang, Y. et al. (2016). High-performance quantum dot light-emitting diodes with hybrid hole transport layer via doping engineering. *Optics Express* 24 (23): 25955–25963.

65 Li, J., Liang, Z., Su, Q. et al. (2018). Small molecule-modified hole transport layer targeting low turn-on-voltage, bright, and efficient full-color quantum dot light emitting diodes. *ACS Applied Materials and Interfaces* 10 (4): 3865–3873.

66 Tuan, N.H., Lee, S., and Dinh, N.N. (2013). Investigation of pure green-colour emission from inorganic–organic hybrid LEDs based on colloidal CdSe/ZnS quantum dots. *International Journal of Nanotechnology* 10 (3–4): 304–312.

67 Kim, D., Fu, Y., Kim, S. et al. (2017). Polyethylenimine ethoxylated-mediated all-solution-processed high-performance flexible inverted quantum dot-light-emitting device. *ACS Nano* 11 (2): 1982–1990.

68 Zhang, Z., Ye, Y., Pu, C. et al. (2018). High-performance, solution-processed, and insulating-layer-free light-emitting diodes based on colloidal quantum dots. *Advanced Materials* 30 (28): e1801387.

69 Kwon, B.W., Son, D.I., Park, D.-H. et al. (2012). Solution-processed white light-emitting diode utilizing hybrid polymer and red–green–blue quantum dots. *Japanese Journal of Applied Physics* 51: 09MH03.

70 Acharya, K.P., Titov, A., Hyvonen, J. et al. (2017). High efficiency quantum dot light emitting diodes from positive aging. *Nanoscale* 9 (38): 14451–14457.

71 Su, Q., Zhang, H., Sun, Y. et al. (2018). Enhancing the performance of quantum-dot light-emitting diodes by postmetallization annealing. *ACS Applied Materials and Interfaces* 10 (27): 23218–23224.

6
Blue Quantum Dot Light-Emitting Diodes

6.1 Introduction

Since the quantum confinement effect of quantum dots (QDs) was discovered more than 30 years ago, the specific applications of QDs have been explored by scientific and industrial circles. Because the radius of QDs is smaller than the Bohr exciton radius, the electronic energy level near Fermi level changes from quasi-continuous to discrete, which makes QDs have unique optical and magnetic properties. In recent years, QDs have been widely used in optoelectronics, light-emitting diodes, and biological fields. Because QDs can adjust the emission color by changing their size and composition, LED devices based on QDs have a wide color gamut from ultraviolet (UV) to infrared. In addition, quantum dot light-emitting diodes (QDLEDs) have many excellent properties, such as wet-heat stability, efficiency close to unit brightness, high color purity, and solution preparation ability. All these characteristics make QDLEDs a new generation of displays with perfect color and realistic brightness.

The development of QDLED is very rapid. Shen et al. reported that the external quantum efficiency (EQE) of double-junction green QDLEDs in solution method exceeded 42%, which is the highest value ever [1]. In 2018, Su et al. demonstrated red QDLEDs in series with EQE exceeding 34% [2]. In addition, the lifetime of green and red QDLEDs has been reported to be over 100 000 hours [3]. These properties make red and green QDLEDs comparable to the most advanced organic light-emitting diodes (OLEDs). However, compared with the sustainable development of green and red QDLEDs, blue QDLEDs are relatively backward in efficiency and life. Zhang et al. adopted the interconnect layer (ICL) structure of ZnMgO/Al/HATCN/MoO_3, and reported that the highest EQE of blue QDLEDs so far is 21.4% [4]. Although this efficiency has reached the requirements for commercialization, the short lifetime and low efficiency of high brightness of blue quantum dots still hinder its application.

The poor performance of blue QDLEDs can be attributed to the following factors:

(1) The photoluminescence quantum yields (PLQYs) of blue QDs are low. Although QDs with high PLQYs have been synthesized in recent years, the PLQY loss of QD films is serious due to the purification of QDs and the energy

transfer during film formation. The low PLQYs of QD films lead to the low recombination efficiency of electrons and holes in the luminescent layer, which leads to the low efficiency of QDLEDs.

(2) The charge transfer efficiency between QDs in solid thin films is low. Generally, QDs include inorganic core/shell structures and surface organic ligands, which are necessary to determine their photoelectric properties. Usually, the surface ligands of colloidal quantum dots choose fatty acids with long hydrocarbon chains [5]. Although these long-chain organic molecules can prevent the aggregation of QDs in organic solvents and passivate the surface defects of QDs, they can also form dense insulating layers to affect the charge injection and carrier transport of QDs [6]. Therefore, choosing appropriate ligands can not only stabilize QDs and maintain their inherent good photoelectric properties but also realize efficient charge transfer between QDs in solid thin films.

(3) The injection of electrons and holes in blue QDLEDs is unbalanced. In the traditional organic–inorganic composite QDLED, there is a large interface barrier between the QD light-emitting layer with deep valence band energy level and the organic hole transport layer (HTL) [7]. At present, quite a few reports have tried to study the transport layer materials with more balanced charge injection [8]. However, the research on how to match blue QDs and charge transport materials with different components and shell thicknesses is not sufficient.

(4) Charge transfer caused by heterojunction between QDs and ZnO in electron transport layer (ETL) and charge accumulation in ETL produce blue QDs with positive charges and ZnO with negative charges, which reduces the radiation recombination efficiency in the emission layer of QDs, which is also the reason for the short life of blue QDLEDs [9].

(5) As we all know, the brightness of the display screen is determined by the luminance of the LED, which is defined as the wavelength-weighted power emitted by the light source per unit area in a specific direction (unit solid angle). The luminous efficiency based on human vision:

$$L_v = 683 I \bar{y}(\lambda)/S \qquad (6.1)$$

where L_v is the brightness in cd m^{-2}, I is the radiation power in W sr^{-1}, and S is the surface area of LED. $\bar{y}(\lambda)$ refers to the average sensitivity of human eyes to light with different wavelengths. Different from its green and red bands, the blue band (440–490 nm) in the visual spectrum has low luminous efficiency, ranging from 0.02 to 0.20, while the green band (520–555 nm) is 0.80–1.00, and the orange/red band (590–640 nm) is 0.20–0.70. Compared with 460 nm blue light, 530 nm green light, and 620 nm red light only need 1/9 and 1/4 radiation power, respectively, to achieve the same visual effect, so blue QDLEDs need a higher radiation power I than that of green light or red light QDLEDs with the same brightness to compensate for the lower luminous efficiency [10]. In addition to brightness, the low efficiency of blue emission also requires the narrow bandwidth and "concise" spectral line shape of QDLEDs to achieve the required saturated blue light.

To sum up, the current QDLEDs must use blue backlight to emit red, green, and blue light through the color conversion layer composed of red and green QDs. Because QDs advance the emitted light to the front of the liquid crystal display (LCD) panel, expensive passive color filters can be replaced. At present, the mainstream QD displays all adopt this mode, so they should be called QD-LCDs, not backlit active matrix QDLED (AM-QDLED) displays. In order to realize flexible ultra-thin AM-QDLED devices, it is very important to obtain blue QDs with high efficiency and long lifetime. However, the blue-emitting nanoparticles (NPs) and their short operational lifetime are the key obstacles in the progression of these devices. In view of this, the performance of blue QDLEDs can be improved by designing nanostructured (core–shell structure) particles with low defects, appropriate bandgap energy levels, and high PLQYs, modifying the surface of semiconductor QDs with appropriate ligands, developing charge transport materials with better energy level matching, and optimizing the interface and device structure of blue devices.

In this chapter, we review the research progress of blue QDs, including the luminescent characteristics and device performance of high-efficiency cadmium-containing blue QDs. The main focus of this chapter is on the blue-emitting QDs used for QDLED applications. To meet ultra-high definition (UHD) TVs, the primary blue color Commission Internationale de l'Elcairage (CIE) (0.131, 0.046) is required according to the color gamut BT. 2020 standard, thus blue QDs emission color with an emission wavelength between 460 and 475 nm is required. We review and analyze the most suitable blue QD materials in terms of their composition, QY, full width at half maximum (FWHM), the lifetime, environmental effects, and QDLED device characterizations. Because of the potential physiological toxicity of cadmium, cadmium-free QDs have become a research hotspot in recent years, and we also introduced this kind of eco-friendly and non-toxic QDs in a certain space. With the continuous optimization of the structure of QD luminescent materials and the deeper understanding of the device mechanism, QDLEDs have developed a brand-new inverted device structure, which provides a new idea for the preparation process of blue QDLEDs. Finally, we will discuss the existing problems and possible solutions in the application of blue QDLEDs.

6.2 Blue Quantum Dot Luminescent Materials

Among all the QD-emitting materials reported in the literature, QD materials with the general formula $Zn_xCd_{1-x}Se$ or $Cd_xZn_{1-x}Se_yS_{1-y}$ can cover the entire visible range. Blue emission is achieved at a small fraction of the Se-containing or very small size (<2 nm) of emission core. Considering the narrow emission bandwidth and the significant improvement of high PLQY, these materials are considered ideal blue emitters. In addition, the core–shell type QDs ZnSe/CdSe and ZnSe/CdS/ZnS also cover the entire visible spectral range from blue to red. The introduction of lower bandgap ZnTe ($E_g = 2.25$ eV) in the core component of ZnSe ($E_g = 2.7$ eV) is a very useful method to extend the emission range of ZnSe-based QDs to the relevant blue color ($\lambda_{em} > 460$ nm), thus extending the emission range of these

6 Blue Quantum Dot Light-Emitting Diodes

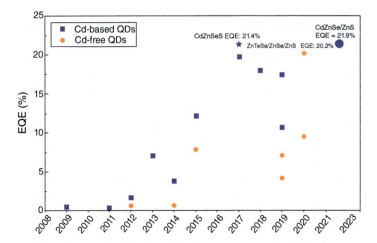

Figure 6.1 Development timeline of external quantum efficiency of blue QDLEDs. Blue dots are cadmium-containing quantum dots, and orange dots are cadmium-free quantum dots.

materials to the blue region. This resulted in the general properties of alloyed core–shell and core-double-shell QDs. Alloy core–shell and core–shell-double-shell QDs with the general compositions ZnSeTe/ZnS and ZnSeTe/ZnS are formed. As mentioned earlier, the research on blue QDLEDs was relatively slow for a long time. On the one hand, it is difficult to find a suitable synthesis method to obtain smaller blue QDs, and on the other hand, it is full of challenges to choose a suitable charge transport layer (CTL) material to match the wide bandgap of blue QDs. After continuous research, the luminous efficiency of blue QDs has made a major breakthrough. The development of external quantum efficiency of blue QDLEDs is shown in Figure 6.1. Through the development of QDs and surface ligands and the optimization of device structure and interface engineering, the external quantum efficiency of cadmium-containing blue QDLEDs has been improved from less than 0.1% to about 20%, which almost meets the requirements of commercialization [4]. In 2020, Samsung Electronics reported that the QDLED with a cadmium-free QD with EQE as high as 20.2% had a lifetime T_{50} of 15 850 hours when the initial brightness was 100 cd m^{-2}, which marked a major breakthrough in cadmium-free blue QDLED [11]. However, how to reduce the cost of materials, obtain higher efficiency, maintain high efficiency during the operation of devices, and the environmental pollution of cadmium-containing QDs are all restrictive factors for the large-scale commercialization of QDLEDs at present.

Here are some of the key breakthroughs in blue QDLED technology development during the years 2007–2023:

1. 2007: First demonstration of blue QDLEDs using cadmium-containing QDs with high color purity and efficiency.
2. 2010: Introduction of nontoxic, cadmium-free QDs for blue QDLEDs.
3. 2012: Development of a hybrid device architecture combining a blue QDLED with a green phosphorescent OLED, achieving high efficiency and brightness.

4. 2014: Achievement of long operational lifetime of blue QDLEDs through improved device structure and materials.
5. 2016: Demonstration of a flexible blue QDLED using solution processing and an ultrathin polymer substrate.
6. 2019: Integration of a blue QDLED with a perovskite solar cell, enabling high efficiency in both light emission and energy conversion.
7. 2023: Introduction of a new generation of blue QDLEDs using advanced synthesis techniques and device architectures, achieving record high efficiency and stability.

These breakthroughs have significantly advanced the development of blue QDLED technology, improving performance and enabling new applications.

6.2.1 Blue Quantum Dots Containing Cadmium

In 2007, Tan et al. first reported blue QDLEDs containing cadmium QDs with bright and saturated blue light emission, which aroused research interest in QDLEDs containing cadmium [12]. By improving the synthesis method of core/shell CdS/ZnS QDs in blue light devices, and using the mixture of oleylamine and octadecene as the solvent for the formation and growth of CdS crystal nuclei, they found that the interaction between oleylamine molecules and Cd^{2+} ions has a profound influence on the growth kinetics of the nuclei, and the emission spectral bandwidth of CdS nuclei is greatly reduced under the condition of properly controlling the dosage of oleylamine. At the same time, the introduction of any free ligand into QD powder will introduce impurities and crack/pinhole defects into the subsequent spin-coated QD films, because the number of molecules on the surface of nanocrystals decreases rapidly during the purification process, and the passivation effect on the surface of QDs is weakened. Therefore, it is necessary to optimize the purification steps to reduce the concentration of free ligands in QD samples and avoid the significant reduction in QD yield.

Starting from pre-nucleated CdSe or ZnSe seeds, a bright blue-emitting $Zn_xCd_{1-x}Se$ NPs were prepared. The "embryonic nuclei-induced alloying process" allowed for the separation of the nucleation of CdSe (or ZnSe) and the subsequent growth of the nanocrystals together with gradual alloying, as shown in Scheme 6.1. Finally, the fabricated device adopts multilayer structure, and the core–shell CdS/ZnS QDs with optimized structures are used in the light-emitting layer. Figure 6.2a shows the EL spectrum of the photoelectric device, and it can be seen that its FWHM is only 20 nm, which ensures the color purity of the emitted light. The spectral characteristics of the device output are similar to PL emission of QD solution, except that there is a small shoulder peak in the UV region due to the residual excitation of poly-TPD (peak value is 410 nm). There is no obvious shift in peak wavelength between EL and PL. The emission intensity of green and red spectral regions is reduced to less than 5% of the total emission of the device, which ensures the purity of the blue region of QDLED emission spectrum. It is worth mentioning that the narrow bandwidth (~20 nm) of blue QDLED reported

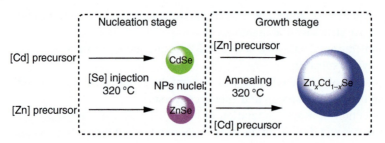

Scheme 6.1 Schematic representation of the synthesis approach proposed in structures of NPs: $Zn_xCd_{1-x}S/ZnS$.

in this paper is equivalent to that of single crystal III–V composite devices obtained by epitaxial growth, which reflects the good crystallinity of CdS/ZnS QDs used as emitting elements of LED devices. This is advantageous compared with OLEDs because the emission spectrum of OLEDs is usually very wide (~60 nm). Figure 6.2b shows the variation curves of current density and brightness with bias voltage, and the turn-on voltage is as low as 2.5 V. Under high brightness, the LED turns white due to the saturation of camera sensor pixels, while the peripheral area turns blue due to backscattering. Under the high brightness of 1600 cd m^{-2}, the emission peak of the device is located at 460 nm. This work makes the low working voltage and high brightness of blue QDLEDs comparable to those of red and green QDs, which lays the foundation for subsequent research.

In order to further improve the efficiency, Anikeeva et al. used QD cores alloyed with ZnCdS and passivated with oleylamine and oleic acid [13]. The fluorescence spectrum shows that the emission peak is located at 460 nm, as shown in Figure 6.2c. The maximum EQE value and the corresponding peak power efficiency of blue QDLEDs are 0.4% and 0.2 lm W^{-1}, respectively. Even if there is a small amount of organic electroluminescent contribution, QDLED pixels based on ZnCdS QDs emit blue light visually. Qian et al. applied CdSe–ZnS core–shell QDs and new ETL material ZnO NPs to QDLEDs in solution treatment [14]. The device consists of indium tin oxide (ITO), poly(ethylene dioxythiophene), polystyrene sulfonate (PEDOT:PSS) (40 nm), poly (N,N'-bis (4-butylphenyl)-N,N'-bis(phenyl) benzidine) (poly-TPD) (45 nm), and CdSe. As shown in Figure 6.2d, the maximum brightness and power efficiency of blue light emitted by the device are 4200 cd m^{-2} and 0.17 lm W^{-1}, respectively. In addition, the turn-on voltage is only 2.4 V, which is lower than the photovoltage of the corresponding device (PL peak: 470 nm, $h\nu = 2.6$ eV). This phenomenon can be attributed to the Auger upconversion mechanism induced by ZnO NPs. It is worth noting that the highest efficiency of QDLED in this work is achieved under high brightness, which is of great significance for practical display and lighting applications. Blue CdSe/ZnS QDs have been proven to be an efficient and stable luminescent material, with their emission peak exceeding 460 nm, and the harmful effects of short-wave blue light on human eyes have been reduced.

In 2017, Wang et al. prepared CdSe/ZnS QDs and QDLEDs optimized for ZnO NPs [15]. CdSe/ZnS QDs have high PLQY due to the limitation of the thick shell on the non-radiative recombination process. QDLED with the best performance shows

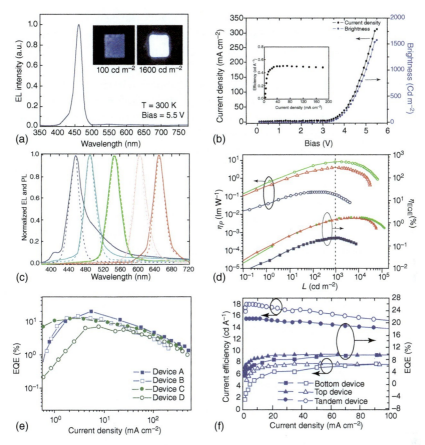

Figure 6.2 (a) Electroluminescence spectrum of QDLED at 5.5 V voltage. Illustration: photos of LED surface taken at the brightness of 100 and 1600 cd m^{-2}. (b) The current density and brightness of blue QDLED are curves of voltage (I–V and L–V). The illustration shows the relationship between lumen efficiency and current density. (c) EL spectrum (solid line) and PL spectrum (dotted line) of RGB QDLEDs when bias voltage is applied. QD monolayer of red and green quantum dots and n-hexane solution of blue quantum dots. (d) The curves of PE, EQE, and luminous brightness of blue light, green light, and orange–red light QDLEDs. (e) The relationship between EQE and current density of devices based on CdSe/ZnS QDs. (f) CE–J–EQE curves of blue CdZnSeS QDLEDs with different structures.

an EL emission peak of 468 nm, with an EQE of 19.8%, and a lumen efficiency of 14.1 cd A^{-1}, as shown in Figure 6.2e. The CIE color coordinates of the device are (0.136, 0.078), which is quite close to the NTSC 1953 standard (0.14, 0.08). Due to the limitations of preparation conditions, the lifetime of 100 cd m^{-2} is only 47.4 hours, which can be further improved by avoiding exposure to air, optimizing the structure of thick-shell QDs, and using better device packaging methods. At present, it is reported that the blue QDs with the highest EQE adopt CdZnSeS as luminescent material [4]. Blue QDLEDs with ZnMgO/Al/HATCN/MoO$_3$ interconnection layers and series structures have achieved extremely high CE (17.9 cd A^{-1}) and EQE

(21.4%), as shown in Figure 6.2f. In addition, high efficiency can be well maintained in a wide range of brightness from 10^2 to 10^4 cd m^{-2}. Even under the high brightness of 20 000 cd m^{-2}, the EQE of blue QDLEDs can still maintain 78% of its maximum value.

Recently, blue QDs with a large core and an intermediate shell structure featuring nonmonotonically-graded energy levels with near-unity photoluminescence quantum efficiency and efficient charge injection were applied in QDLEDs (Scheme 6.2). This strategy significantly reduces surface-bulk coupling and increases the device's efficiency. With CIE coordinates of (0.127, 0.081) and (0.135, 0.063), at an initial luminance of 1000 cd m^{-2}, the devices exhibit T_{95} operational lifetimes of 106 and 75 hours at $L_0 = 1000$ cd m^{-2}, respectively, significantly surpassing the existing records [16]. Enlarging the core size results in a more confined exciton wavefunction, while a non-monotonically graded shell reduces surface-bulk coupling. This results in a blueshift of the emission wavelength without compromising charge injection, as well as an increase in device efficiency and lifetime.

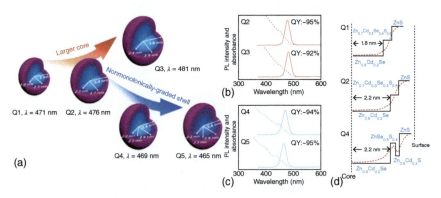

Scheme 6.2 Blue quantum dots with reduced surface-bulk coupling. (a). The core/shell/shell structures for Q1–Q5. The radial element distribution is obtained using high-resolution TEM and ICP-OES. (b, c). PLQE, PL spectra, and absorbance spectra of Q2–Q5. (d). Schematic diagram of the conduction band minimums of Q1, Q2, and Q4. The electronic energy levels of the sub-shells are estimated by linearly combining the electronic energy levels of the isolated materials according to the measured radial compositional profiles.

Blue QDLED devices based on cadmium-containing QDs have shown significant progress in recent years. These devices offer several advantages over traditional organic-based LEDs, including high color purity, narrow emission spectra, and excellent photostability.

However, the use of cadmium in QDs raises concerns about their potential toxicity and environmental impact. To address these concerns, researchers have been exploring the use of alternative materials, such as cadmium-free QDs, to reduce the environmental impact while maintaining the desirable properties of cadmium-containing QDs.

Despite these challenges, the development of blue QDLED devices based on cadmium-containing QDs has made significant strides, with improvements in device efficiency, lifetime, and color purity. Ongoing research is focused on addressing the environmental concerns associated with cadmium-containing QDs while maintaining their superior optical properties. In conclusion, while the development of blue QDLED devices based on cadmium-containing QDs has faced challenges, they continue to show promise and represent an exciting area of research in the field of optoelectronics.

6.2.2 Cadmium-Free Quantum Dots

Due to the potential physiological toxicity of cadmium and the increasingly strict environmental protection policies around the world, the development of cadmium-containing QDs has been limited to some extent. Therefore, QDLEDs without cadmium will be extremely important in AM-QDLED display and lighting in the future, including QDLEDs based on InP, ZnSe, Cu, and AlSb.

6.2.2.1 Quantum Dots Based on InP

With the fine control of InP core size distribution and the optimization of multi-shell heterostructure engineering, the luminescence performance of InP QDs devices has been greatly improved in PL broadening and PLQY [17]. Previously, due to the small bandgap (1.35 eV) of InP QDs, it was difficult to realize the controllable synthesis of small InP core (<3 nm) and the controllable epitaxial growth of shell. The performance of InP QDs-related devices has been poor, and there are few related reports [18]. Shen et al. reported the LED device based on blue InP QDs for the first time in 2017, which uses InP/ZnS small core–shell tetrahedral QDs as the luminescent material, as shown in Figure 6.3a [19]. The absolute PLQY of InP/ZnS QDs can reach 76%, and the FWHM is as narrow as 44 nm. At the same time, the stability of InP/ZnS QDs is improved, and it can be stored in air for more than 1000 hours. However, the performance of the blue QDLED device is poor, and the maximum brightness is only 90 cd m^{-2}, as shown in Figure 6.3b. This may be due to the low PLQY of QD-assembled luminescent films and the low carrier injection and transmission. Although the extra-thick ZnS shell can enhance PLQY and ultimately suppress the non-radiative Förster resonance energy transfer process (FRET) between small-sized QDs, the lattice mismatch between InP and ZnS will lead to QD structural defects [20].

In order to reduce the influence of lattice mismatch, Shen et al. introduced a thin gap bridge layer between InP core and ZnS shell by "low temperature nucleation and high temperature growth" [21]. They synthesized InP/GaP/ZnS/ZnS QDs with high PLQY (81%) and large size (~7.0 ± 0.9 nm), and the corresponding QDLEDs showed record brightness and external quantum efficiency, which were 3120 cd m^{-2} and 1.01%, respectively, as shown in Figure 6.3c. Due to the weakening of quantum confinement effect, the energy level of highest occupied molecular orbital (HOMO) state increases in the case of continuous ZnS coating. Therefore, the increase in VBM confirmed by experiments is beneficial to injecting holes from TFB into the light-emitting layer, as shown in Figure 6.3d. At the same time, for electron injection, the increase in lowest unoccupied molecular orbital (LUMO) energy level

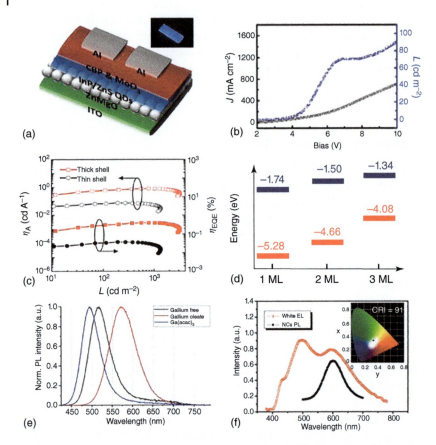

Figure 6.3 (a) Shows the multilayer structure of InP/ZnS QDLED, and the illustration shows a QDLED device with blue light emission. (b) J–V–L curve of InP/ZnS QDLED device. (c) The curves of CE and EQE of InP/GaP/ZnS QDs based on thin and thick ZnS shells with brightness. (d) With the increase in ZnS shell thickness, the HOMO and LUMO energy levels in InP/GaP/ZnS quantum dots change. (e) The normalized PL spectrum of In(Zn)P QDs synthesized by adding 0.03 mmol gallium oleate or 0.025 mmol Ga(acac)$_3$ without adding gallium. (f) PL spectrum of InP/ZnS QDs and EL spectrum of white QDLEDs.

reduces the injection efficiency, so the influence of unbalanced carrier injection can be reduced. Large-scale density functional theory (DFT) calculations of QDs with thousands of atoms show that the thicker shell layer is beneficial to more balanced carrier injection in the QD film. At the same time, the resonance peak between tightly filled QDs is suppressed, which promotes the performance improvement of blue light devices.

Wegner et al. explored the introduction of gallium into colloidal InP QDs in order to adjust their light emission characteristics [22]. As shown in Figure 6.3e, QDs will emit red light (precursor: gallium oleate) or blue light (precursor: gallium acetylacetonate, Ga(acac)$_3$) according to the properties of gallium precursors. The addition of Ga(acac)$_3$ changed the nucleation and growth kinetics of In(Zn)P QDs, which can be inferred from the narrowing of the size distribution of the sharper first exciton peak.

Further X-ray diffraction patterns show a slight but obvious transition to a large angle, which indicates that the lattice constant decreases in the presence of gallium. Yang et al. reported that high-quality InP/ZnS core–shell QDs were synthesized by a simple one-pot solvothermal method [23]. The obtained QDs have high quantum yield (above 60%), wide spectral tunability, narrow emission spectrum, and good light stability. Using high-quality InP/ZnS QDs, the white QDLED device shows a color rendering index as high as 91, as shown in Figure 6.3f. InP/ZnSeS/ZnS QDs were synthesized by Kim et al. using the aminophosphine ligand of $P(DMA)_3$ [24]. With the increase in Se content in the inner shell of ZnSe from 0.1 to 1.5 mmol, the spectrum was obviously redshifted, because the quantum confinement effect gradually weakened as electrons leaked to the adjacent inner shell domain. Then, in the presence of gallium iodide (GaI_3), InP QDs were subjected to cation-exchange reaction at a lower temperature. By changing the amount of GaI_3 (0.68–1.35 mmol), the alloying degree of Ga can be well controlled, and finally a series of double-shell InGaP/ZnSeS/ZnS QDs were synthesized. With the increase in Ga alloy degree, the photoluminescence peak shows a blueshift from 475 to 465 nm, while maintaining a high fluorescence quantum yield of 80–82%. The blue QLED device prepared with InGaP/ZnSeS/ZnS QDs has a maximum luminance of 1038 cd m^{-2}, a current efficiency of 3.8 cd A^{-1}, and an external quantum efficiency of 2.5%. Zhang et al. introduced a thin GaP bridging layer into the core–shell structure to effectively reduce the lattice mismatch between the InP core and the ZnS shell and successfully grew InP/GaP/ZnS QDs with a thick ZnS shell [21]. These thick-shell QDs have a large particle size (7.0 ± 0.9 nm) and exhibit high stability and high fluorescence quantum yield (~81%). The blue QLED based on thick-shell QDs has a record brightness of 3120 cd m^{-2}, which is 35 times higher than previous reports, and the peak external quantum efficiency is as high as 1.01%.

Although some progress has been made in the research of InP QDs, the performance of blue-light QDLEDs based on InP is still poor, and In as a precious metal, is expensive. Therefore, efforts are still needed in the selection of luminescent materials, the optimization of device structure, and the in-depth understanding of device operation mechanisms.

6.2.2.2 Quantum Dots Based on ZnSe

As mentioned earlier, the lattice mismatch between CdSe and ZnS easily leads to the formation of defects and low PLQY, which is a major problem in the study of CdSe-containing QDs [25]. In addition, the electroluminescent spectrum of light-emitting diodes based on CdSe QDs generally has a wide FWHM of ~30 nm, and at the same time, because the defect state participates in optical radiation, it has a long background radiation wavelength [13]. In recent years, zinc-based QDs, especially ZnSe QDs, have attracted more and more attention because of their wide bandgap energy of 2.8 eV and their luminescence range between UV and blue light [26]. Xiang et al. prepared efficient cadmium-free ZnSe/ZnS core–shell QDs with PL emission peak at 420 nm and FWHM of 16 nm, indicating that the quality of ZnSe QDs is better than most blue cadmium QDs [27]. They also found that the EQE increased with the increase in the thickness of the QD layer. When the

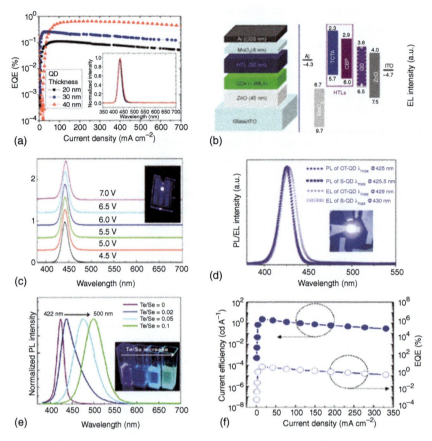

Figure 6.4 (a) EQE–CE curves with different QDs thicknesses. The illustration shows the EL spectrum of the corresponding device. (b) Left: schematic diagram of QDLED structure, and right: schematic diagram of material energy levels of each functional layer. (c) EL spectrum of QDLED at different voltages. The illustration shows a picture of QDLED running at 6.0 V. (d) Normalized PL spectra of ZnSe/ZnS core/shell quantum dots with octylthiol (OT-QD) and S-ODE(S-QD) as S sources and EL spectra of corresponding ultraviolet quantum dot light-emitting diodes. Illustration: Electroluminescent image at 11 V. (e) PL spectra and fluorescence images of ZnSeTe/ZnS core/shell quantum dots with varying Te/Se molar ratios (illustration). (f) The relationship curve of current efficiency and EQE with current density.

thickness of the QD layer was 40 nm, the EQE of the device was 0.65%, and the efficiency rolled off very little below the current density of 500 mA cm^{-2}, as shown in Figure 6.4a. With the increase in the thickness of QD layer, the distance between the light-emitting region and HTL/QD interface increases, which leads to the decrease in nonradiative recombination of interface excitons and the improvement in device efficiency. However, when the thickness of the QD layer is greater than 40 nm, the device's performance is poor because of the uneven thickness of the spin coating. Ji et al. reported the deep blue LED with inverted device structure based on ZnSe/ZnS core/shell QDs [28]. The EL emission peak of QDLED is 441 nm with FWHM of 15.2 nm, which is obviously smaller than that of blue LEDs containing cadmium QDs. As shown in Figure 6.4b, the device structures are ITO/ZnO (35 nm), ZnSe/ZnQDs (~3 monolayers [MLs]), HTL (45 nm), MoO$_3$ (8 nm) and Al (200 nm). The prepared QDLEDs also have ultra-high color purity with CIE color coordinates of (0.16, 0.015), maximum brightness of 1170 cd m^{-2}, and peak current efficiency of 0.51 cd A^{-1}, as shown in Figure 6.4c. Notably, no EL emission of organic matter or surface defect state of QDs was observed. In order to further improve PLQY and photochemical stability, Shen et al. proposed a method of synthesizing ZnSe/ZnS core–shell QDs by low-temperature implantation and high-temperature growth, which is different from most traditional methods of high-temperature nucleation and low-temperature growth [29]. They synthesized nearly monodisperse ZnSe/ZnS core/shell QDs with high PLQYs (up to 80%), high color purity (FWHM of 12–20 nm), and good spectral adjustability in the UV range (400–455 nm). Importantly, this kind of ZnSe/ZnS core/shell QDs shows very good photochemical stability under UV irradiation and repeated purification, which is very important for the further application of QDLEDs. The maximum brightness of high-brightness violet QDLEDs is 2632 cd m^{-2} and the peak EQE is 7.83%, as shown in Figure 6.4d. Considering the photoluminescence function, the devices based on ZnSe/ZnS QDs show a competitive advantage over the best blue–violet QDLEDs containing cadmium. However, its EL wavelength (437 nm) is beyond the expected range of long-wave blue light (440–460 nm). In order to obtain ZnSe/ZnS QDs with PL peak of 460 nm, the size of ZnSe/ZnS QDs should be large enough to reduce the quantum confinement effect. Similar to the previously mentioned InGaP QDs, the feasible way for ZnSe QDs to realize blue light emission is to alloy with low-bandgap ZnTe (2.25 eV). Yang et al. developed a thermal synthesis method for ZnSeTe QDs [30]. By changing the Te/Se ratio, the PL peak value of the shelled ZnS became adjustable in the range of 422–500 nm. The double-shell ZnSeTe/ZnSe/ZnS QDs with the best Te/Se ratio and ZnSe inner shell thickness have blue light emission with a peak of 441 nm, a high PLQY of 70%, and a narrow FWHM of 32 nm, as shown in Figure 6.4e. By using this QD, they prepared high-performance ZnSeTe QDLEDs by solution method, with a peak brightness of 1195 cd m^{-2} and an EQE of 4.2%, as shown in Figure 6.4f. These perfect properties can be attributed to the relatively thick double outer shell and ZnSe inner shell inhibiting Forst energy transfer, Auger recombination, and other non-radiation processes. However, under the constant current density of 200 cd m^{-2} and 9.7 mA cm^{-2}, the lifetime (T_{50}) of the device is only five minutes. In practical display and lighting applications, the service life of cadmium-free blue light diodes will be the most critical issue, which needs further study in the future.

6.2.2.3 Quantum Dots Based on Cu

For a long time, Cu-based QDs have been used as a supplement to red or green cadmium-containing QDs, and the research on blue Cu-based QDs is less. Recently, the production of quaternary Zn-Cu-In-S (ZCIS) or Cu-In-Ga-S (CIGS) QDs by alloying Cu-In-S (ZCIS) or Cu-In-Ga-S (CIGS) with high bandgap materials of ZnS or Cu-Ga-S (CGS) can achieve excellent spectra. In 2014, Yang et al. synthesized a series of CIGS QDs with varying In/Ga ratios, PL peak wavelength of 479–578 nm, and PLQYs of 20–85% [31]. However, the sky-blue CIGS QD device shows a maximum

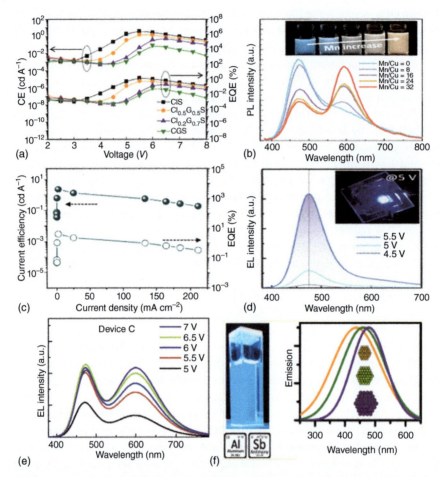

Figure 6.5 (a) Voltage curves of CE and EQE of four CIGS/ZnS QDLEDs with different compositions. (b) Photoluminescence spectrum of double-shell doped ZCGS quantum dots, and the illustration shows the color change of QDs solution with the change of Mn doping concentration. (c) Current density correlation curves of QDLED devices CE and EQE doped with Mn/Cu = 16 concentration. (d) EL spectrum changes under different voltages. Illustration: QDLED equipment image collected at 5 V. (e) The voltage-dependent EL spectrum change of the two-color white QDLED device C. (f) Left: photo of QD solution under ultraviolet irradiation. Right: PL spectrum of colloidal AlSb quantum dots dispersed in n-hexane solution.

CE value of only 1.65 cd m^{-2} and a low EQE of 0.6%, as shown in Figure 6.5a. The poor performance may be due to the low PL QY of CIGS QDs (about 20%). After that, there are some reports about the synthesis of CIGS QD, but there are no related device performance characterizations.

After that, on the basis of ZGS/ZnS/ZnS double-shell QDs, Yang et al. doped Mn with different Mn/Cu concentrations into ZGS matrix through surface adsorption and lattice diffusion [32]. Mn doping into the host QD leads to the characteristic transition emission of $^4T_1 \rightarrow {}^6A_1$ of Mn^{2+}, which is a direct sign of Mn effective doping into the lattice. By changing the concentration of Mn to control the emission specific gravity of Mn in the whole PL, the emission color of doped QDs can be easily adjusted from blue and white to red and white, as shown in Figure 6.5b. All undoped and doped QDs show a relatively high QY of 74.79%, which indicates that Mn doping and its high-concentration doping have not formed a non-radiation process. The QDLED doped with Mn/Cu = 16 produces a maximum brightness of $1352\,cd\,m^{-2}$, a current efficiency of $2.3\,cd\,A^{-1}$, and a relatively high EQE of 4.2%, as shown in Figure 6.5c. Unlike Mn-doped QDLEDs containing cadmium, there is almost no Mn transition radiation in the EL of Zn-Cu-Ga-S: Mn/ZnS/ZnS-based QDLEDs, which may be due to the fact that HTL preferentially injects holes into the QD emitter layer, and then quickly captures holes in the acceptor. Their team demonstrated the preparation of a high-efficiency blue (475 nm) QDLED by full solution method, showing that the maximum EQE is 7.1% and the peak current efficiency is $11.8\,cd\,A^{-1}$, as shown in Figure 6.5d [33]. In addition, an interesting voltage-dependent spectral change is that at high voltage, the yellow region of EL spectrum has a relatively obvious increase in luminous intensity compared with the blue region, as shown in Figure 6.5e, which may be attributed to the difference in Auger de-excitation process between CIS and ZGCQDs due to different excited state lifetimes.

6.2.2.4 Quantum Dots Based on AlSb

Tetrahedral coordination effect in III–V covalent compounds is stronger, crystallization is slower, reaction temperature is higher, and growth time is longer, so the size of III–V QDs does not increase obviously [34]. In addition, the use of two separate monoatomic precursors and the lack of suitable precursors limit the control of colloid nucleation and growth [35]. For the synthesis of AlSb, the main challenge is the poor reactivity of antimony precursors. Therefore, there is less research on AlSb in III–V semiconductors. In 2019, Jalali et al. synthesized colloidal AlSb QDs (PLQY up to 18%) with exciton transition in UV-A region and tunable band-edge emission in blue light spectrum, as shown in Figure 6.5f [36]. QDs show bright light emission in the blue spectral region. At the same time, with the size increase of QDs, the quantum confinement effect weakens, and the wavelength of the first exciton peak and PL spectrum redshift. Although there is no report on the device performance of QDLEDs of AlSb QDs, nontoxic QDs with bright emission have high potential for biological and photoelectric applications, so AlSb QDs may be the next research hotspot.

6.3 Optimization of Charge Transport Layer (CTL)

Due to the large bandgap and low HOMO level of blue QDs, and the high HOMO of HTL material, there is a large energy barrier between QD emitter and HTL, which makes it difficult to inject holes from HTL to QDs. The imbalance between electron and hole injection leads to the excessive accumulation of electrons in the QD emission layer, and the non-radiative Auger recombination process is enhanced. In addition, redundant electrons that are not recombined may leak into HTL, resulting in parasitic emission and affecting the spectral purity of the device. In order to solve this problem, it is necessary to improve the CTL to achieve charge balance, such as by improving hole injection or reducing electron injection. According to the influence of different exciton injections, we will divide them into HTLs and ETLs to explain the latest progress in this field.

Scheme 6.3 shows the energy level arrangement of commonly used hole transfer materials and electron transfer materials in QDLED devices. There is a relatively higher energy level barrier from the HTL (HOMO: −5.4 to −6.3 eV) injected into the emitting layer of Cd-based QDs (VBM: −6.6 eV); therefore, high conductivity of the hole injection/transport layer is needed to be well considered in QDLEDs, especially for blue devices. Consequently, there is a need to design highly conductive and good-hole injection/transport materials with sufficiently high energy levels. For ETM, ZnO-based ETLs with a deeper VBM facilitate the trapping of holes in the QD luminescent layer. And the low CBM forms an effective barrier against electron injection. To control the electron injection barrier, various ZnO ETLs with doped structures (e.g. ZnMgO) have been proposed.

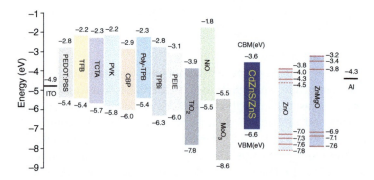

Scheme 6.3 Energy-level alignment of organic HTM/inorganic ETM, blue QDs, and electrodes.

6.3.1 Hole Transport Layer

In order to improve hole injection, the HOMO energy level of HTLs needs to be lower, thus reducing the hole injection energy barrier of the device. Poly-TPD has been widely used in the preparation of HTL of organic light-emitting devices by solution method and as the organic solvent for the subsequent deposition of QD layer because of its low process temperature and good resistance to organic solvents [14]. However, due to its poor energy level matching with ZnSe/ZnS QDs layer, Xiang et al. used PVK instead of poly-TPD layer [27]. When HOMO is −5.8 eV, the hole injection barrier of poly-TPD device decreases from 1.4 to 0.8 eV of PVK device, which leads to the enhancement of charge balance and the increase of hole current, as shown in Figure 6.6a [37]. The enhancement of hole injection ability also leads to the reduction of the turn-on voltage from 4.4 to 3.5 V. When the voltage is 8 V, the optical power density of PVK device is 4.2 mW cm^{-2}, which is obviously higher than 6.4×10^{-4} mW cm^{-2} of poly-TPD device, as shown in Figure 6.6b. In addition, the bandgap of poly-TPD is not enough to effectively limit excitons in

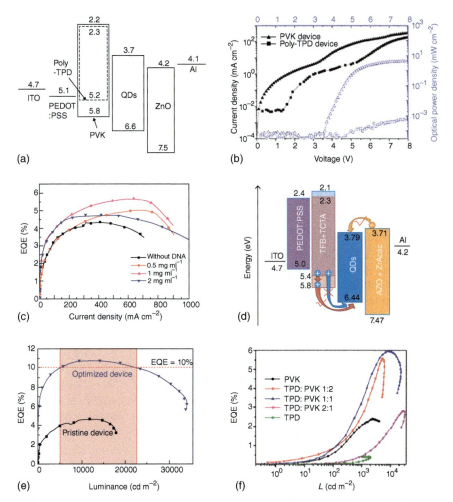

Figure 6.6 (a) Energy band diagram of multilayer QDLED device with poly-TPD or PVK as HTL. The unit of energy level is eV. (b) The characteristic curves of current density–voltage (black) and optical power density–voltage (blue) of devices with Poly-TPD and PVK as HTL. (c) EQE–current density curves of the device under different DNA concentrations. (d) Energy level diagram of QDLEDs optimized by doped CTLs. (e) EQE–L curve of QDLEDs optimized by doping CTLs. Red light region: the curve part with EQE greater than 10%. (f) EQE–L characteristic curve of QDLEDs based on composite HTL.

the QD layer. Excitons can be transferred from QD layer to poly-TPD layer, which leads to the decrease of quantum efficiency of QD transition radiation. Compared with poly-TPD, PVK has higher exciton energy, which can provide better exciton confinement [16].

Tan et al. reported a high-performance blue QDLED by inserting a thin buffer layer of deoxyribonucleic acid (DNA) between poly-TPD HTL and ZnCdS/ZnS core/shell QD emission layers [38]. The deep HOMO level and shallow LUMO level of DNA not only effectively enhance the hole injection, but also confine the injected electrons to the luminescent layer, thus ensuring the charge balance of the QD layer. The optimized device shows high brightness of 16 655 cd m^{-2} and current efficiency of 2.3 cd A^{-1}. The enhancement of this performance is attributed to the better energy level matching on the HTL-QDs interface after the introduction of the DNA intermediate layer, as shown in Figure 6.6c. The key to improve the charge balance of light-emitting layer is to inhibit redundant electron injection, improve hole injection, and introduce additional functional layers (electron blocking layer [EBL] or double HTLs) and doped CTL, which are two effective strategies to adjust the charge transport mobility and reduce the charge injection barrier [39, 40]. Although the addition of functional layer can improve the performance of the device, it makes the manufacturing process of the device more complicated [41]. In contrast, CTL doping is a simpler and more effective method. Wang et al. chose tris(4-carbazoyl 9-ylphenyl) amine (TCTA) as HTL dopant to reduce the hole injection barrier [42]. The electron mobility of AZO NPs on the cathode side can also be controlled by doping metal complex ZrAcac, as shown in Figure 6.6d. High-performance blue QDLEDs were achieved by CTL doping, with the maximum brightness of 34 874 cd m^{-2} and EQE of 10.7%. What is more striking is that the optimized device EQE can keep above 10% in the brightness range of 500 022 000 cd m−2, as shown in Figure 6.6e. Therefore, CTL doping not only improves the charge balance but also improves the stability of blue QDLEDs. Li et al. mixed TPD into PVK as HTL and also tried to construct TPD/PVK double-layer structure [43]. The peak EQEs of QDLEDs optimized by TPD modification can reach 13.40%, which is 3–4 times higher than that of pure PVK QDLEDs, as shown in Figure 6.6f. This work provides a general but very simple and effective method to realize full-color QDLEDs with low turn-on voltage, brightness, and high efficiency by effective optimization of CTLs.

Core–shell structured QD materials help to confine the charge/exciton to the core due to the introduction of a wide bandgap shell (e.g. ZnS), thus eliminating various defects and increasing the photoluminescence (PL) quantum yields of blue QDs. On the other hand, the wide bandgap shell inevitably increases the barrier to the hole injection, especially in blue QDLEDs, where the valence band (VB) is already very low. To balance the charge injection, efficient blue QDLEDs with an EQE of more than 16% were achieved by inserting thin layer of poly(methyl methacrylate) (PMMA) or tert-butyldimethylchlorosilane modified poly(p-phenylene dioxole) (TBS-PBO) to hinder electron injection for balancing charge injection [44, 45]. To achieve efficient blue QDLEDs, HTLs with deeper HOMO energy levels are often used to balance charge injection. For example, Wu et al. developed CNPr-TFB hole transport polymers with deeper HOMO energy levels [46], which exhibited superior hole conductivity with more holes being transported into the QD luminescent layer. Recently, Se has been applied to the whole ZnCdSe/ZnSe QD and/or cadmium-doped zinc sulfide (CdZnS) as the outermost shell. It reduces the barrier of hole injection and enhances the hole conductivity. The obtained blue QDLED achieves an excellent brightness of up to 62 600 cd m^{-2} [47]. Poly-N-vinylcarbazole (PVK) with deep HOMO energy level of around −5.8 eV is the most popular HTL for the blue QD. However, the scene of electron trapping behavior in PVK HTL decreases the probability of exciton formation in the blue QDs (Figure 6.7), resulting in low device efficiency in the blue QDLEDs.

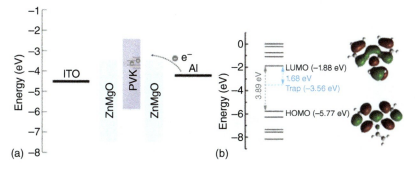

Figure 6.7 The electrical properties of electron-only devices of PVK. (a) Energy level diagram of the electron-only device containing PVK layer; (b) The LUMO/HOMO and the trap levels of PVK.

Ji et al. introduced an interlayer of ZnSe/ZnS I-QDs between PVK and QD EML to mitigate the electron traps caused by PVK [47]. The final device exhibits a maximum EQE of 20.6% and a current efficiency of 19.1 cd A^{-1}, an increase of ≈35% compared to using PVK alone as the HTM. This improvement is mainly attributed to the introduction of the I-QDs interlayer, which not only limits the electron trapping caused by PVK but also confines more charge carriers to the QDs, thus increasing the probability of exciton formation/recombination within the EML. In addition, a stepped energy level arrangement is formed between PVK and E-QDs, which facilitates enhanced hole injection and improves the balance of charge injection. Figure 6.8 shows the schematic structure of the multilayered QDLEDs, consisting of ITO (≈100 nm)/PEDOT:PSS (≈32 nm)/PVK (≈29 nm)/I-QDs (≈8 nm)/E-QDs (≈12 nm)/ZnMgO (35 nm)/Al. The cross-sectional transmission electron microscope (TEM) image of the device, and the corresponding energy levels of materials. The device results suggest that the insertion of I-QD layers limits the electron trapped, which not only increases the probability of exciton recombination within the emissive layer but also facilitates the hole injection, improving the charge balance in blue devices.

To decrease the energy barrier for hole injection from HTM polymers to QDs, making the holes to be easily injected into the QD EMLs and enabling a balanced recombination of charge carriers in QDLEDs, a new strategy was developed by the Chen group [48], in which cadmium-doped zinc sulfide (CdZnS) is used as the outermost shell to synthesize core/triple shell structures with the blue QDs structure of CdZnSe/ZnSeS/ZnS/CdZnS (Figure 6.9). QDLEDs made by the new blue QDs exhibit peak EQE of 8.4%.

Due to the energetic disorder of polymer organic hole transport layers (HTLs, such as TFB) and the size discrepancy between QDs and amorphous HTL segments in blue QDLEDs, electron leakage from QDs to HTLs leads to low device efficiency. The Wang group proposed a new strategy to design polymer HTLs (PF8CZ) with shallower LUMO levels and reduced energy disorder to manufacture blue QDLEDs, resulting in devices with extremely high EQE of 21.9% (Figure 6.10). More importantly, the blue QDLED has a lifetime of 4400 hours at an initial brightness of 100 cd m^{-2}, representing the best-performing solution-processed blue QDLED [49].

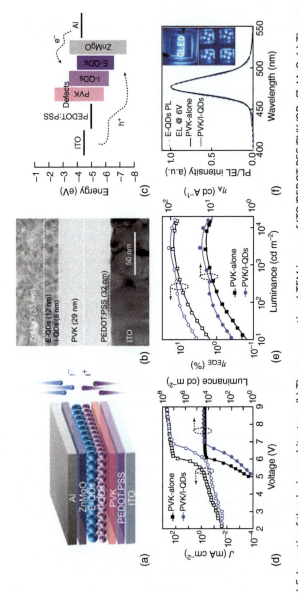

Figure 6.8 (a) Schematic of the device architecture. (b) The cross-sectional TEM image of ITO/PEDOT:PSS/PVK/QDs/ZnMgO. (c) The energy level diagram of each layer used in this work. (d) Photoelectrical properties of QDLEDs. Current density–voltage–luminance ($J-V-L$). (d) and (e) EQE–luminance–current efficiency (ηEQE–L–ηA) for these two blue QDLEDs. (f) Normalized PL spectra of E-QDs film and EL spectra of two devices. The insets are photographs of PVK/I-QDs-based devices under driving voltage of 6 V.

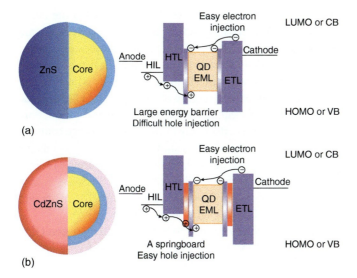

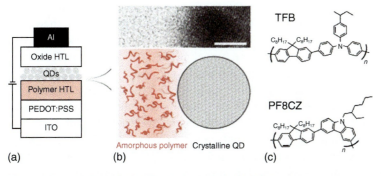

Figure 6.9 (a) A typical type-II core/shell structure of a QD and an energy diagram in a conventional QDLED; (b) The design strategy used the outermost cadmium-doped zinc sulfide (CdZnS) shell as a springboard to make hole injection easier (from the core to the outermost shell, the layers with yellow, blue, and red colors represent the core, ZnS shell, and outermost CdZnS shell, respectively).

Figure 6.10 (a) Typical QDLED structure; (b) HTL/QD interface and cross-sectional transmission electron microscopy image of the interface (top right; scale bar, 5 nm); Chemical structure of hole transport polymers: TFB and PF8CZ.

6.3.2 Electron Transport Layer

In addition to improving hole injection, reducing electron injection is also an effective method to promote the balance of charge injection and improve the recombination efficiency of QD layer. ZnO NPs are widely used as ETL in QDLEDs because of their good electron transport performance. However, due to the close conduction band energy level of QDs-ZnO, electrons in the luminescent layer can spontaneously transfer to the adjacent ZnO NPs, which will lead to serious exciton shift and reduce the proportion of radiation recombination and device efficiency. Sun et al. studied Al-doped ZnO NPs as a substitute for ZnO [50]. By adjusting the concentration of Al doping, the energy band structure of Al-doped ZnO (AZO) was changed. With the increase in Al doping content, the work function and conduction band edge of ZnO gradually increase, thus effectively inhibiting the charge transfer at QDs–ETL interface. QDLEDs with AZO as ETLs show current density–voltage characteristics similar to those of reference devices, as shown in Figure 6.11a, but the maximum CE and EQE of QDLEDs doped with 10% Al are 59.7 cd A^{-1} and 14.1%, which are 1.8 times higher than those of devices without ZnO NPs, as shown in Figure 6.11b, and show that doping improves the device's performance.

Another method to optimize electron injection is to insert an EBL between QDs and ETL. Qu et al. inserted lithium fluoride (LiF) between ZnO and QDs layers, which improved the efficiency and stability of blue QDLEDs [51]. LiF interface layer promotes electron injection into QDs through electron tunneling effect and inhibits exciton quenching at QD/ZnO interface. Compared with PL spectrum, EL spectrum has a redshift of 8 nm, which is mainly due to FRET and quantum confinement Stark effect. The maximum EQE and CE of the blue QDLED device are 9.8% and 7.9 cd A^{-1}, respectively, which are 1.45 times and 1.39 times that of the control device, as shown in Figure 6.11c. The working life of the device is also increased by 2 times, as shown in Figure 6.11d. Zhao et al. tried to explain the electron tunneling effect. If there is a potential barrier E1 at the heterojunction interface, when the LiF interface layer is inserted, the potential barrier will be reduced to E2. In addition, the inserted LiF interface layer will generate an additional barrier E3. When the thickness of LiF is E2 + E3 < E1, carriers will easily tunnel through the heterojunction interface, thus enhancing electron injection. In blue QDLEDs, there is a large potential barrier at the ZnO/LiF interface. By optimizing the thickness of the LiF interface layer (2 nm), strong electron tunneling effect can be generated, and the injection of electrons into the blue QD layer can be enhanced [52]. In addition, Lin et al. inserted PMMA interface layer between blue QD layer ZnCdSe/ZnS/ZnS and zinc oxide ETL to prevent electric injection. The optimized blue device showed the maximum EQE of 16.2%, which was 65% higher than that of traditional devices [39].

Nanostructure engineering is also an effective way to tune the electron mobility of ZnO NPs. For example, ZnO-ETLs are modified to adjust the trap density of ZnO NPs. The post-annealing process of ZnO thin films is a typical treatment method that has a significant impact on the device's performance. Although the annealing temperature has no effect on the bandgap of ZnO thin films, the defect state density of ZnO thin films decreases with the increase in annealing temperature, indicating that the electron mobility of ZnO NPs decreases. In addition, Al:Al$_2$O$_3$ can be introduced as cathode to reduce the potential barrier between QDs and cathode and improve exciton confinement [53]. Under the joint action of annealed ZnO-ETL and Al:Al$_2$O$_3$ cathode, when the annealing temperature of ZnO is 80 °C, the high brightness of 27 753 cd m^{-2} and high EQE of 8.92% are achieved. This is attributed to the

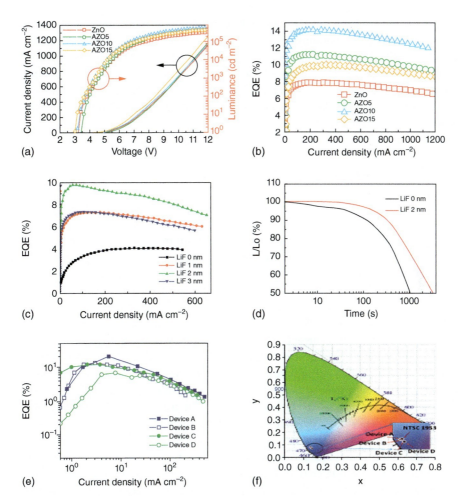

Figure 6.11 (a) QDLEDs current density and brightness versus voltage characteristic curves based on ZnO, AZO5, AZO10, and AZO15 NPs. (b) EQE–current density characteristic curve of QDLEDs based on ZnO, AZO5, AZO10, and AZO15 NPs. (c) EQE–current density curves of blue QDLED devices with different LiF thicknesses. (d) Working life curves of device A (LiF 0 nm) and device C (LiF 2 nm), with initial brightness of 1000 cd m^{-2}. (e) EQE–current density curves of ZnO NPs devices with different sizes. (f) CIE color coordinates based on small-sized ZnO NPs devices.

annealed ZnO with decreased electron mobility beneficial to the charge injection balance between holes and electrons. Wang et al. synthesized two types of ZnO NPs and studied the effects of different sizes of ZnO NPs on device performance [15]. The device fabricated by using small-sized ZnO NPs achieves a CE of 14.1 cd A^{-1} and a maximum EQE of 19.8%, as shown in Figure 6.11e. The CIE color coordinate of the device is (0.136, 0.078), which is quite close to the NTSC 1953 standard (0.14, 0.08). As shown in Figure 6.11f, its excellent color purity makes it an ideal blue light source. In conclusion, high-performance blue QDLEDs are fabricated based on ZnO NP ETL and Al:Al$_2$O$_3$ cathode. Therefore, the electron mobility in ZnO NP films can be adjusted for better electron–hole balance by controlling the annealing temperature, which promotes the decomposition of Zn(OH)$_2$ and allows more defects in ZnO crystal to be filled, leading to lower gap state density, thus lowering the mobility of ZnO NP films. The gap states can decrease the hopping barrier for electrons to some degree. This simple way to adjust charge injection balance via annealing temperature requires no additional modification of device architecture, making it applicable for fabrication of different structured QDLEDs.

6.4 Device Structure

In the traditional QDLED structure, an orthogonal solvent must be selected to prevent physical damage during solution deposition. With the deeper understanding of the physical mechanism of devices, the device with inverted structure can solve the limitation of solvent selection. In addition, since the bottom transparent cathode can be directly connected to the N-channel metal oxide or amorphous silicon thin film transistor (TFT) backplane, QDLED with inverted structure is very beneficial for display applications. Kwak et al. reported that ZnO-NP thin film treated with solution was used as ETL, and CBP and MoO$_3$ evaporated in vacuum were used as HTL/hole injection layer (HIL) inverted blue devices, which had a high efficiency of 1.7% and a low turn-on voltage of 3.0 V, as shown in Figure 6.12a [54]. Among them, the ZnO NP film in the inverted device structure has three important advantages: (i) As a universal EIL/ETL emitted by RGB QDs, it has efficient electron injection and transmission characteristics; (ii) It provides a stable platform for continuous QD deposition; (iii) Using conventional organic materials with excellent properties, the thermoluminescence materials are systematically engineered. The remarkable improvement in performance is due to the fact that ZnO and CBP are more effective in injecting charges into QDs, which leads to more direct and effective carrier recombination in QD layers. In addition, the devices with conventional structures show obvious brightness attenuation after continuous operation for several hours at an initial brightness of 500 cd m^{-2} and a constant current density. In contrast, the half-life of QDLEDs with inverted structure is about 600 hours, which is better than that of conventional devices, as shown in Figure 6.12b. The stability of the device is significantly improved, which is consistent with low turn-on voltage and high EQE. The HIL with systematic design has lower HOMO, which enhances the charge balance in the device. Therefore, it is possible to independently optimize charge injection, transmission, and luminescence to achieve efficient QDLEDs.

With the optimization of device structure and the structural engineering improvement of QD luminescent materials, the device performance of QDLEDs has been further improved. By inserting ultra-thin PEI between ITO layer and ZnO layer to reduce the oxygen vacancy defect of zinc oxide and the electron accumulation at the interface between zinc oxide and QDs, Zhong et al. proposed a blue QDLEDs with inverted structure [55]. The color coordinates of the prepared blue QD device are (0.14, 0.04), which is close to the standard blue coordinate, and the peak EQE increases from 3.5% to 5.5%, as shown in Figure 6.12c. By PEI modification of

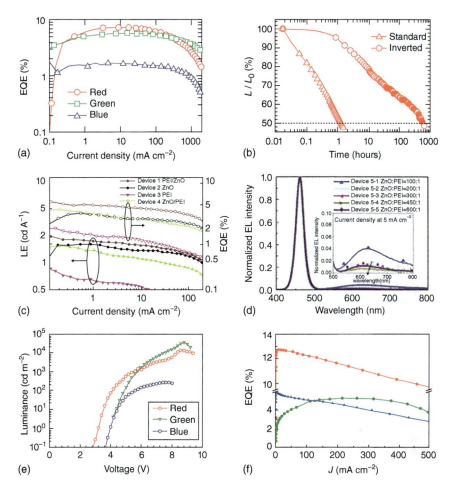

Figure 6.12 (a) Current density EQE characteristic curve of inverted QDLEDs with CBP as HTL. (b) Working life characteristic curves of devices with standard and inverted structures. Measure the brightness of QDLEDs when operating continuously at a constant current density with an initial brightness of 500 cd m^{-2}. Both QDLEDs are made of the same QDs and characterized under the same conditions (i.e. packaging conditions, initial brightness, humidity, and temperature). (c) QDLEDs EQE–J–LE characteristic curves with different types of ETLs. (d) The normalized EL spectra of different ETL devices were used at 5 mA cm^{-2} current density. The illustration shows the amplified EL spectrum. (e) The brightness voltage characteristic curves of inverted structure QDLED prepared by solution method for red, green, and blue emission. (f) EQE–J characteristic curve of RGB QDLEDs using double-layer PVK/TFB materials.

ZnO NPs, the surface defects of zinc oxide can be reduced and the broadband red emission of zinc oxide surface defects can be suppressed, as shown in Figure 6.12d. In addition, reducing the barrier height of ITO/ZnO and QDs/ZnO can reduce the electron accumulation at the interface between ZnO and QD and the radiation of defect state on ZnO surface. The research results can provide an effective way to prepare pure blue QDLEDs with high efficiency. However, vacuum evaporation is a high-cost technology, which is not suitable for large-scale applications. Castan et al. reported for the first time the inverted blue QDLED prepared by solution method, which promoted the deposition of PEDOT:PSS on hydrophobic polymer hole transport materials by using polyoxyethylene tridecyl ether (PTE) [56]. The maximum brightness of red, green, and blue QDLED devices is 12 510, 32 370, and 249 cd m^{-2}, respectively, as shown in Figure 6.12e, and the turn-on voltages are 2.8, 3.6, and 3.6 V, respectively. However, the current efficiency of the blue light device is only 0.06 cd A^{-1}, which is due to the influence of PTE on reducing the current flowing through the device.

In order to further solve the solvent erosion of HTL on QD light-emitting layer, Liu et al. used polar solvent 1,4-dioxane as the solvent of PVK-HTL, and hydrated phosphomolybdic acid (PMAH) dissolved in isopropanol as HIL [57]. The maximum brightness of the prepared blue QDLED is 1280 cd m^{-2}, and the peak EQE is 4.69%. In order to further improve the electroluminescent performance of devices, they also developed a double-layer thermoluminescent material PVK/TFB, which is used to provide stepped hole injection, promote hole injection, and achieve carrier balance. The second layer of thermoluminescent material TFB is dissolved in p-xylene to prevent PVK layer from physical damage. The results show that the peak LE of red and blue QDLEDs is 22.1 and 1.99 cd A^{-1}, respectively, and the maximum EQE is 12.7% and 5.99%, respectively, as shown in Figure 6.12f.

However, the reported EQE of red and green QDLEDs is usually over 20%, while it is still rare in blue QDLEDs. One way to achieve high efficiency and long life is to use a series structure, which has been widely used in OLEDs. As we mentioned earlier, the blue QDLED device with the highest EQE reported at present uses a series structure [4]. The device uses ICL with ZnMgO/Al/HATCN/MoO$_3$ structure, which has high transparency, efficient charge generation/injection capability, and high stability against solvent damage during the deposition of the upper functional layer. Therefore, the prepared blue QDLEDs show record EQE and current efficiency of 21.4% and 17.9 cd A^{-1}, respectively, and have high brightness of 26 800 cd m^{-2}. These performances make blue QDLEDs an ideal choice for the next generation of full-color displays and solid-state lighting markets.

6.5 Summary

In recent decades, QDLEDs have made great progress in material chemistry, charge transfer materials, nanostructure engineering, and device structure. The efficiency and working life of blue QDLEDs have been greatly improved, but they still lag behind red and green QDLEDs. In addition, due to the difficulties in the preparation of blue-light cadmium-free QDs, the development of cadmium-free quantum dots is relatively slow compared with cadmium-containing QDs. The existing problems with blue QDLEDs listed below are the key to accelerating their industrialization and commercialization.

1. The efficiency of QDLEDs containing cadmium is relatively low

    On the one hand, the low efficiency of QDLEDs containing cadmium is due to the low PLQYs of blue QDs. In addition, there is a large energy level difference between QDs and traditional organic HTLs, leading to unbalanced charge injection and low hole mobility, which is another main reason for low device efficiency. In order to develop efficient blue QDs and related QDLEDs, it is necessary to conduct in-depth research on the composition and crystal structure of blue QDs with appropriate energy level structure, and high-performance HTL materials with high working function and hole mobility.

2. Cadmium-free QDLEDs performance

    Although the performance of cadmium-containing QDLEDs is better than that of cadmium-free QDLEDs, they contain toxic element Cd, so they will be strictly restricted in consumer products and environmental protection. Synthesis of cadmium-free QDs, such as QDs based on ZnSe, InP, AlSb, and Cu, is the focus of current research. However, the device performance based on cadmium-free QDs shows a huge gap with that of QDLEDs containing cadmium. Therefore, there is an urgent need to optimize the synthesis method, synthesize cadmium-free QDs with few defects and uniform size, improve the composition and structure of device functional layers to balance the charge transmission in the device, and improve the efficiency and working life of the device.

Despite the challenges and competition, QDLED has a broad application prospect in the display field, even competing with OLED and micro-LED. We look forward to the success of QDLED displays in daily life in the near future, adding luster to our lives.

Blue QDLEDs have made significant progress in recent years, but they still face several challenges that need to be addressed before their commercialization.

One of the main challenges for blue QDLEDs is their relatively low efficiency compared to red and green QDLEDs. This is due to several factors, including the low PLQY of blue QDs, the large energy level difference between QDs and traditional organic HTLs, and the unbalanced charge injection caused by low hole mobility. To improve the efficiency of blue QDLEDs, researchers need to conduct in-depth studies on the composition and crystal structure of blue QDs, as well as develop high-performance HTL materials with high working functions and hole mobility.

Another major challenge for blue QDLEDs is the difficulty in developing cadmium-free QDs, which are necessary for consumer product applications due to the toxicity of cadmium. While significant progress has been made in the synthesis of cadmium-free QDs, such as QDs based on ZnSe, InP, AlSb, and Cu, the performance of devices based on cadmium-free QDs still lags behind those based on cadmium QDs. Future research will focus on optimizing the synthesis methods to produce cadmium-free QDs with few defects and uniform size, improving the composition and structure of the device functional layers to balance the charge transmission, and enhancing the efficiency and lifetime of the device.

In addition to improving the efficiency of blue QDLEDs and developing cadmium-free QDs, future research directions for blue QDLEDs may include improving color purity, achieving higher brightness and better stability, and developing more efficient and cost-effective device fabrication techniques. Overall, the development of blue QDLEDs still has a long way to go, but with continued research and development, they have the potential to become a commercially viable technology for display applications.

References

1 Shen, P., Cao, F., Wang, H. et al. (2019). Solution-processed double-junction quantum-dot light-emitting diodes with an EQE of over 40%. *ACS Applied Materials and Interfaces* 11 (1): 1065–1070.

2 Su, Q., Zhang, H., Xia, F. et al. (2018). 73-4: Tandem red quantum-dot light-emitting diodes with external quantum efficiency over 34%. *SID Symposium Digest of Technical Papers* 49 (1): 977–980.

3 Dai, X., Zhang, Z., Jin, Y. et al. (2014). Solution-processed, high-performance light-emitting diodes based on quantum dots. *Nature* 515 (7525): 96–99.

4 Zhang, H., Chen, S., and Sun, X.W. (2018). Efficient red/green/blue tandem quantum-dot light-emitting diodes with external quantum efficiency exceeding 21%. *ACS Nano* 12 (1): 697–704.

5 Wang, R., Shang, Y., Kanjanaboos, P. et al. (2016). Colloidal quantum dot ligand engineering for high performance solar cells. *Energy and Environmental Science* 9 (4): 1130–1143.

6 Anikeeva, P., Madigan, C.F., Halpert, J.E. et al. (2008). Electronic and excitonic processes in light-emitting devices based on organic materials and colloidal quantum dots. *Physical Review B* 78 (8): 085434.

7 Zhang, Y., Zhang, F., Wang, H. et al. (2019). High-efficiency CdSe/CdS nanorod-based red light-emitting diodes. *Optics Express* 27 (6): 7935–7944.

8 Xie, L., Xiong, X., Chang, Q. et al. (2019). Inkjet-printed high-efficiency multilayer QLEDs based on a novel crosslinkable small-molecule hole transport material. *Small* 15 (16): 1900111.

9 Chen, S., Cao, W., Liu, T. et al. (2019). On the degradation mechanisms of quantum-dot light-emitting diodes. *Nature Communications* 10 (1): 765.

10 Steckel, J.S., Zimmer, J.P., Coesullivan, S. et al. (2004). Blue luminescence from (CdS)ZnS core-shell nanocrystals. *Angewandte Chemie* 43 (16): 2154–2158.

11 Kim, T., Kim, K.-H., Kim, S. et al. (2020). Efficient and stable blue quantum dot light-emitting diode. *Nature* 586 (7829): 385–389.

12 Tan, Z., Zhang, F., Zhu, T. et al. (2007). Bright and color-saturated emission from blue light-emitting diodes based on solution-processed colloidal nanocrystal quantum dots. *Nano Letters* 7 (12): 3803–3807.

13 Anikeeva, P., Halpert, J.E., Bawendi, M.G., and Bulovic, V. (2009). Quantum dot light-emitting devices with electroluminescence tunable over the entire visible spectrum. *Nano Letters* 9 (7): 2532–2536.

14 Qian, L., Zheng, Y., Xue, J., and Holloway, P.H. (2011). Stable and efficient quantum-dot light-emitting diodes based on solution-processed multilayer structures. *Nature Photonics* 5 (9): 543–548.

15 Wang, L., Lin, J., Hu, Y. et al. (2017). Blue quantum dot light-emitting diodes with high electroluminescent efficiency. *ACS Applied Materials and Interfaces* 9 (44): 38755–38760.

16 Chen, X., Lin, X., Zhou, L. et al. (2023). Blue light-emitting diodes based on colloidal quantum dots with reduced surface-bulk coupling. *Nature Communications* 14 (1): 284.

17 Hahm, D., Chang, J.H., Jeong, B.G. et al. (2019). Design principle for bright, robust and color-pure InP/ZnSe$_x$S$_{1-x}$/ZnS heterostructures. *Chemistry of Materials* 31 (9): 3476–3484.

18 Tamang, S., Lincheneau, C., Hermans, Y. et al. (2016). Chemistry of InP nanocrystal syntheses. *Chemistry of Materials* 28 (8): 2491–2506.

19 Shen, W., Tang, H., Yang, X. et al. (2017). Synthesis of highly fluorescent InP/ZnS small-core/thick-shell tetrahedral-shaped quantum dots for blue light-emitting diodes. *Journal of Materials Chemistry C* 5 (32): 8243–8249.

20 Chen, F., Lin, Q., Shen, H., and Tang, A. (2020). Blue quantum dot-based electroluminescent light-emitting diodes. *Materials Chemistry Frontiers* 4 (5): 1340–1365.

21 Zhang, H., Ma, X., Lin, Q. et al. (2020). High-brightness blue InP quantum dot-based electroluminescent devices: the role of shell thickness. *Journal of Physical Chemistry Letters* 11 (3): 960–967.

22 Wegner, K.D., Pouget, S., Ling, W.L. et al. (2019). Gallium – a versatile element for tuning the photoluminescence properties of InP quantum dots. *Chemical Communications* 55 (11): 1663–1666.

23 Yang, X., Zhao, D., Leck, K.S. et al. (2012). Full visible range covering InP/ZnS nanocrystals with high photometric performance and their application to white quantum dot light-emitting diodes. *Advanced Materials* 24 (30): 4180–4185.

24 Kim, K.-H., Jo, J.-H., Jo, D.-Y. et al. (2020). Cation-exchange-derived InGaP alloy quantum dots toward blue emissivity. *Chemistry of Materials* 32 (8): 3537–3544.

25 Wood, V. and Bulovic, V. (2010). Colloidal quantum dot light-emitting devices. *Nano Reviews* 1 (1): 5202.

26 Reiss, P., Quemard, G., Carayon, S. et al. (2004). Luminescent ZnSe nanocrystals of high color purity. *Materials Chemistry and Physics* 84 (1): 10–13.

27 Xiang, C., Koo, W.H., Chen, S. et al. (2012). Solution processed multilayer cadmium-free blue/violet emitting quantum dots light emitting diodes. *Applied Physics Letters* 101 (5): 053303.

28 Ji, W., Jing, P., Xu, W. et al. (2013). High color purity ZnSe/ZnS core/shell quantum dot based blue light emitting diodes with an inverted device structure. *Applied Physics Letters* 103 (5): 053106.

29 Wang, A., Shen, H., Zang, S. et al. (2015). Bright, efficient, and color-stable violet ZnSe-based quantum dot light-emitting diodes. *Nanoscale* 7 (7): 2951–2959.

30 Jang, E., Han, C., Lim, S. et al. (2019). Synthesis of alloyed ZnSeTe quantum dots as bright, color-pure blue emitters. *ACS Applied Materials and Interfaces* 11 (49): 46062–46069.

31 Kim, J.H., Lee, K.H., Jo, D.Y. et al. (2014). Cu–In–Ga–S quantum dot composition-dependent device performance of electrically driven light-emitting diodes. *Applied Physics Letters* 105 (13): 133104.

32 Kim, J., Kim, K., Yoon, S. et al. (2019). Tunable emission of bluish Zn–Cu–Ga–S quantum dots by Mn doping and their electroluminescence. *ACS Applied Materials and Interfaces* 11 (8): 8250–8257.

33 Yoon, S., Kim, J., Kim, K. et al. (2019). High-efficiency blue and white electroluminescent devices based on non-Cd I–III–VI quantum dots. *Nano Energy* 63: 103869.

34 Kim, S.W., Sujith, S., and Lee, B.Y. (2006). $InAs_xSb_{(1-x)}$ alloy nanocrystals for use in the near infrared. *Chemical Communications* 46: 4811–4813.

35 Choi, H., Kim, K. et al. (2015). Synthesis of colloidal InSb nanocrystals *via in situ* activation of $InCl_3$. *Dalton Transactions: An International Journal of Inorganic Chemistry* 44 (38): 16923–16928.

36 Jalali, H.B., Sadeghi, S., Sahin, M. et al. (2019). Colloidal aluminum antimonide quantum dots. *Chemistry of Materials* 31 (13): 4743–4747.

37 Wang, E., Li, C., Peng, J., and Cao, Y. (2007). High-efficiency blue light-emitting polymers based on 3,6-silafluorene and 2,7-silafluorene. *Journal of Polymer Science Part A* 45 (21): 4941–4949.

38 Wang, F., Jin, S., Sun, W. et al. (2018). Enhancing the performance of blue quantum dots light-emitting diodes through interface engineering with deoxyribonucleic acid. *Advanced Optical Materials* 6 (21): 1800578.

39 Lin, Q., Wang, L., Li, Z. et al. (2018). Nonblinking quantum-dot-based blue light-emitting diodes with high efficiency and a balanced charge-injection process. *ACS Photonics* 5 (3): 939–946.

40 Conaghan, P.J., Menke, S.M., Romanov, A.S. et al. (2018). Efficient vacuum-processed light-emitting diodes based on carbene-metal-amides. *Advanced Materials* 30 (35): 1802285.

41 Zhang, Z., Ye, Y., Pu, C. et al. (2018). High-performance, solution-processed, and insulating-layer-free light-emitting diodes based on colloidal quantum dots. *Advanced Materials* 30 (28): 1801387.

42 Wang, F., Sun, W., Liu, P. et al. (2019). Achieving balanced charge injection of blue quantum dot light-emitting diodes through transport layer doping strategies. *Journal of Physical Chemistry Letters* 10 (5): 960–965.

43 Li, J.H., Liang, Z., Su, Q. et al. (2018). Small molecule-modified hole transport layer targeting low turn-on-voltage, bright, and efficient full-color quantum dot light emitting diodes. *ACS Applied Materials and Interfaces* 10 (4): 3865–3873.

44 Jin, X., Chang, C., Zhao, W. et al. (2018). Balancing the electron and hole transfer for efficient quantum dot light-emitting diodes by employing a versatile organic electron-blocking layer. *ACS Applied Materials and Interfaces* 10 (18): 15803–15811.

45 Su, Q., Zhang, H., Sun, Y. et al. (2018). Enhancing the performance of quantum-dot light-emitting diodes by postmetallization annealing. *ACS Applied Materials and Interfaces* 10 (27): 23218–23224.

46 Wu, W., Chen, Z., Zhan, Y. et al. (2021). An efficient hole transporting polymer for quantum dot light-emitting diodes. *Advanced Materials Interfaces* 8 (15): 2100731.

47 Wang, F., Hua, Q., Lin, Q. et al. (2022). High-performance blue quantum-dot light-emitting diodes by alleviating electron trapping. *Advanced Optical Materials* 10 (13): 2200319.

48 Liu, B., Guo, Y., Su, Q. et al. (2022). Cadmium-doped zinc sulfide shell as a hole injection springboard for red, green, and blue quantum dot light-emitting diodes. *Advanced Science* 9 (15): 2104488.

49 Deng, Y., Peng, F., Lu, Y. et al. (2022). Solution-processed green and blue quantum-dot light-emitting diodes with eliminated charge leakage. *Nature Photonics* 16 (7): 505–511.

50 Sun, Y., Wang, W., Zhang, H. et al. (2018). High-performance quantum dot light-emitting diodes based on Al-doped ZnO nanoparticles electron transport layer. *ACS Applied Materials and Interfaces* 10 (22): 18902–18909.

51 Qu, X., Zhang, N., Cai, R. et al. (2019). Improving blue quantum dot light-emitting diodes by a lithium fluoride interfacial layer. *Applied Physics Letters* 114 (7): 071101.

52 Zhao, J.M., Zhang, S.T., Wang, X.J. et al. (2004). Dual role of LiF as a hole-injection buffer in organic light-emitting diodes. *Applied Physics Letters* 84 (15): 2913–2915.

53 Cheng, T., Wang, F., Sun, W. et al. (2019). High-performance blue quantum dot light-emitting diodes with balanced charge injection. *Advanced Electronic Materials* 5 (4): 1800794.

54 Kwak, J., Bae, W.K., Lee, D. et al. (2013). Bright and efficient full-color colloidal quantum dot light-emitting diodes using an inverted device structure. *Nano Letters* 12 (5): 2362–2366.

55 Zhong, Z., Zou, J., Jiang, C. et al. (2018). Improved color purity and efficiency of blue quantum dot light-emitting diodes. *Organic Electronics* 58: 245–249.

56 Castan, A., Kim, H., and Jang, J. (2014). All-solution-processed inverted quantum-dot light-emitting diodes. *ACS Applied Materials and Interfaces* 6 (4): 2508–2515.

57 Liu, Y., Jiang, C., Song, C. et al. (2018). Highly efficient all-solution processed inverted quantum dots based light emitting diodes. *ACS Nano* 12 (2): 1564–1570.

7

Near-Infrared Quantum Dots (NIR QDs)

7.1 Introduction of Near-Infrared Quantum Dots

NIR QDs have potential applications in night vision, biomedical imaging, optical communication, computing, and friend-foe identification systems. At present, commercial NIR light-emitting diodes (LEDs) are mainly grown on crystal substrates by III–V-type semiconductors and processed in special clean rooms. This limits their integration into various photoelectric devices and also increases the cost. Therefore, the research and development of efficient and low-cost NIR materials has attracted extensive attention. Among them, organic light-emitting diode (OLED), perovskite light-emitting diode (PeLED), and quantum dot light-emitting diode (QDLED) are mainly studied. It is difficult for OLED to extend into the NIR spectral range because organic molecules usually emit light with a wavelength (λ) less than 1 μm [1]. In addition, for organic semiconductors, according to the energy gap law, the vibration coupling increases and then the excited state is quenched rapidly [2, 3]. The quantum yield of luminescence decreases with the increase of emission wavelength. On the contrary, simply changing the composition and size of perovskite and quantum dots (QDs) can easily adjust the emission spectrum from visible light to NIR range, which can perfectly match the "transparent window" of silica fiber. It is worth noting that the external quantum efficiency (EQE) of perovskite materials is the best, but the adjustment of their emission spectrum in NIR range is limited. For undoped perovskite materials, their emission wavelength is limited to 800 nm, and the stability of perovskite materials is still the biggest stumbling block in their commercialization. In a word, QDLED has better comprehensive performance in the NIR emission wavelength range.

In the research process of QDLED materials, compared with the development of visible colloidal quantum dots (CQDs), the development process of NIR QDs is relatively slow due to the lack of effective synthesis methods and characterization techniques. NIR QDs are mainly classified as follows: IV (Si, Ge, Sn) [4], IV–VI (PbS, PbSe, PbTe) [5], III–V (InAs, InSb) [6], II–VI (HgCdTe, HgSe, HgTe) [7], I–VI (Ag_2S, Ag_2Se, Ag_2Te) [8], and I–III–VI ($CuInS_2$, $CuInSe_2$, $AgBiS_2$, $AgInSe_2$) [9]. The emerging NIR QDs also include metal halide perovskite nanocrystals, such as $CsSnI_3$, $CsSn_xPb_{1-x}I_3$, $FAPbI_3$, and $Cs_xFA_{1-x}PbI_3$. As shown in Figure 7.1, these NIR QDs have wide adjustability in different spectral regions [10]. Among them, the

Colloidal Quantum Dot Light Emitting Diodes: Materials and Devices, First Edition. Hong Meng.
© 2024 WILEY-VCH GmbH. Published 2024 by WILEY-VCH GmbH.

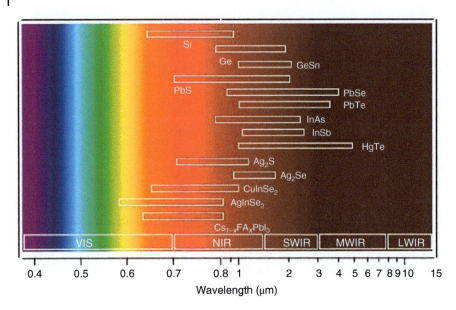

Figure 7.1 Schematic diagram of emission wavelength range of representative NIR QDs.

traditional NIR QDs (PbS, CdHgTe, etc.) have narrow bandgaps, but their further application is limited due to the inherent toxicity of heavy metal ions. Although $CuInS_2$ and $AgInSe_2$ NIR QDs with low toxicity were synthesized in organic phase, the complicated synthesis steps, harsh reaction conditions, and the need for post-treatment of ligand exchange and phase transfer limited their further application to some extent. Among them, lead QDs are considered as promising candidates for developing NIR LEDs because of their narrow bandgap (0.28 eV), large Bohr radius (46 nm), and high extinction coefficient ($105\,m^{-1}\,cm^{-1}$). This chapter mainly introduces three kinds of NIR QD materials, which have been studied more widely, and then expounds on the optimization schemes of these materials, including ligand engineering, core/shell engineering, and matrix engineering. Finally, the research on NIR–QDLED is prospected.

7.2 Near-Infrared Quantum Dot Materials

At present, in the field of LED, IV–VI, II–VI, IV QDs and perovskite QDs are the most widely studied because of their excellent characteristics, and since 2014, a lot of research has been conducted on perovskite QDs because of their excellent photoelectric properties and relatively simpler solution preparation methods. Because there is a monograph on NIR perovskite, the diagram will not be discussed in this chapter. Figure 7.2 shows the great performance breakthrough of NIR LED in recent years. These performance breakthroughs are all reflected in NIR QD materials, including ligand engineering, core/shell engineering, and matrix engineering.

7.2 Near-Infrared Quantum Dot Materials

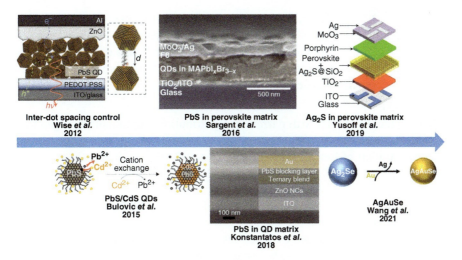

Figure 7.2 Development of PbX QDs based on NIR–QDLEDs in recent years. Source: Gong et al. [11]/Springer Nature; Maria Vasilopoulou et al. [12]/Springer Nature.

In 2012, Wise et al. adjusted the distance between QDs by using bifunctional connecting molecules with different lengths, thus achieving the balance between carrier injection and radiation recombination in the QD layer [13]. Finally, correspondingly, the radiation intensity of their devices and EQE were increased by eight times and two times, respectively, compared with the best reports at that time. During the synthesis of CQDs, a large number of defects will be formed on the surface, which will lead to an increase in non-radiation recombination. In 2015, Vladimir et al. synthesized PbS/CdS QDs with core/shell structure by cation-exchange method to reduce the annihilation of photoluminescence (PL) in PbS core [14]. The EQE and power conversion efficiency (PCE) of the device prepared by QDs with core/shell structure are more than twice those recorded at that time. In order to improve the efficiency and stability of devices, some researchers tried to add QDs to organic substances, such as poly [2-methoxy-5-(2′-ethylxyloxy)-1,4-phenyl vinyl] (MEH–PPV) [15, 16]. Although this method improves the performance of devices, it increases energy loss and causes the turn-on voltage and working voltage to rise. Considering that the hot perovskite materials have good carrier mobility, the Sargent group took the lead in using organic and inorganic perovskite as the matrix material of PbS QDs in 2015. After that, the EQE of the system increased continuously by optimizing the lattice matching between perovskite and QDs and using multi-quantum-well perovskite materials [11, 17, 18]. In 2019, Abd et al. added Ag_2S QDs to ternary cationic perovskite containing Cs with few defects and higher mobility, and at the same time introduced an interface layer to optimize the carrier injection balance in the device structure, obtaining a high EQE of 17% [12]. Different from the previous chemical passivation of organic and perovskite substrates, in 2018, Gerasimos et al. used ZnO QDs as the substrate to reduce the density of defect states by remote charge passivation [19]. This method is simpler and more efficient, and 7.9% EQE is obtained. In order to realize the application of NIR QDs in biological sciences, it is essential to have high PLQY and good biocompatibility. In 2021, the Wang group synthesized Ag–Au–Se (AgAuSe) QDs by alloying strategy and obtained 65.3% PLQY records.

Infrared QDLED development has been an active research area over the past two decades, with many important breakthroughs and advancements made. Here are some key developments and device performance data:

Year: 2000–2005
- Development of CQDs with narrow size distribution and tunable bandgap for IR applications.
- Demonstration of first IR QDLED based on PbS QDs, with a peak emission wavelength of 1.5 μm, EQE of 0.03%, and a lifetime of 20 hours.

Year: 2005–2010
- Development of new synthetic routes for high-quality IR QDs with improved stability and brightness.
- Introduction of new device architectures, such as the use of nanocrystal superlattices and multiple emissive layers, to improve performance.
- Demonstration of IR QDLEDs with peak emission wavelengths up to 2.3 μm, EQE of 1%, and lifetimes of several hundred hours.

Year: 2010–2015
- Introduction of new materials systems, such as HgTe and PbTe, for mid-IR applications.
- Development of novel surface passivation techniques to reduce surface trap states and improve carrier injection and transport.
- Demonstration of IR QDLEDs with peak emission wavelengths up to 3.8 μm, EQE of 3%, and lifetimes exceeding 1000 hours.

Year: 2015–2023
- Development of new device architectures, such as the use of plasmonic structures and hybrid structures, to further improve performance.
- Demonstration of IR QDLEDs with peak emission wavelengths up to 4.4 μm, EQE of 8%, and lifetimes exceeding 5000 hours.

7.2.1 Chalcogenide Lead Quantum Dots

Among all types of NIR QD that have been discovered, lead QDs (PbX; PbX = PbS, PbSe, PbTe) have shown great prospects in basic scientific research and technical application because of their narrow bandgaps (0.41, 0.278, and 0.31 eV at room temperature), broadband absorption, large exciton Bohr radius (due to high dielectric constant and small effective carriers), and the generation of multiple excitons. In addition, their bandgaps can be widely adjusted from about 0.3 eV to more than 2.0 eV. These remarkable characteristics make PbX QD an ideal candidate for NIR photoelectric applications [20]. In recent years, great progress has been made in the field of PbS QD NIR LEDs, as shown in Table 7.1. These advances have already been introduced in Section 7.2 and will not be repeated here.

In order to obtain ideal colloidal NIR QDs, it is very important to choose appropriate synthesis methods and control reaction conditions. For PbX QDs, thermal injection synthesis with rapid nucleation and continuous growth is the most commonly

Table 7.1 Research progress of lead sulfide quantum dots near-infrared light-emitting diodes.

Year	Group	EL Wavelength (nm)	Peak EQE (%)	Radiance (WSr^{-1} m^{-2})
2012	Wise Frank W.	1232	2.0	6.4
2015	Vladimir Bulovic	1242	4.3	0.75
2016	Sargent Edward H.	1391	5.2	2.6
2017	Yang Xuyong	1500	4.12	6.04
2018	Konstantatos G.	1400	7.9	9

used synthesis method because of its low cost, simple preparation, and high yield. PbX QDs are generated by the reaction of lead and sulfur precursors with solvents and ligands in a synthetic system. At present, the most commonly used solvents in the synthesis process are 1-octadecene (ODE) and oleic acid (OA). OA is also the main choice of ligand for NIR QDs, which is used to stabilize QDs and passivate dangling bonds on the surface of NIR QDs.

Lead oleate is one of the common precursors produced by the reaction of lead salts with non-coordination solvents such as OA (ODE and diphenyl ether), such as $PbCl_2$, PbO, and $Pb(CH_3COO)_2$. In some cases, oleylamine is also used as a ligand. There are many selenium sources that can be used to prepare selenium precursors, such as N-trioctylphosphine selenide (Top-Se) and BIS (trimethylsilyl) selenium (TMS-Se), SeO_2, and selenium urea. As for sulfur precursors, BIS (trimethylsilyl) sulfide (TMS-S) is also a common sulfur precursor in the synthesis of PbS QDs. In 2003, Hines et al. injected TMS-S into lead oleate precursor at high temperature to obtain PbS QDs with PLQY of 20–30% and narrow particle size dispersion (10–15%) [21]. There are other materials that can be used as sulfur sources for the synthesis of PbS QDs, such as sulfur powder, thioacetamide, sodium sulfide, hydrogen sulfide gas, and thiourea. However, in many sulfur-containing compounds, the obtained QDs PLQY are not high and the adjustable range is smaller than that synthesized by TMS-S. Trioctyl tellurium precursor needed to synthesize PbTe QDs was obtained by dissolving Te powder in TOP. In 2006, Murphy et al. synthesized PbTe QDs by injecting TOP-Te into lead oleate generated by PbO and OA reactions in ODE solvent [22].

Because PbSe QDs synthesized by TOP-Se, PbS QDs synthesized by TMS-S, and PbTe QDs synthesized by TOP-Te are highly toxic and unstable in the air, many researchers focus on finding alternative and environmentally friendly precursors. Although these other chalcogenide precursors can also be used to synthesize high-quality QDs, they still cannot completely replace TOP-Se, TMSS, and TOP-Te due to the requirements of reactions or the quality of QDs.

7.2.2 Chalcogenide Cadmium Quantum Dots

Recently, based on transition metal complexes (for example, iridium and platinum), the peak EQE of NIR light sources of OLED and perovskite PeLED has reached as high as >9% and 10% at 721 nm, even more than 20% (20.7% at 803 nm and 21.6% at 800 nm) [23–26]. However, for the NIR LED based on transition-metal complex, high cost, relatively scarce resources, and efficiency decline under high brightness are still great challenges for its large-scale and long-term application [27]. In NIR, organic dyes and semiconductor polymers often suffer from low-light photoluminescence quantum yield (PLQY) [11]; and perovskite semiconductor often suffers from serious trap-mediated non-radiation loss, which has been identified as the main efficiency limiting factor of LED [26]. In addition, its highly unstable characteristics still hinder its industrialization.

As an alternative material, II–VI semiconductor QDs show their unique advantages as NIR emitters due to their high PLQY, easily adjustable size-dependent emission, low-cost solution processability, and scalable high-quality QD production [28–32]. The latest development of visible light LED based on II–VI QD (especially type-I structure) has met the requirements of display and solid-state lighting [33–38]. In particular, the recent new strategy of optimizing shell materials to obtain better energy levels matching with the highest occupied molecular orbital (HOMO) of hole transport layer has promoted its industrialization [34, 35, 37]. However, simply adjusting the size of high-quality type-I QDs cannot modulate their emission wavelength to deep red or NIR region, which limits their application in the NIR field. In view of the fact that CdTe/CdSe II QD can be easily extended to the NIR region, it has high PLQY and good stability, because the valence state and conduction band energy of the core are higher than that of the shell [39, 40], which makes them the most promising NIR materials.

7.2.3 Silicon Quantum Dots

Nanocrystals have attractive EQE because of their size-adjustable emission characteristics, compatibility with solution treatment, and recently demonstrated high-quality devices. Up to now, almost only group II–VI, III–V, and IV–VI semiconductor nanocrystals have been studied. Particularly noteworthy are II–VI and IV–VI nanocrystals, which have been used to prove that they have tunable electroluminescence (EL) in both visible and infrared parts of the electromagnetic spectrum [4–10, 13–16].

Although there is great interest in the study of II–VI and IV–VI colloidal semiconductor nanocrystals, little attention has been paid to class IV systems (including silicon). Although bulk silicon is characterized by indirect bandgap, silicon nanocrystals (SiNCs) with a diameter of less than 5 nm have been proven to have exceptionally high PL efficiency. SiNCs are also attractive because of their potential low toxicity and high natural abundance. Therefore, an attempt has been made to integrate SiNC into a hybrid light-emitting device [20, 41].

In 2010, Cheng et al. synthesized SiNCs by nonthermal plasma process, and a hybrid nanocrystal–organic light-emitting device (NC-OLED) was prepared, which was usually composed of inorganic nanocrystal light-emitting layers placed between organic charge transport layers as shown in Figure 7.3 [41]. The low efficiency previously observed for silicon NC-OLED may also reflect the low and limited injection

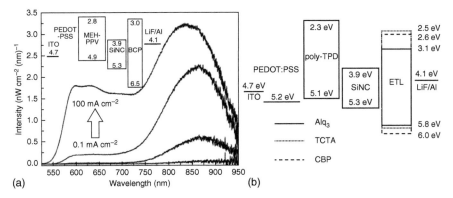

Figure 7.3 (a) Structure and emission spectrum of mixed nanocrystal–organic light-emitting device, (b) energy level diagram of double heterojunction device.

efficiency of charge carriers in SiNC layer. The use of optimized wide bandgap HTL and ETL materials solves these problems by creating a double heterostructure, which allows charge carriers and excitons to be confined in the nanocrystal emission layer. Therefore, in 2011, Cheng et al. achieved the best EQE of 8.6%.

7.3 Optimization of Near-Infrared Quantum Dot Materials

7.3.1 Regulation of Near-Infrared Quantum Dots by Ligand Engineering

The synthesis of PbS QDs uses long-chain alkyl groups as ligands (such as OA or oleylamine), but these long-chain insulating ligands block the injection and transmission of carriers between QDs, which greatly reduces the performance of optoelectronic devices. Therefore, for optoelectronic devices, it is necessary to replace the original long-chain alkyl group with short ligands (such as 3-mercaptopropionic acid (MPA) and 8-mercaptooctanoic acid (MOA)) to improve the carrier transport between QDs, thus improving the performance of devices. There are two main methods of organic ligand exchange: solid ligand exchange and solution ligand exchange.

The process of solid-state ligand exchange is shown in Figure 7.4a. QDs with long-chain alkyl ligands are deposited on the substrate by spin coating, and then the film is covered with a solution containing short-chain organic ligands. After a short soaking time, the solution was removed by spin coating, and then the proton solvent was dropped on the substrate to clean the exchanged ligands and excess new short ligands. In addition, the concentration of QDs needs to be carefully adjusted to ensure the complete exchange of ligands, and the thickness of each layer is usually 10–30 nm.

The process of ligand exchange in solution is shown in Figure 7.4b, in which QDs of long-chain alkyl ligands are dispersed in nonpolar solvents and short-chain

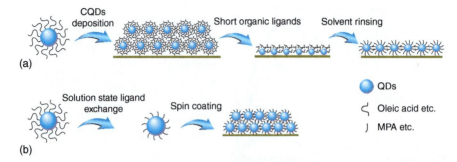

Figure 7.4 (a) Solid ligand exchange, (b) solution ligand exchange.

hydrophilic organic ligands are dissolved in polar solvents. QDs are wrapped by short-chain hydrophilic ligands and then transferred to polar solvents. The QD solution after ligand exchange can be used as ink, and the thin film can be prepared by one-step spin-coating method, which can be directly spin coated or sprayed without other exchange steps.

At present, most LEDs, solar cells, and photodiodes use solid-state ligand exchange. The method is simple in process and can produce smooth and compact QD films, the long-chain alkyl ligands are basically removed, and the strong combination between sulfhydryl ligands and QDs enables rapid and effective exchange.

7.3.2 Control of Near-Infrared Quantum Dots by Core/Shell Structure

In order to improve the stability of QDs, a shell material with larger bandgap and better chemical stability is grown on the QD core to passivate the well state, which can limit the holes and electrons in the core and protect the core from being exposed to air.

QDs with shell structure are usually divided into two types: type-I, where the wave functions of electrons and holes are confined to the same region (core or shell) of QDs; type-II, where the wave functions of electrons and holes are separated into two regions of QDs. The bandgap energy of type-II QDs is smaller than that of the core and shell of QDs, so that the fluorescence emission spectrum of QDs is red-shifted. There are two ways to form core/shell structure: one is cation exchange and the other is epitaxial shell growth. The coated shell can exist in three cases: uniform shell, composition gradient shell, or alloy shell.

Cation exchange is the most commonly used method to form the QD shell of PbX because Pb^{2+} cations near the surface can easily exchange with other metal ions in the solution, such as Cd^{2+} cations, which can be proved by the blue-shift emission generated. Compared with the QD using only the core, the QD of the core/shell will have blue-shift emission because it reduces the size of PbX QD. Pietryga et al. first produced PbSe/CdSe and PbS/CdS core/shell QDs by cation-exchange method in 2008 [42]. These QDs are stable to spectral attenuation and spectral shift.

7.3.3 Quantum Dots in the Matrix

Colloidal quantum dots (CQDs) have gradually become promising infrared luminescent materials due to their adjustable bandgap, high quantum efficiency, and solvability. However, the introduction also mentions that QDs will have strong luminescence annihilation in the state of solid thin films. The methods to solve the self-annihilation of QDs include growing protective shell structures, coating insulating organic ligands, and introducing polymer matrix.

Since 2002, organic and inorganic materials have been used as the matrix materials of QDs. In 2002, Tessler et al. used MEH-PPV as the matrix material [16]. Compared with the polymer in visible light spectrum, the PL intensity of QD polymer composites is obviously weakened. This difference proves that the energy of polymer body is transferred to InAs-ZnSe NCs. Although the EQE of the device is about 0.5%, the turn-on voltage is quite high (about 15 V). The method of mixing with polymer is only suitable for core/shell structural materials, while PbX, a NIR material with good application prospects, is a single structural material. In 2005, the Sargent group embedded a single structure PbS QD after ligand exchange into MEH–PPV and improved PLQY to increase the EQE of the device by more than 30 times [15].

In order to obtain bright luminescence, high voltage is needed to inject enough current, thus increasing power loss. To sum up, currently available CQD thin films compromise between luminous efficiency and charge transport, which leads to unacceptably high-power consumption [43]. Considering that perovskite has a good diffusion length, some researchers have overcome this problem by embedding CQD into high-mobility perovskite matrix (QDiP) [11, 12, 17, 18, 44].

As shown in Figure 7.5a, QDiP system has two heterogeneous material structures, but it is not enough to produce epitaxial bonding between them, and energy must also be considered. The epitaxial interface structure between perovskite matrix and CQDs will provide excellent passivation performance and improve PLQY. At present, QDiP adopts a heteroepitaxial structure with I-type energy band arrangement, as shown in Figure 7.5b, which is suitable for luminescent materials through spatially confined enhanced radiation recombination. Using type-I arrangement can make electrons and holes effectively leak out of the matrix and be confined in CQD [11]. The synthesis of QDiP mainly includes two steps: the preparation of CQD and the preparation of CQD and perovskite hybrid films by reprecipitation [11, 12, 17, 18].

In 2015, the Sargent group proposed that QDs grow perovskite heteroepitaxy to form the structure of QDs in perovskite matrix [18]. In 2016, better lattice matching was obtained by adjusting the composition of halogen ions in perovskite, as shown in Figure 7.5c, and the device performance with EQE of 5.2% was obtained [11]. Considering the fast and unbalanced carrier dynamics and a large number of phase separations in metal halide perovskite, they reported a QDiP material, as shown in Figure 7.5d. The matrix used in this material is a perovskite with reduced dimensions, and the EQE of 8.1% is achieved at 980 nm when the radiation is as high as 7.4 W Sr^{-1} m^{-2}. Similarly, Yusoff et al. embedded Ag_2S @ SiO_2 CQDs in perovskite matrix and optimized the charge injection balance of the device. The EQE at 1397 nm was 16.98%, and the power conversion efficiency was 11.28% [12].

7 Near-Infrared Quantum Dots (NIR QDs)

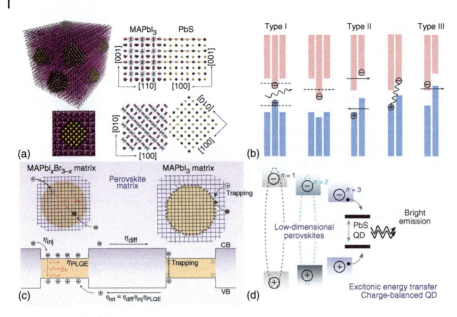

Figure 7.5 (a) QDiP structure simulation; (b) heteroepitaxial structures with different energy band arrangements; (c) QDiP structure with matched lattice parameters; and (d) QDiP system with quasi-two-dimensional perovskite as substrate.

7.4 Summary and Prospect

This chapter begins with the introduction of three kinds of NIR QDs, including PbX, CdX, and Si QDs. The structure, synthesis methods, and research progress of these three kinds of materials are described accordingly. Then, the main factors affecting the EQE of devices are summarized, and the corresponding understanding of these scientific problems and the existing solutions is put forward.

The research after NIR QDs will focus on environment-friendly lead-free and cadmium-free QD materials with long life and stable performance.

There has been significant progress in the development of NIR QDs for LEDs. These materials have several advantages over conventional organic and inorganic semiconductors, including their narrow emission spectra, high PL quantum yields, and tunable emission wavelengths.

One of the most promising NIR QD materials is lead sulfide (PbS) QDs, which have demonstrated high PL quantum yields and efficient EL in LED devices. Other materials, such as CdS, CdSe, and Si QDs, have also shown potential for use in NIR LEDs. Despite the progress made, there are still some challenges that need to be addressed.

One issue is the toxicity of lead and cadmium in some of these materials, which limits their potential for widespread commercial use. To overcome this, there is a growing focus on developing lead-free and cadmium-free alternatives that are more environmentally friendly. Additionally, there is a need to improve the efficiency and stability of NIR QD-based LEDs, particularly in terms of their lifetime and stability under operating conditions.

Future research in this area is likely to focus on the development of new materials with improved properties, as well as the optimization of device structures and fabrication processes. There will also be a need for continued investigation into the fundamental physics of QD materials and their interaction with other components in the LED devices, in order to further improve their performance and efficiency. Overall, the outlook for NIR QD-based LEDs is promising, with the potential to enable new applications in sensing, imaging, and lighting.

References

1 Yang, X., Ren, F., Wang, Y. et al. (2017). Iodide capped PbS/CdS core-shell quantum dots for efficient long-wavelength near-infrared light-emitting diodes. *Scientific Reports* 7 (1): 14741.
2 Siebrand, W. (1967). Radiationless transitions in polyatomic molecules. I. Calculation of Franck—Condon factors. *The Journal of Chemical Physics* 46 (2): 440–447.
3 Wilson, J.S., Chawdhury, N., Al-Mandhary, M.R. et al. (2001). The energy gap law for triplet states in Pt-containing conjugated polymers and monomers. *Journal of the American Chemical Society* 123 (38): 9412–9417.
4 Wheeler, L.M., Anderson, N.C., Palomaki, P.K. et al. (2015). Silyl radical abstraction in the functionalization of plasma-synthesized silicon nanocrystals. *Chemistry of Materials* 27 (19): 6869–6878.
5 Hendricks, M.P., Campos, M.P., Cleveland, G.T. et al. (2015). A tunable library of substituted thiourea precursors to metal sulfide nanocrystals. *Science* 348 (6240): 1226–1230.
6 Franke, D., Harris, D.K., Chen, O. et al. (2016). Continuous injection synthesis of indium arsenide quantum dots emissive in the short-wavelength infrared. *Nature Communications* 7 (1): 12749.
7 Lei, W., Antoszewski, J., and Faraone, L. (2015). Progress, challenges, and opportunities for HgCdTe infrared materials and detectors. *Applied Physics Reviews* 2 (4): 041303.
8 Zhang, Y., Hong, G., Zhang, Y. et al. (2012). Ag_2S quantum dot: a bright and biocompatible fluorescent nanoprobe in the second near-infrared window. *ACS Nano* 6 (5): 3695–3702.
9 Li, L., Daou, T.J., Texier, I. et al. (2009). Highly luminescent $CuInS_2$/ZnS core/shell nanocrystals: cadmium-free quantum dots for in vivo imaging. *Chemistry of Materials* 21 (12): 2422–2429.
10 Lu, H., Carroll, G.M., Neale, N.R., and Beard, M.C. (2019). Infrared quantum dots: progress, challenges, and opportunities. *ACS Nano* 13 (2): 939–953.
11 Gong, X., Yang, Z., Walters, G. et al. (2016). Highly efficient quantum dot near-infrared light-emitting diodes. *Nature Photonics* 10 (4): 253–257.

12 Vasilopoulou, M., Kim, H.P., Kim, B.S. et al. (2020). Efficient colloidal quantum dot light-emitting diodes operating in the second near-infrared biological window. *Nature Photonics* 14 (1): 50–56.

13 Sun, L., Choi, J.J., Stachnik, D. et al. (2012). Bright infrared quantum-dot light-emitting diodes through inter-dot spacing control. *Nature Nanotechnology* 7 (6): 369–373.

14 Supran, G.J., Song, K.W., Hwang, G.W. et al. (2015). High-performance shortwave-infrared light-emitting devices using core-shell (PbS-CdS) colloidal quantum dots. *Advanced Materials* 27 (8): 1437–1442.

15 Konstantatos, G., Huang, C., Levina, L. et al. (2005). Efficient infrared electroluminescent devices using solution-processed colloidal quantum dots. *Advanced Functional Materials* 15 (11): 1865–1869.

16 Tessler, N., Medvedev, V., Kazes, M. et al. (2002). Efficient near-infrared polymer nanocrystal light-emitting diodes. *Science* 295 (5559): 1506–1508.

17 Gao, L., Quan, L.N., García de Arquer, F.P. et al. (2020). Efficient near-infrared light-emitting diodes based on quantum dots in layered perovskite. *Nature Photonics* 14 (4): 227–233.

18 Ning, Z., Gong, X., Comin, R. et al. (2015). Quantum-dot-in-perovskite solids. *Nature* 523 (7560): 324–328.

19 Pradhan, S., Di Stasio, F., Bi, Y. et al. (2019). High-efficiency colloidal quantum dot infrared light-emitting diodes via engineering at the supra-nanocrystalline level. *Nature Nanotechnology* 14 (1): 72–79.

20 Liu, H., Zhong, H., Zheng, F. et al. (2019). Near-infrared lead chalcogenide quantum dots: synthesis and applications in light emitting diodes. *Chinese Physics B* 28 (12): 128504.

21 Hines, M.A. and Scholes, G.D. (2003). Colloidal PbS nanocrystals with size-tunable near-infrared emission: observation of post-synthesis self-narrowing of the particle size distribution. *Advanced Materials* 15 (21): 1844–1849.

22 Murphy, J.E., Beard, M.C., Norman, A.G. et al. (2006). PbTe colloidal nanocrystals: synthesis, characterization, and multiple exciton generation. *Journal of the American Chemical Society* 128 (10): 3241–3247.

23 Cao, Y., Wang, N., Tian, H. et al. (2018). Perovskite light-emitting diodes based on spontaneously formed submicrometre-scale structures. *Nature* 562 (7726): 249–253.

24 Graham, K.R., Yang, Y., Sommer, J.R. et al. (2011). Extended conjugation platinum (II) porphyrins for use in near-infrared emitting organic light emitting diodes. *Chemistry of Materials* 23 (24): 5305–5312.

25 Kim, D.-H., D'aléo, A., Chen, X.-K. et al. (2018). High-efficiency electroluminescence and amplified spontaneous emission from a thermally activated delayed fluorescent near-infrared emitter. *Nature Photonics* 12 (2): 98–104.

26 Xu, W., Hu, Q., Bai, S. et al. (2019). Rational molecular passivation for high-performance perovskite light-emitting diodes. *Nature Photonics* 13 (6): 418–424.

27 Wang, S., Yan, X., Cheng, Z. et al. (2015). Highly efficient near-infrared delayed fluorescence organic light emitting diodes using a phenanthrene-based

charge-transfer compound. *Angewandte Chemie International Edition* 54 (44): 13068–13072.

28 Chen, O., Zhao, J., Chauhan, V.P. et al. (2013). Compact high-quality CdSe–CdS core–shell nanocrystals with narrow emission linewidths and suppressed blinking. *Nature Materials* 12 (5): 445–451.

29 Kwak, J., Bae, W.K., Lee, D. et al. (2012). Bright and efficient full-color colloidal quantum dot light-emitting diodes using an inverted device structure. *Nano Letters* 12 (5): 2362–2366.

30 Qin, H., Niu, Y., Meng, R. et al. (2014). Single-dot spectroscopy of zinc-blende CdSe/CdS core/shell nanocrystals: nonblinking and correlation with ensemble measurements. *Journal of the American Chemical Society* 136 (1): 179–187.

31 Shirasaki, Y., Supran, G.J., Bawendi, M.G., and Bulović, V. (2013). Emergence of colloidal quantum-dot light-emitting technologies. *Nature Photonics* 7 (1): 13–23.

32 Zhang, H., Hu, N., Zeng, Z. et al. (2019). High-efficiency green InP quantum dot-based electroluminescent device comprising thick-shell quantum dots. *Advanced Optical Materials* 7 (7): 1801602.

33 Dai, X., Zhang, Z., Jin, Y. et al. (2014). Solution-processed, high-performance light-emitting diodes based on quantum dots. *Nature* 515 (7525): 96–99.

34 Li, X., Lin, Q., Song, J. et al. (2020). Quantum-dot light-emitting diodes for outdoor displays with high stability at high brightness. *Advanced Optical Materials* 8 (2): 1901145.

35 Shen, H., Gao, Q., Zhang, Y. et al. (2019). Visible quantum dot light-emitting diodes with simultaneous high brightness and efficiency. *Nature Photonics* 13 (3): 192–197.

36 Song, J., Wang, O., Shen, H. et al. (2019). Over 30% external quantum efficiency light-emitting diodes by engineering quantum dot-assisted energy level match for hole transport layer. *Advanced Functional Materials* 29 (33): 1808377.

37 Yang, Y., Zheng, Y., Cao, W. et al. (2015). High-efficiency light-emitting devices based on quantum dots with tailored nanostructures. *Nature Photonics* 9 (4): 259–266.

38 Zhang, H., Chen, S., and Sun, X.W. (2018). Efficient red/green/blue tandem quantum-dot light-emitting diodes with external quantum efficiency exceeding 21%. *ACS Nano* 12 (1): 697–704.

39 Kim, S., Fisher, B., Eisler, H.-J., and Bawendi, M. (2003). Type-II quantum dots: CdTe/CdSe (core/shell) and CdSe/ZnTe (core/shell) heterostructures. *Journal of the American Chemical Society* 125 (38): 11466–11467.

40 Shea-Rohwer, L.E., Martin, J.E., Cai, X., and Kelley, D.F. (2012). Red-emitting quantum dots for solid-state lighting. *ECS Journal of Solid State Science and Technology* 2 (2): R3112.

41 Cheng, K.-Y., Anthony, R., Kortshagen, U.R., and Holmes, R.J. (2011). High-efficiency silicon nanocrystal light-emitting devices. *Nano Letters* 11 (5): 1952–1956.

42 Pietryga, J.M., Werder, D.J., Williams, D.J. et al. (2008). Utilizing the lability of lead selenide to produce heterostructured nanocrystals with bright, stable infrared emission. *Journal of the American Chemical Society* 130 (14): 4879–4885.

43 Cheng, K.-Y., Anthony, R., Kortshagen, U.R., and Holmes, R.J. (2010). Hybrid silicon nanocrystal-organic light emitting devices for infrared electroluminescence. *Nano Letters* 10 (4): 1154–1157.

44 Ishii, A. and Miyasaka, T. (2020). Sensitized Yb^{3+} luminescence in $CsPbCl_3$ film for highly efficient near-infrared light-emitting diodes. *Advanced Science* 7 (4): 1903142.

8

White QDLED

8.1 Generation of White Light

In 1879, Edison manufactured the world's first practical tungsten incandescent lamps, which sparked an emerging revolution in electric lighting technology. To date, humans have been using electrical energy for lighting to expand productive activities for more than 100 years. Electricity is an important factor for sustainable economic development, and lighting accounts for about 20% of the world's energy consumption, with only 30% of the electrical energy used for lighting is actually used to emit light; the rest is dissipated in the form of heat. In order to cope with the huge demand for electrical energy and reduce carbon dioxide emissions, many countries have announced that they will phase out energy-intensive incandescent lamps [1]. It has always been an important goal of human scientific research to develop new and efficient electric light sources. The light-emitting diode (LED) is a common light-emitting semiconductor device that can efficiently convert electrical energy into light energy by releasing energy from an electron–hole complex.

In the past fifty years, many groups conducting research on LED have put great effort into preparing a variety of light-emitting materials and devices with excellent photovoltaic properties. Due to the advantages of LED's long lifespan and high efficiency, it is expected to replace incandescent lamps [2]. The mainstream solution for the preparation of white LEDs is the combination of yellow rare earth phosphors (YAG: Ce^{3+}) and blue LEDs [3]. However, the red light-emitting part of YAG: Ce^{3+} is missing, mainly concentrated in the yellow–green light region, resulting in high color correlation temperature (CCT > 4500 K) and low color rendering index (CRI < 80) of white LEDs [3, 4]. The particles of rare earth phosphors are generally at the micron level, and larger sizes can lead to light scattering, reducing the efficiency of light-emitting devices.

Therefore, there is an urgent need to explore materials for the preparation of high-efficiency and high-color rendering white LEDs. Among the LED materials with different properties, quantum dots (QDs) have attracted more attention due to their high color purity, tunable emission wavelengths, high photoluminescence quantum yield (PLQY), and excellent intrinsic stability. Additionally, QDs have potential applications in various research areas, including biomedicine,

Colloidal Quantum Dot Light Emitting Diodes: Materials and Devices, First Edition. Hong Meng.
© 2024 WILEY-VCH GmbH. Published 2024 by WILEY-VCH GmbH.

photodetection, new energy, and information display. The high PLQY potential of QDs has attracted the attention of many researchers in these research areas.

8.2 Quantum Dots for White LEDs

QDs, which are invisible to the naked eye, are extremely small semiconductor nanocrystals with a particle size of less than 10 nm. They have the advantages of adjustable wavelength size components, high luminous efficiency, small particle size, and almost zero light scattering. These advantages have led to widespread attention and research in the field of display lighting [5–8]. A quantum dot light-emitting diode (QDLED) is a thin layer of quantum dots used as the light-emitting layer in LED devices. Electrons and holes converge on both sides of the quantum dot layer to form photons, which are then recombined to emit light. Since the first appearance of QDLEDs in 1994 [9], research on white QDLEDs has developed rapidly. Researchers have focused on optimizing the optical properties of quantum dots, improving their stability, enhancing compatibility, and reducing toxicity. White LEDs based on blue LEDs with CdSe quantum dots have been successfully prepared, with a CRI of about 91, which is better than commercial YAG:Ce^{3+} white LEDs [10]. However, the prospect of CdSe quantum dots as a color conversion material for white LEDs is unclear due to the toxicity of the heavy metal element Cd, as well as the severe self-absorption and energy transfer caused by its small Stokes shift. The LED efficiency may be reduced due to the small Stokes shift [11]. Compared to conventional white LEDs, QDLEDs with quantum dots as down-conversion materials occupy more of the red light spectrum and can cover a wider range of the spectrum than conventional white LEDs. In the visible region, the more red light components present, the lower the color temperature of the white light produced. This is more comfortable and can help delay visual fatigue when used for indoor lighting, while also being less irritating to the human eye. Therefore, there is an urgent need to develop white light quantum dot LED materials with low toxicity, high efficiency, and a large Stokes shift.

The white LED based on quantum dot photoluminescence mechanisms includes three ways to achieve white light emission: yellow–blue composite white light quantum dots, quantum dots that can directly emit white light upon excitation, and composite quantum dots made of the three primary colors. These three methods can be fundamentally divided into two categories: single light-emitting white light devices and composite light-emitting white light devices. Both have their advantages and disadvantages. Single-emitting white light devices require the synthesis of special materials as a single-emitting layer of white light, which may be expensive and difficult to achieve. On the other hand, mixing two or more groups of quantum dots usually leads to different attenuation rates between different emission wavelengths, resulting in the deterioration of device performance over time. Therefore, the stability of the quantum dot white light emitting diode (WLED) devices made is not satisfactory [12]. The following subsections will show the general methods and recent advances for making white quantum dot light-emitting diodes (WQDLEDs).

8.2.1 Yellow–Blue Composite White Light Quantum Dots

The most important application of yellow QDLEDs, which emit at 570–590 nm, is to combine blue light-emitting chips to make white QDLEDs. The preparation of blue and yellow high-efficiency devices (or phosphors) is a crucial cornerstone of such white light devices and has great potential commercial value. The current commercial yellow phosphors mainly use precious rare earth elements, which are costly and polluting, and there is an urgent need to find alternatives. In addition, if high-efficiency yellow light devices can be prepared in one step, compared with using red and green light to synthesize yellow light, the efficiency can be significantly increased. Therefore, the study of high-performance yellow light QDLED devices has potential research vitality. At present, the reported yellow light-emitting materials mainly include cadmium-containing QDs, such as CdTe and CdSe, and cadmium-free QDs, such as InP, InGaN, cesium chalcogenide quantum dots, and doped carbon quantum dots.

8.2.1.1 Cadmium-Containing Yellow Light Quantum Dots

Among cadmium-containing yellow light QD materials, in 2008, the Tan group reported white light QDLED devices with CdSe-ZnS core/shell structured quantum dots as the emitting layer [3]. They created a new white QDLED device structure with a blue-emitting polymer poly(N,N'-bis(4-butylphenyl)-N,N'-bis(phenyl)benzidine) (poly TPD) film as the substrate, on which a layer of yellow-emitting CdSe-ZnS core/shell quantum dots was deposited, creating a simple bilayer light-emitting device that produces binary complementary white light. By optimizing the thickness of the QD layer, they achieved stable, bright-white QDLED devices at a wide range of bias voltages. The devices start at voltages as low as 3.15 V and have a maximum brightness of 2600 cd/m². The CIE coordinates of the LED color shift only slightly from (0.32, 0.36) to (0.33, 0.37). This work demonstrates that compound yellow and blue light devices can yield highly efficient white light devices but with significant improvements in the device structure. The double-layer light-emitting devices do not have a hole transport layer and an electron transport layer, so the exciton transport efficiency is greatly reduced, making the efficiency of exciton compounding much less efficient than that of multilayer structured devices. In addition, the intrinsic mechanism needs to be further investigated to regulate the devices performance by regulating the thickness of the light-emitting layer.

In 2014, Yin et al. reported CdTe quantum dots and YAG hybrid phosphors for white LEDs as shown in Figure 8.1 [4]. By combining CdTe quantum dots and YAG hybrid phosphors with blue light-emitting chips, the white LED showed white light areas with excellent coordinates ($x = 0.30$, $y = 0.29$) and a CRI value of 75 at 20 mA. This finding confirms the high CRI value of white LEDs by adding a composite QD of CdTe and YAG phosphor. However, this work did not succeed in synthesizing a high-performance, low-cost yellow phosphor.

8.2.1.2 Cadmium-Free Yellow Light Quantum Dots

In 2013, the Jang group reported a simple solvothermal synthesis of InP/ZnS core/shell QDs using a safer and cheaper phosphorus precursor than the most

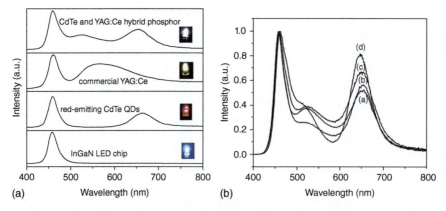

Figure 8.1 (a) Photoluminescence spectra of InGaN LED chips with red-emitting CdTe quantum dots, commercial YAG:Ce and CdTe and YAG:Ce hybrid phosphors. (b) Scale YAG:Ce:CdTe quantum dots of (a) 1, (b) 1.25, (c) 1.5, and (d) 1.75, respectively. Inset shows photos of individual LEDs at an operating current of 20 mA. Source: Yin et al. [4]/John Wiley & Sons.

commonly used tris-(trimethylsilyl)phosphine (P[TMS]$_3$) and tris-(dimethylamino) phosphine P(N[CH$_3$]$_2$)$_3$ [13]. The bandgap of InP quantum dots can be easily controlled by varying the solvothermal growth time (4 and 6 hours) at 150 °C, and the InP/ZnS core–shell structured quantum dots showed green and yellow colors with 41–42% emission quantum yield after continuous heating at 220 °C for 6 hours. In addition, a mixture of green and yellow InP/ZnS quantum dots was applied to produce white light with wider spectral coverage, leading to enhanced CRI of 76 and luminescence efficiency of 32.1 lm W^{-1} at 20 mA. This work advances the synthesis of cadmium-free yellow quantum dot devices, but is less environmentally friendly as phosphorus-containing quantum dots remain toxic.

In 2015, the Yang group reported white LEDs based on InGaN blue quantum wells and green–yellow quantum dots (Figure 8.2) [8]. The phosphorus-free white LEDs, consisting of four layers of InGaN/GaN quantum dots and four layers of quantum wells were grown by metal–organic chemical vapor deposition. By mixing the green–yellow light from the quantum dots and the blue light from the quantum wells, white emission was shown under electrical injection. The peak electroluminescence wavelengths of the quantum dots and quantum wells are 548 and 450 nm, respectively, at an injection current of 5 mA, and the CRI Ra is 62. This work uses the composite of quantum wells and quantum dots to realize the preparation of white light devices, which is relatively novel, but the CRI is slightly inferior and can only be applied to the fields where color discrimination is not required, which is a big limitation.

Since the Park group reported high brightness yellow–green emitting CuInS$_2$ colloidal quantum dots as shown in Figure 8.3. They coated CuInS$_2$ quantum dots with multilayer ZnS quantum dots [6]. With the formation of the first and second ZnS shell layers, the peaks in the PL spectral band shifted to shorter wavelengths (670 → 559 nm), and the quantum efficiency increased significantly from 31.7% for the core quantum dots to 80.0% for the CSS quantum dots. Bright white LEDs

Figure 8.2 Schematic diagram of white LED samples based on InGaN blue quantum dots and green–yellow quantum dots. Source: Yang et al. [8]/with permission of Elsevier.

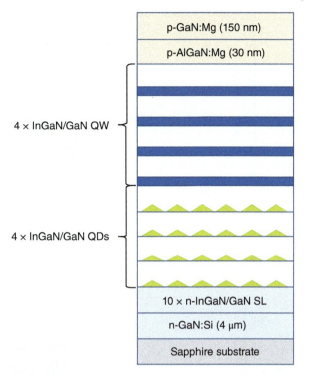

were prepared by coating CIS/ZnS/ZnS-CSS quantum dots on blue LEDs, which show bright natural white light [(CIEx, CIEy) = (0.3229, 0.2879), η_L = 80.3 lm W^{-1}, Tc = 6140 K, Ra = 73. The prepared white LEDs were also relatively stable with the increase in forward current. This result indicates that CIS/ZnS/ZnS quantum dots are a promising material for the preparation of white light QDLED devices with significantly lower toxicity compared to Cd-containing QDs.

In 2016, the Wang group reported multicolor fluorescent LEDs based on cesium-lead halide chalcogenide quantum dots-a series of monochromatic LED devices made by combining all-inorganic cesium-lead halide chalcogenide quantum dots and blue LED chips (Figure 8.4) [7]. The study designed a liquid-phase color-changing layer to maintain the high quantum efficiency of the quantum dots while protecting them from environmental factors such as water and oxygen, and further demonstrated that this structure can suppress thermal effects on the device surface. The EQE of yellow QDLEDs was confirmed to be 12.4% when the luminous efficiency of 63.4 lm W^{-1} was achieved. In addition, these devices also show good color stability at increasing operating currents, and the CRI of liquid warm white QDLEDs can reach 86 at a color temperature of 2890 k. This work explores new directions for the preparation and synthesis of calcium titanite quantum dots in the field of high-performance yellow and white light devices.

In 2018, the He group reported the high-purity yellow light emission from calixarene CsPb (Br$_x$I$_{1-x}$)$_3$ quantum dots and its application in yellow LEDs (Figure 8.5) [5]. This work systematically investigates quantum yield of 50% yellow

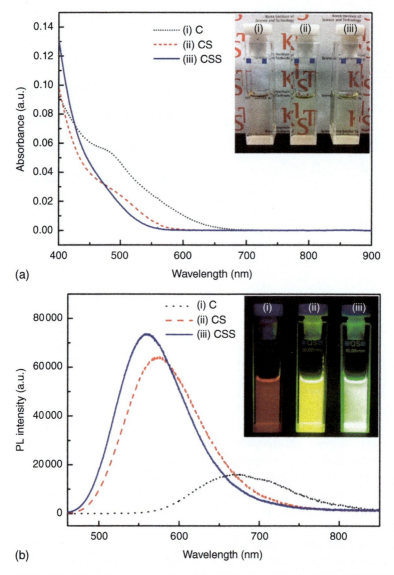

Figure 8.3 Absorption spectra (a) and (b) PL spectra of (i) CIS core (C), (ii) CIS/ZnS core/shell (CS), and (iii) CIS/ZnS/ZnS core/shell/shell (CSS) quantum dots. (a) and (b) insets show the photographs of CIS, CIS/ZnS, and CIS/ZnS/ZnS solutions taken under room light and UV light, respectively. Source: Park et al. [6]/American Chemical Society.

light QDLEDs and designs a novel thermal insulation structure for yellow light QDLEDs. The chromaticity coordinates of the yellow light QDLEDs hardly change in the operating range of the driving current from 5 to 150 mA. In addition, the devices luminous efficiency reaches 13.51 mW^{-1} at 6 mA. These results indicate that CsPb(Br$_x$I$_{1-x}$)$_3$ calcium titanite quantum dots are a potential candidate for high-purity yellow LEDs.

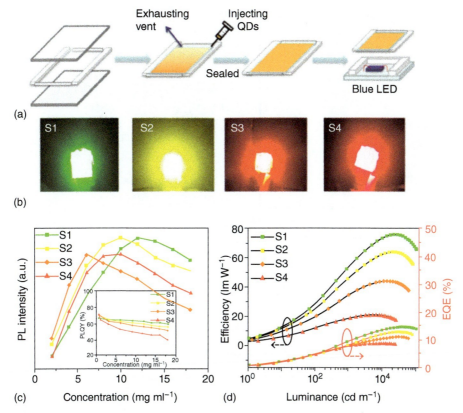

Figure 8.4 (a) Process flow of liquid-type QDLEDs; (b) pictures of each LED sample with different luminescence colors operating at 40 mA under dark conditions; (c) output intensity of each monochromatic light (green (S1), yellow (S2), red (S3), and red (S4) as the QD concentration in the solution increases; inset shows the PLQYs versus the QD of each monochromatic emission concentration; (d) luminous efficiency and EQE as a function of luminance of green (S1), yellow (S2), red-orange (S3), and red (S4) QDLEDs. Source: Wang et al. [7]/AIP Publishing LLC.

In 2020, Wang et al. reported a controlled synthesis of highly efficient red, yellow, and blue carbon nanodots for photoluminescent devices as shown in Figure 8.6 [10]. While most of the reported CDs are hydrophilic, Wang et al. reported an oil-soluble quantum dot, which is critical for enhancing device efficiency. In this work, red, yellow, and blue emitting carbon nanodots with high quantum yields were successfully prepared by the solvent and reaction temperature mastery using the solvothermal method. CNDs&PVP composite phosphors and composite films were successfully prepared, and red, yellow, and blue CNDs&PVP-based QDLEDs were obtained with CIE coordinates of (0.29, 0.33). This work provides a new idea to deal with the luminescence mechanism and application of multi-color doped CNDs composites.

8.2.2 Three-Base Color Quantum Dot Composite

Quantum dot materials can be classified according to their fluorescence emission peak range as blue [11], green [9], and red [14] quantum dots, which are also known

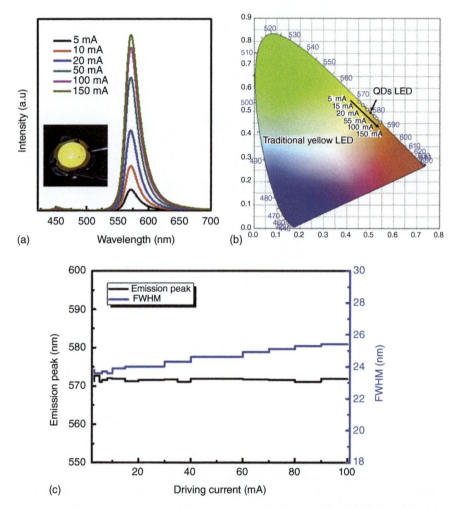

Figure 8.5 Performance of QDLEDs with different driving currents. (a) EL spectrum; (b) chromaticity coordinates of the yellow quantum dot LED; (c) variation of peak wavelengths and FWHMs with driving current. Source: He et al. [5]/Elsevier.

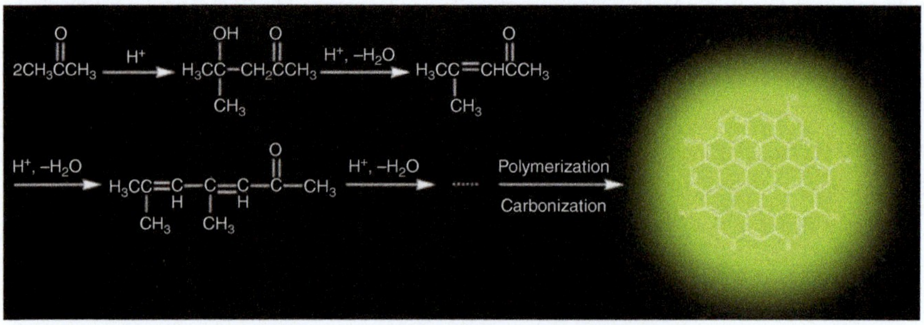

Figure 8.6 Schematic diagram of the synthetic route of oil-soluble co-doped carbon quantum dots.

as three-color quantum dot materials, and the research progress in obtaining white light by compounding three-color quantum dots is relatively slow. This structure was grown on an InP substrate by molecular beam epitaxy. By controlling the deposition time of the CdSe layer, the size and emission wavelength of each CdSe quantum dot can be precisely tuned. As a result, a combination of three stacked CdSe quantum dot layers corresponding to red, green, and blue emission, respectively, was mixed and electrically driven to obtain white light. However, important parameters of the white light source, including CRI, color coordinates, and color correlation temperature (CCT), are not given in the report. The three stacked active layers of this electrically driven device can be replaced by a single layer consisting of a CdSe/ZnS-QD composite corresponding to each of the three base colors, thus simplifying the device structure. By controlling the Fórster energy transfer and charge trapping mechanisms between the different emitting components, the single-emitting layer device emits white light with a luminance of $1050\,cd\,m^{-2}$ and a color coordinate of (0.32, 0.45) [12]. In 2006, Li et al. used CdSe/ZnS emitting blue (490 nm), green (540 nm), and red (618 nm). Core–shell quantum dots were used to prepare three-color white LEDs [15]. By mixing different sizes of three-color CdSe/ZnS quantum dots as the light-emitting layer, a white light with CIE coordinates of (0.32, 0.45) was obtained with a maximum brightness of $1050\,cd/m$ at $58\,mA/cm^2$ and a turn-on voltage of 6 V in air.

Anikeeva et al. also reported the use of a mixture of three quantum dots as a monolayer emitter in an electrically driven structure, and the device had an external quantum efficiency (EQE) of 0.36%, a color coordinate of (0.35, 0.41), and a CRI of 86 [16]. However, the organic surface ligands around the quantum dots had charge injection problems, which passivated the quantum dot surface and created a large potential barrier between the quantum dots and the charge transport layer. However, the charge injection problem of the organic surface ligands around the quantum dots makes the surface of the quantum dots passivated and creates a large potential barrier between the quantum dots and the charge transport layer, which makes it difficult to construct quantum dots with better performance. Although not plagued by the charge injection problem, quantum dot-based WLEDs have been realized in the past decade. Typically, by mounting a combination of red-emitting quantum dots and green rare earth phosphors or green-emitting quantum dots and yellow rare earth phosphors on a blue GaN chip, the resulting WLEDs can display high CRI values. In addition to the combination of quantum dots and rare earth phosphors, mixtures of quantum dots with different emission peak wavelengths also offer new options for WLED integration. Chen et al. reported a hybrid integration of InGaN-based blue-emitting chips with red- and green-emitting $CuInS_2$ QDs with a WLED CRI of 95 and CCT of 4600–5600 K [17]. By using blue or purple GaN chips in the appropriate ratio to excite a mixture of CdZnS/ZnS and CdZnS/ZnSe quantum dots with different emission wavelengths, white light with a CRI greater than 90 can also be obtained [18].

A special structure of WQDLED was developed by Zhang et al. as shown in Figure 8.7 [19]. By mixing red (R), blue (B), and green (G) quantum dots together,

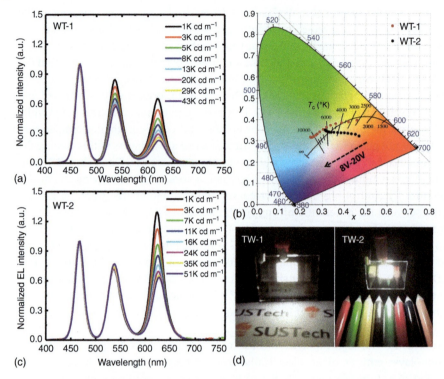

Figure 8.7 (a) Normalized EL spectrum of TW-1; (b) CIE coordinates of TW-1 and TW-2 at different driving voltages; (c) normalized EL spectrum of TW-2; (d) photographs of TW-1 and TW-2 at 15 V operation. Source: Zhang et al. [19]/John Wiley & Sons.

the optimal peak current efficiency (CE) was 22.54 cd A^{-1} and the EQE was 7.07%. To solve the non-radiative Forster resonance energy transfer (FRET) problem between B, G-, and R-QDs, they developed tandem WQDLEDs. By superimposing blue and yellow QDLEDs, the peak CE of two cells tandem WQDLEDs was 30.3 cd A^{-1}, and the maximum EQE was 15.2%. To further improve color stability and efficiency, a three-unit R/G/B tandem WQDLED was developed with a CE value of 55.06 cd A^{-1}, an EQE value of 23.88%, and a luminance value of 65 690 cd m^{-2}. In addition, the tandem WQDLEDs have high color stability with pure white CIE coordinates of (0.33, 0.34) and a CRI of 80, demonstrating that the tandem WQDLEDs have the advantages of high efficiency, pure white, high color stability, and high color rendering.

Jang et al. synthesized multishell green and red InP/ZnSeS/ZnS quantum dots with 82% and 80% PLQYs from cheap and safe P(N[CH3]$_2$)$_3$ aminophosphines (Figure 8.8) [20], and then formed QD silica composites by an ATPMS-based, catalyst-free, aqueous sol–gel reaction. The slow supply of water vapor and the lack of catalyst were found to be effective in preventing damage to the quantum dots, thus maximizing their original PL retention during the silica reaction. Monochromatic and bichromatic QD silica composites were combined as color

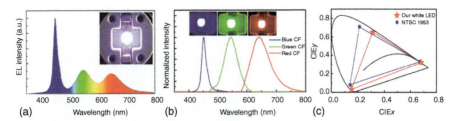

Figure 8.8 (a) Typical EL spectra and EL images (60 mA) of a tricolor white QDLED encapsulated with green/red InP quantum dot-silica composite; (b) normalized blue, green, and red spectra and corresponding images; (c) individual color filters with (b) (marked with red stars) and NTSC standard for the three primary colors (marked with blue circles) radiation corresponding to the CIE color coordinates. Source: Jang et al. [20]/Royal Society of Chemistry.

converters with a blue LED in an on-chip configuration. The device stability of two- and three-color QDLEDs fabricated in this manner was evaluated over a sustained operating time at 60 mA, and after 100 hours of operation, the initial QD emission maintained a significant advantage of up to 93–94%, demonstrating the effectiveness of silica-embedded QD passivation.

A polymer-mediated QD assembly strategy was developed to prepare tricolor QDs@Psi powders with microsphere morphology and white emission by a hybrid method as shown in Figure 8.9 [21]. The prepared B-CQDs@Psi, (Y-CuInS$_2$ @ZnS) @Psi, and (R-CuInS$_2$ @ZnS) @Psi are far superior to the previously reported quantum dots or QD-based composites. The tricolor QDs@Psi exhibit excellent photostability, which is significantly better than the corresponding QD aqueous solutions, and high thermal stability. The fabricated LEDs display solar spectrum simulated emission with luminous efficiency as high as 127.5 lm W^{-1}, CIE coordinates of (0.37,0.37), and CCT of 4500 K. Moreover, compared with commercial white LEDs, the UV light is completely absorbed by the conjugated structure in B-CQDs due to the spectral absorption of CuInS$_2$ @ZnS, and the blue light emission is weak, so it is a relatively healthy LED light source. In addition, this QD-based LED has a recorded CRI of 97, which can be used in professional fields that require high CRI.

8.2.3 Quantum Dots with Direct White Light Emission

There are two types of QD energy spectroscopy: one is direct electroluminescence based on quantum dots, and the other is using quantum dots as nanophosphors and optically exciting them. Narrow-band emission-based electroluminescent QDLEDs are capable of reproducing high-purity colors, and direct white light-emitting QDLEDs are a hot topic of current research. Direct white light is photoluminescence in which only one type of QD layer emits white light, and this white light can be directly excited by the light source of UV LEDs. This type of white LED can overcome some common drawbacks of devices made from multiple quantum dots or fluorescent layers due to self-absorption, scattering, and reabsorption.

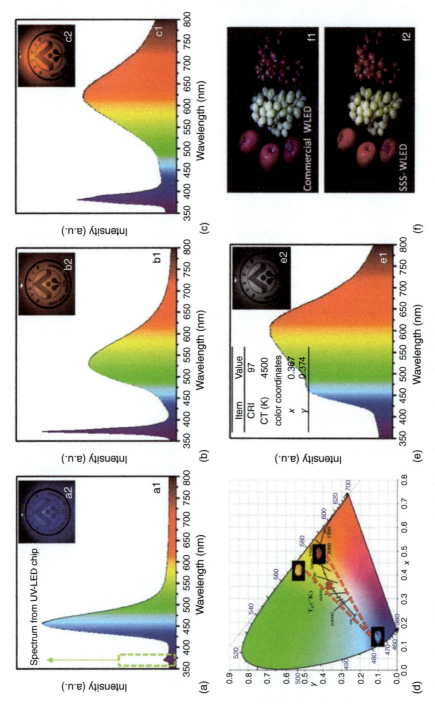

Figure 8.9 (a–c) Luminescence spectra of individual luminescent LED-coated B-CQDs@Psi, (Y-CuInS$_2$@ZnS) @Psi, (R-CuInS$_2$@ZnS) @Psi powders; (d) CIE; (e) luminescence spectra of white LED-coated tri-color QDs@Psi. (f1) fruit color under commercial WLED, (f2) tricolor WLED based on QDs@psi. Source: Hu et al. [21]/John Wiley & Sons.

White LED phosphors with quantum dots that do not contain heavy metals are relatively environmentally friendly and more sustainable. For example, a very interesting white LED based on CZIS quantum dots was proposed by Zhang et al. [22]. White light-emitting QDs can be prepared using commercially available low-toxicity precursors (copper acetate, zinc acetate, indium acetate, and sulfur powder). In addition, these quantum dots have tunable luminescence spectra in the visible to near-infrared band (520 to 750 nm) and have high QY (>70%) without covering any wide bandgap shell material. Compared with other colloidal quantum dots, CZIS quantum dots prepared by this method have long luminescence lifetime, large Stokes shift, and good chemical and thermal stability. A simple illumination device was fabricated using a commercial blue–white LED covered with CZIS QD film, and the color of the LED changed from cold blue to warm yellow light when the light emitted from the chip was transmitted through the QD film.

Chen et al. demonstrated the preparation of white-light LEDs directly using white-light-emitting ZnSe quantum dots, showing their great potential for lighting applications [23]. They synthesized direct white-light-emitting ZnSe quantum dots using ZnO and Se powders as precursors using a colloidal chemistry method. The PL of the samples exhibited strong white emission in the visible range (FWHM of about 200 nm) under the set environmental conditions. Based on this, the near-UV InGaN chip was used as the excitation source to fabricate white LEDs with CIE chromaticity coordinates of (0.38, 0.41). In addition, Bowers et al. successfully obtained a CdSe QDLED emitting cool white light [24], prepared CdSe quantum dots (~1.5 nm) with strong luminous intensity and broadband emission (420–710 nm), covering almost the entire visible region, achieving a relatively balanced white light emission with CIE chromaticity coordinates of (0.32, 0.37). Subsequently, researchers have also tried various methods to improve the brightness of direct white light-emitting CdSe quantum dots.

However, in most cases, the efficiency does not meet the demand, which to some extent limits its further application in white LEDs. In order to improve the LE, new modification methods are urgently needed. By varying the doping concentration of the prepared quantum dots, different white lights can be successfully prepared. The Hickey group developed a method to produce Mn and Cu co-doped ZnSe quantum dots (Cu-Mn-ZnSe quantum dots) by a versatile hot injection colloidal synthesis method [14], and the resulting quantum dots have high quality in both colloidal solution and solid powder of white light emission, as shown in Figure 8.10a with 17% PLQY. Different emission colors were observed from the doped quantum dots by different amounts of copper precursors, as shown in Figure 8.10b–d. The PL decay characteristics of different emission centers in Mn-ZnSe and Cu-Mn-ZnSe quantum dots are shown in Figures 8.10e,f. Compared with the Mn-ZnSe QD sample (264 ms), the lifetime of Cu-Mn-ZnSe quantum dots can reach 324 ms. This double-doped QD can be synthesized in a large-scale and environmentally friendly way and is expected to be the future direction of white light-emitting LEDs. Although the preparation process of doped white light-emitting quantum dots is relatively complicated, it offers the possibility of replacing toxic element-based QDLED.

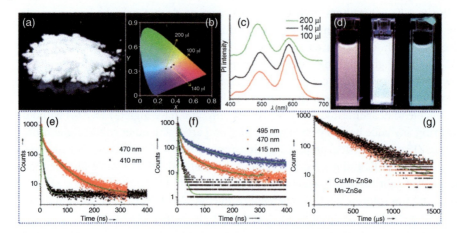

Figure 8.10 (a) Cu-Mn-ZnSe doped QD powder showing white light emission under UV lamp with 365 nm excitation, (b) CIE coordinates, and (c) emission spectra of quantum dots with different Cu precursor contents, (d) photographed under UV lamp at 365 nm, (e) Mn-ZnSe doped quantum dots, (f) Cu-Mn-ZnSe doped quantum dots at different peak positions, and (g) Cu-Mn-ZnSe and Mn-ZnSe doped quantum dots with PL decay trajectories at 585 nm. Source: Panda et al. [14]/John Wiley & Sons.

8.3 Summary Outlook

The development of white and orange–yellow QDLEDs has been an active area of research in recent years. Some of the key breakthrough technology developments in this area from 2000 to 2023 are:

1) Core–shell QDs: The use of core–shell quantum dots with a CdSe core and a ZnS shell has significantly improved the stability and efficiency of white QDLEDs. These core–shell QDs can be further engineered to emit different colors by varying the shell thickness and composition.
2) High-efficiency orange–yellow QDLEDs: In 2015, researchers demonstrated an orange–yellow QDLED with an EQE of 17.2%, which is among the highest reported for QDLEDs. This was achieved by optimizing the QD size and concentration, as well as the device architecture.
3) Hybrid QDLEDs: Hybrid QDLEDs, which combine colloidal QDs with organic materials, have shown promise for achieving high-efficiency white emission. In 2020, researchers reported a hybrid QDLED with an EQE of 22.1% and a CRI of 91, which is close to the ideal value of 100.
4) Color-tunable QDLEDs: The ability to tune the color of QDLEDs by changing the QD size and composition has been an important breakthrough in the field. In 2018, researchers reported a color-tunable QDLED that can cover the entire visible spectrum with high color purity and efficiency.
5) Solution-processed QDLEDs: Solution-processing techniques have been developed to fabricate QDLEDs with low cost and high efficiency. In 2022, researchers reported a white QDLED with an EQE of 16.7% that was fabricated using a simple solution processing technique.

Device performance data for white and orange-yellow QDLEDs in recent years are summarized as follows:

- White QDLEDs:
  - CIE value: (0.33, 0.33)
  - Emission wavelength: 450–650 nm
  - FWHM: <30 nm
  - Efficiency: up to 16.7% EQE (2022)
  - Lifetime: up to 1000 hours (2019)
- Orange-yellow QDLEDs:
  - CIE value: (0.42, 0.50)
  - Emission wavelength: 580–605 nm
  - FWHM: <35 nm
  - Efficiency: up to 17.2% EQE (2015)
  - Lifetime: up to 700 hours (2015)

In the early stage of QDLED development, the only core CdSe quantum dots as emitters result in poor optical performance and high cost, and low profit. Due to the surface defects and hanging bonds, it makes the exciton radiation-free leap loss unusually significant, which is the most central problem limiting QD light-emitting devices. To solve this problem, researchers have proposed a number of feasible strategies. First, surface passivation modification of quantum dots was first found to be effective. As the research on QD light-emitting materials continued to develop, researchers found that the problem of radiation-free jumping was significantly improved by covering CdSe cores, such as (CdS, ZnS, and ZnSe), with broad-banded shells. The core–shell structure has become the preferred strategy for modifying QD luminescent materials, resulting in a narrow luminescence range (<25 nm) and more than 90% PLQY [11]. However, the core–shell structure is more complex and costly to synthesize compared to the normal structure of quantum dots, and further improvement is needed. The second strategy is to modulate the particle size of quantum dots. The quantum dots are assembled into a fine film, the quantum dot spacing is reduced, and the exciton is more easily captured by a QD with defects [9]. This leads to a decrease in energy transfer efficiency. Therefore, increasing the particle size of the shell layer becomes an optional strategy. For example, the Lim group found that the fluorescence lifetime of CdSe/ZnS QD film was significantly lower than that of QD solution [12], and the efficiency of QD film gradually recovered with increasing the particle size of quantum dots, indicating that the reduction of energy transfer efficiency caused by inter-dot micro-motion could be suppressed by increasing the shell thickness. Therefore, the exploration of multishell quantum dots has become a new research direction. On the other hand, modification edits of ligands are gradually proposed. Replacing unstable ligands with strongly bound stable ligands (e.g. thiol, oleic acid, etc.) allows quantum dots to maintain their initial properties even after multiple purifications [25].

In addition to the low energy efficiency caused by radiation-free leaps, the other biggest problem is that high-efficiency QD light-emitting devices generally contain Cd elements, which are very toxic and must be gradually replaced, and exploring high-efficiency Cd-free QD light-emitting materials remains an irrevocable research mainstream in the future. quantum dots for direct-emission white LEDs can avoid the different emission wavelengths between multiple layers that bring different attenuation rates, leading to deterioration of device performance over time, and can emit narrow-band high-purity white light. The current research is difficult, but there is still great research potential due to where its unique advantages lie. Three base color composite white LEDs with quantum dots compared to the direct emission of white light and yellow–blue composite luminescence, the research progress is relatively slow because the nontoxic Zn class quantum dots such as ZnS (Se) emission peak position are mainly concentrated in the blue light region. It is difficult to change the size of the Cd group QDs to get three base color quantum dots, and then composite to get white light. Yellow quantum dot light-emitting devices are currently the most active and most widely used in the composite with blue devices, the preparation of high-efficiency white light devices. Therefore, the future direction of development is mainly toward white light devices.

From the material point of view, quantum dots have a large room for improvement mainly from the aspects of preparation, surface passivation, and ligand modification. How to prepare quantum dots with core–shell structure or alloy structure rapidly and at low cost is the main direction of research on synthesis methods. Since the particle size has also become one of the core elements to improve the inter-dot burst, the regulation and optimization of the quantum dot particle size are also factors to be considered for the preparation. As for ligand modification, it is crucial to find strong binding and stable ligands to ensure the purification and lifetime of quantum dots.

From the device point of view, the device needs to be further optimized for the transport layer. The selection of the appropriate hole transport layer and electron transport layer will significantly improve the luminescence performance compared to the traditional double-layer structure. In principle, the selection of the transport layer, the matching of energy levels, the suppression of exciton loss, the optical properties of the transport layer itself, and stability will be considered.

WQDLEDs have received significant attention due to their potential applications in high-quality displays and solid-state lighting. In general, white QDLEDs can be realized through two main approaches: (i) using a mixture of blue, green, and red QDs and (ii) employing a single-color QD with broad emission that covers the entire visible spectral range.

Significant progress has been made in both approaches in recent years. For the first approach, the efficiency and color purity of each individual color QD has been significantly improved, leading to high-efficiency and color-saturated white QDLEDs. In addition, the use of multilayered structures and advanced device architectures, such as tandem QDLEDs, has further improved the performance of white QDLEDs. For the second approach, researchers have made great efforts to synthesize new types of quantum dots that have broad emission spectra, such as alloyed semiconductor QDs and halide perovskite QDs. These QDs have exhibited high quantum yields, wide emission spectra, and color purity and have shown great potential for use in high-quality white QDLEDs.

Despite the progress, there are still several challenges in the development of white QDLEDs. One of the key challenges is to improve the efficiency and stability of the blue QDs, which are typically the least efficient and least stable among the three primary colors. Another challenge is to achieve high color stability over a wide range of operating conditions, such as temperature and current density. In addition, the development of high-performance charge transport materials and device architectures that can efficiently balance charge injection and transport remains an important area for future research.

In terms of future research directions, there is a growing interest in the development of sustainable and environmentally friendly white QDLEDs. This includes the use of nontoxic and low-cost quantum dots, such as silicon-based QDs, and the development of device architectures that can further improve the efficiency and stability of white QDLEDs. Furthermore, the integration of white QDLEDs with advanced technologies, such as flexible and transparent substrates, is expected to further expand the potential applications of white QDLEDs in various fields.

References

1 Shirasaki, Y., Supran, G.J., Bawendi, M.G., and Bulović, V. (2013). Emergence of colloidal quantum-dot light-emitting technologies. *Nature Photonics* 7 (1): 13–23.
2 Kastner, M.A., Klein, O., Lyszczarz, T.M. et al. (1994). *Artificial Atoms*. Research Laboratory of Electronics (RLE) at the Massachusetts Institute of Technology (MIT).
3 Tan, Z., Hedrick, B., Zhang, F. et al. (2008). Stable binary complementary white light-emitting diodes based on quantum-dot/polymer-bilayer structures. *IEEE Photonics Technology Letters* 20 (23): 1998–2000.
4 Yin, Y., Wang, R., and Zhou, L. (2014). CdTe quantum dots and YAG hybrid phosphors for white light-emitting diodes. *Luminescence* 29 (6): 626–629.
5 He, Y., Gong, J., Zhu, Y. et al. (2018). Highly pure yellow light emission of perovskite CsPb $(Br_xI_{1-x})_3$ quantum dots and their application for yellow light-emitting diodes. *Optical Materials* 80: 1–6.
6 Park, S.H., Hong, A., Kim, J.-H. et al. (2015). Highly bright yellow-green-emitting $CuInS_2$ colloidal quantum dots with core/shell/shell architecture for white light-emitting diodes. *ACS Applied Materials & Interfaces* 7 (12): 6764–6771.
7 Wang, P., Bai, X., Sun, C. et al. (2016). Multicolor fluorescent light-emitting diodes based on cesium lead halide perovskite quantum dots. *Applied Physics Letters* 109 (6): 063106.
8 Yang, D., Wang, L., Lv, W.-B. et al. (2015). Growth and characterization of phosphor-free white light-emitting diodes based on InGaN blue quantum wells and green–yellow quantum dots. *Superlattices and Microstructures* 82: 26–32.
9 Kagan, C., Murray, C., Nirmal, M., and Bawendi, M. (1996). Electronic energy transfer in CdSe quantum dot solids. *Physical Review Letters* 76 (9): 1517.

10 Zheng, K., Li, X., Chen, M. et al. (2020). Controllable synthesis highly efficient red, yellow and blue carbon nanodots for photo-luminescent light-emitting devices. *Chemical Engineering Journal* 380: 122503.

11 Dabbousi, B.O., Rodriguez-Viejo, J., Mikulec, F.V. et al. (1997). (CdSe) ZnS core-shell quantum dots: synthesis and characterization of a size series of highly luminescent nanocrystallites. *The Journal of Physical Chemistry B* 101 (46): 9463–9475.

12 Lim, J., Jeong, B.G., Park, M. et al. (2014). Influence of shell thickness on the performance of light-emitting devices based on CdSe/$Zn_{1-x}Cd_xS$ core/shell heterostructured quantum dots. *Advanced Materials* 26 (47): 8034–8040.

13 Jang, E.-P. and Yang, H. (2013). Utilization of solvothermally grown InP/ZnS quantum dots as wavelength converters for fabrication of white light-emitting diodes. *Journal of Nanoscience and Nanotechnology* 13 (9): 6011–6015.

14 Panda, S.K., Hickey, S.G., Demir, H.V., and Eychmuller, A. (2011). Bright white-light emitting manganese and copper co-doped ZnSe quantum dots. *Angewandte Chemie International Edition* 50 (19): 4432–4436.

15 Li, Y., Rizzo, A., Cingolani, R., and Gigli, G. (2006). Bright white-light-emitting device from ternary nanocrystal composites. *Advanced Materials* 18 (19): 2545–2548.

16 Anikeeva, P.O., Halpert, J.E., Bawendi, M.G., and Bulović, V. (2007). Electroluminescence from a mixed red-green-blue colloidal quantum dot monolayer. *Nano Letters* 7 (8): 2196–2200.

17 Chen, B., Zhong, H., Wang, M. et al. (2013). Integration of $CuInS_2$-based nanocrystals for high efficiency and high colour rendering white light-emitting diodes. *Nanoscale* 5 (8): 3514–3519.

18 Li, F., You, L., Li, H. et al. (2017). Emission tunable CdZnS/ZnSe core/shell quantum dots for white light emitting diodes. *Journal of Luminescence* 192: 867–874.

19 Zhang, H., Su, Q., Sun, Y., and Chen, S. (2018). Efficient and color stable white quantum-dot light-emitting diodes with external quantum efficiency over 23%. *Advanced Optical Materials* 6 (16): 1800354.

20 Jang, E.-P., Jo, J.-H., Lim, S.-W. et al. (2018). Unconventional formation of dual-colored InP quantum dot-embedded silica composites for an operation-stable white light-emitting diode. *Journal of Materials Chemistry C* 6 (43): 11749–11756.

21 Hu, G., Sun, Y., Zhuang, J. et al. (2020). Enhancement of fluorescence emission for tricolor quantum dots assembled in polysiloxane toward solar spectrum-simulated white light-emitting devices. *Small* 16 (1): 1905266.

22 Zhang, J., Xie, R., and Yang, W. (2011). A simple route for highly luminescent quaternary Cu-Zn-In-S nanocrystal emitters. *Chemistry of Materials* 23 (14): 3357–3361.

23 Chen, H.S., Wang, S.J.J., Lo, C.J., and Chi, J.Y. (2005). White-light emission from organics-capped ZnSe quantum dots and application in white-light-emitting diodes. *Applied Physics Letters* 86 (13): 131905.

24 Bowers, M.J., McBride, J.R., and Rosenthal, S.J. (2005). White-light emission from magic-sized cadmium selenide nanocrystals. *Journal of the American Chemical Society* 127 (44): 15378–15379.

25 Shen, H., Cao, W., Shewmon, N.T. et al. (2015). High-efficiency, low turn-on voltage blue-violet quantum-dot-based light-emitting diodes. *Nano Letters* 15 (2): 1211–1216.

9

Non-Cadmium Quantum Dot Light-Emitting Materials and Devices

9.1 Introduction

Quantum dots have attracted a lot of attention from researchers because of their excellent intrinsic properties. Their high quantum efficiency, high color purity, low-cost solution handling capability, and easily adjustable emission wavelength contribute to the excellent performance of quantum dot light emitting diodes (QDLEDs). Many attempts have been made by scientists to enhance the electroluminescence performance of red, green, and blue-related QDLEDs. As of 2020, the external quantum efficiency (EQE) of QDLEDs in all three primary colors has been recorded to exceed 20% [1, 2]. With better color purity and simpler material synthesis than commercial OLED technology, QDLEDs are expected to be the next generation of display and lighting technology.

However, the toxicity of cadmium elements in typical cadmium-containing quantum dots such as CdS remains a major challenge for the large-scale application of QDLED technology. The presence of cadmium in quantum dot materials can cause serious damage not only to the environment but also to human health. The European Union's Restriction of Hazardous Substances (RoHS) rules already prohibit the presence of these materials in any consumer electronics product that contains more than trace amounts of the corresponding heavy metal elements. Therefore, scientists have put a lot of effort to achieve high-performance cadmium-free quantum dots.

Table 9.1 lists some of the performance parameters of cadmium-free QDLEDs. Indium phosphide (InP) quantum dots were first synthesized by A.J. Nozik's group back in 1995. Subsequently, a series of Class I-III-VI quantum dots were synthesized sequentially, among which, the most studied were mainly CuInS class quantum dots. However, it was not until the second half of 2011 that scientists produced the first QDLED with InP quantum dots as the light-emitting layer, and although its EQE was only 0.008%, it kicked off the development of QDLED devices. Other cadmium-free QDLEDs have been introduced since then. Currently, the highest EQEs of red, green, and blue QDLEDs are recorded at 21.4%, 6.3%, and 7.83%, respectively. In this chapter, we focus on the development history of some traditional green QDLED materials. The cadmium-free quantum dots mentioned in this paper not only do not contain cadmium but also do not contain other toxic heavy metals such as Pb and As. In the later part, we mainly introduce the development

Colloidal Quantum Dot Light Emitting Diodes: Materials and Devices, First Edition. Hong Meng.
© 2024 WILEY-VCH GmbH. Published 2024 by WILEY-VCH GmbH.

history of QDLED technology for cadmium-free quantum dots InP, Class I-III-VI, and ZnSe, discuss some methods to improve the performance of QDLED, and make an outlook.

Here are some key breakthroughs in non-cadmium colloidal quantum dot (CQD) QDLED technology development from 2000 to 2023, along with some device performance data:

- 2005: Development of lead sulfide (PbS) CQDLEDs with EQE of 0.5%, peak emission at 1000 nm, and full width at half maximum (FWHM) of 200 nm.
- 2009: Introduction of cadmium-free CQDs made of InP with EQE of 3.3% and peak emission at 630 nm.
- 2012: Demonstration of all-inorganic $CsPbBr_3$ perovskite CQDLEDs with EQE of 0.1% and peak emission at 530 nm.
- 2015: Development of $CsCuInS_2$/ZnS CQDs with EQE of 6.2% and peak emission at 545 nm.
- 2017: Improvement of green-emitting CQDLEDs made of InP/ZnS with EQE of 19.3% and peak emission at 550 nm.
- 2019: Introduction of quasi-2D all-inorganic $CsPbI_2Br$ CQDs with EQE of 14.3% and peak emission at 525 nm.
- 2021: Demonstration of high-efficiency blue-emitting CQDLEDs made of $CsCuInS_2$/ZnS with EQE of 16.1%, peak emission at 455 nm, and FWHM of 27 nm.
- 2023: Development of nontoxic and environmentally friendly CQDs made of copper iodide (CuI) with EQE of 5.8%, peak emission at 625 nm, and FWHM of 40 nm.

It is worth noting that the performance of non-cadmium CQDLEDs has been steadily improving over the years, with EQE reaching double digits and emission wavelengths covering the entire visible spectrum. The use of perovskite and all-inorganic CQDs has shown great promise for achieving high efficiency and stable devices.

9.2 Quantum Dots and QDLED

9.2.1 InP

Indium phosphide quantum dots are one of the star materials in cadmium-free quantum dots. With its low toxicity and wide and tunable emission wavelength, it has been recognized as an ideal material to replace heavy metal-free in commercial quantum dot displays and solid-state lighting. Currently, InP holds the record for cadmium-free QDLED performance with its 21.4% EQE, which is close to reaching the theoretical limit [19].

As early as 1995, A. J. Nozik assembled InP quantum dots and optimized the synthesis conditions. They varied the concentration of precursors in the solution and the reaction temperature to obtain InP quantum dots with different particle sizes, and the synthesized quantum dots emitted red/green light [25]. Over the next

Table 9.1 Structure and performance of some cadmium-free QDLED devices.

Quantum dots	Emission peak (nm)	EQE (%)	Device structure	References
ZnSe/ZnS	~470	0.65	PEDOT:PSS/PVK/QDs/ZnO/Al	[3]
ZnSe/ZnS	~441	—	Al/MoO3/CBP or TCTA/QDs/ZnO	[4]
ZnSe/ZnS	400–455	7.83	PVK/QDs/ZnO	[5]
ZnSeTe/ZnSe/ZnS	441	4.2	PEDOT:PSS/PVK/QDs/ZnMgO/Al	[6]
$CuInS_2$/ZnS	580	1.1	ITO/PEDOT:PSS/PVK/QDs/ZnO/Al	[7]
Cu:Zn-In-S	580	—	ITO/PEDOT:PSS/poly-TPD/QDs/TPBI/LiF/Al	[8]
$CuInS_2$/ZnS	579	2.19	ITO/PEDOT:PSS/PVK/QDs/$Zn_{0.9}Mg_{0.1}$O/Al	[9]
$CuInS_2$/ZnS	~580	3.22	ITO/ZnO/QDs/CBP/TCTA/MoO3	[10]
$CuInS_2$/ZnS	573	0.63	ITO/PEDOT:PSS/PVK/QDs/ZnO/Al	[11]
Zn-Cu-Ga-S/Cu-In-S	475	7.1	ITO/PEDOT:PSS HIL/PVKHTL/QDs/ZnMgO/Al	[12]
Ag-In-Zn-S/ZnS	550	0.39	ITO/PEDOT:PSS/PVK/QDs/ZnO/Al	[13]
Cu-In-Zn-Se-S	~600	—	ITO/PEDOT:PSS/TFB/QDs/ZnO/Al	[14]
$CuInS_2$/ZnS	645	~3	ITO/ZnO/QDs/CBP/CBP/MoO_3/Al	[15]
InP/ZnSeS	532	0.008	PEDOT:PSS/poly TPD/InP/$TPBi_3$	[16]
InP/ZnSeS	~520	3.46	ITO/ZnO/PFN/QDs/TCTA/MoO_3/Al	[17]
InP/GaP/ZnS//ZnS	530	6.3	ITO/PEDOT:PSS/TFB/QDs/ZnO/Al	[18]
InP/ZnSe/ZnS	~630	21.4	ITO/PEDOT:PSS/TFB/QDs/ZnMgO/Al	[19]
InP/GaP/ZnS//ZnS	488	1.01	ITO/PEDOT:PSS/TFB/QD/ZnO/Al	[20]
InP/ZnSe/ZnS	538	15.0	ITO/PEDOT:PSS/PVK/QDs/PO-T2T/Al	[21]
InP/$ZnSe_{0.7}S_{0.3}$/ZnS	532	15.2	ITO/PEDOT:PSS/TFB/QDs/ZnO/Al	[22]
InP/ZnSe	621	22.52	ITO/PEDOT:PSS/TFB/QDs/ZnMgO/Al	[23]
In(Zn)As/In(Zn)P/GaP/ZnS	1006	13.3	ITO/ZnO/PVK/PEIE/QDs/TPD/MoO3/Ag	[24]

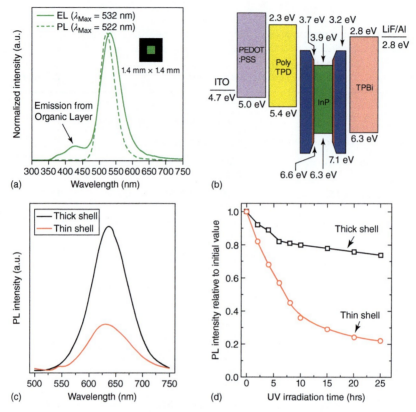

Figure 9.1 (a) Luminescence peaks of InP quantum dots and device luminescence peaks; (b) energy level diagram of each layer of the device; (c–d) effect of different shell layer thicknesses of quantum dots on fluorescence intensity and stability.

two decades, scientists improved the synthesis method to achieve control over the structure of InP quantum dots, including size, shell–core structure, and alloying [26–32]. In the next two decades, scientists improved their synthesis methods to control the structure of InP quantum dots, including size, shell–core structure, and alloying. Until 2011, Seonghoon Lee's group synthesized InP/ZnSeS-like quantum dots and fabricated the first QDLED device with InP quantum dots as the light-emitting layer in the structure of PEDOT: PSS/poly TPD/InP/TPBi3 [16] as shown in Figure 9.1a,b. Their synthesized materials have a core–shell structure, and the shell layer components exhibit gradient uniformity along the radius direction, which helps to enhance the stability of the quantum dots, effectively limits the excitonic wave function, and reduces the oxide and non-radiative binding generated by photo-oxidation or ligand exchange at the material surface traps. Although the device had a very low EQE of 0.008% and the presence of spurious peaks within the device's luminescence spectrum, the study opened the way for research on InP-like QDLEDs. In 2012, Sun's group used a thick ZnS shell to effectively passivate the surface defects of InP nanocrystals, resulting in improved optical properties including high quantum yields, as shown in Figure 9.1c,d, and reducing the half-peak width

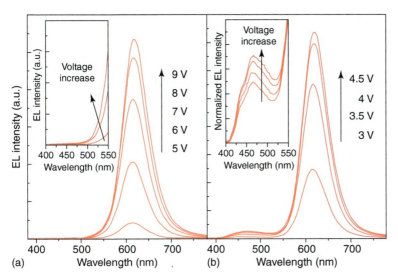

Figure 9.2 Effect of (a) PVK and (b) TFB as hole-transporting materials on the luminescence spectra of the devices.

of the emission spectrum [33]. However, due to the imperfect energy level matching of the device, the emission spectrum of the device contains parasitic emission peaks of poly-TPD in the short wavelength band in addition to the luminescence peaks of quantum dots, which is harmful to the color purity of the display device.

In 2016, in order to remove the luminescence peak of the organic layer in the InP-like QDLED luminescence spectrum and improve the color purity of the device, Kookheon Char's group tried two different hole transport materials (HTL): TFB and PVK [34]. The experimental results in Figure 9.2 show that, unlike TFB, the devices using PVK as HTL have no parasitic luminescence peaks. They propose that this is due to the fact that the lowest unoccupied molecular orbital (LUMO) of PVK is higher than that of TFB, which inhibits electrons from continuing to cross the luminescent layer to bind to holes in the HTL layer, thus effectively removing the emission peak from this organic layer. However, the experiment also showed that the TFB device (EQE 2.5%) outperformed the PVK device (EQE 1.4%) in terms of device efficiency because the TFB had a higher mobility, which resulted in a higher current efficiency for the QDLED.

As another important factor is selection of the electron transport materials (ETM), Liu's group investigated their InP quantum dot material and compared the effect of two different ETL materials, ZnMgO and ZnO, on the material properties [35]. The experimental device structures and results are shown in Figure 9.3. They pointed out that the former can balance the charge binding of the luminescent layer using the higher highest occupied molecular orbital (HOMO), which enhances the device's performance.

There are many studies on the structure of InP quantum dots, which show that the core–shell structure has a better performance. While the core–shell structure has the problem of lattice matching between layers, which will produce more defects and

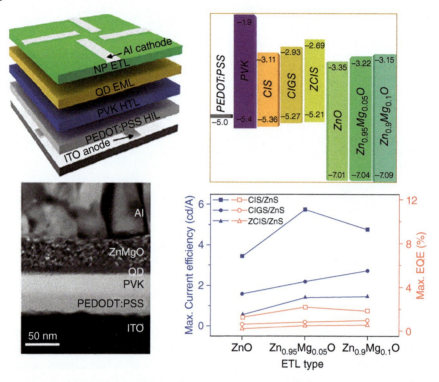

Figure 9.3 Energy level matching diagram and device structure schematic for different electron transport layers ZnMgO and ZnO.

reduce the composite efficiency of excitons, the researchers continue to optimize the shell layer, such as by introducing an intermediate layer GaP, whose presence minimizes the lattice mismatch and successfully reduces the interface defects. Quantum dot structure, device structure, and experimental results are shown in Figure 9.4. QDLEDs based on InP quantum dots using this type of multilayer structure can achieve current efficiencies of up to 13.7 cd/A and EQE of 6.3% [18].

In 2019, Eunjoo Jang's group has specifically optimized QDLEDs with InP/ZnSe/ZnS with a maximum EQE of nearly 21.4% [19]. They synthesized and optimized this class of quantum dots by more complex experimental operations. After preparing the InP core, they added HF solution at the early stage of ZnSe shell growth to prevent defects due to InP reoxidation, such as the survival of $InPO_x$ or In_2O_3. Then they grew another layer of ZnS shell on InP/ZnSe quantum dots by sequentially adding relevant precursors to reduce the material after preparation into devices due to the Osher effect and Förster resonance energy transfer (FRET) efficiency loss after device preparation. The obtained InP (3.3 nm)/ZnSe (1.9 nm)/ZnS (0.3 nm) QDs produced red luminescence with a quantum yield of 98%. However, the shape of the material is triangular rather than the typical round shape of quantum dots, and the researchers increased the growth temperature of the shell to 340 °C to optimize the uniform shape of the QD and successfully increased the PLQY to 100%. Also, charge transport in QDLEDs is influenced by

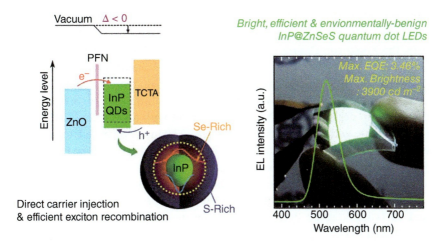

Figure 9.4 Schematic diagram of the device energy level and quantum dot structure after PFN insertion, the electroluminescence spectrum of the device, and the corresponding performance values.

surface ligands. In order to eliminate the blocking of electron or hole carriers caused by the long alkyl chain of the ligand oleic acid (OA) on the surface, the ligand exchange method was used to reduce the alkyl chain length by changing the surface ligand to hexanoic acid (HA). Finally, they designed and prepared the corresponding device with the structure shown in Figure 9.5 as ITO/PEDOT:PSS (35 nm)/TFB (25 nm)/QD (20 nm)/ZnMgO (40 nm)/Al (100 nm) with an EQE of 21.4% and a lifetime of 4 300 h at 1 000 cd m^{-2} brightness (T_{75}). This red light device is comparable to the best performance achieved by Cd-containing QDLEDs and represents an important milestone in the history of InP-based QDLEDs.

Theoretically, the color of quantum dots can be changed by adjusting their size to change the bandgap width to achieve color change [36]. Usually, the smaller the diameter of quantum dots, the more blueshift in emission and absorption spectra will occur. However, blue and green InP quantum dots are generally smaller in size than red InP quantum dots, thus increasing the difficulty of synthesis, and their

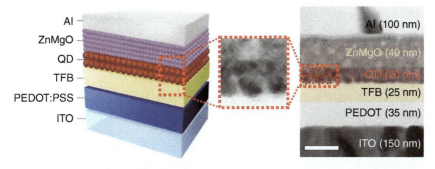

Figure 9.5 Device structure of the highest-efficiency cadmium-free QDLED available.

wider bandgap also imposes higher requirements on the charge transport layer, which severely limits the performance of green and blue InP-based QDLEDs.

In 2013, the Kookheon Char group used PFN layers to improve charge balance in green InP-class QDLED devices [17]. The thick ZnSeS heterostructure shell in their quantum dots also plays an important role in improving the EQE of QDLEDs. Their device achieved an EQE of 3.46% and a maximum luminance of 3900 cd m^{-2}. Feng Teng's group designed an InP-like quantum dot with an InP/GaP/ZnS//ZnS core–shell structure with a PLQY of about 70% [18]. The insertion of GaP interfacial layer effectively reduces the lattice mismatch and interfacial defects. The thick ZnS shell is designed to suppress the performance loss due to FRET between closely packed quantum dots after spin coating. The EQE and current efficiency of the green electroluminescent devices fabricated on InP core/shell quantum dots are 6.3% and 13.7 cd A^{-1}, respectively, which is the most efficient InP-based green QDLED achieved so far.

As the most difficult of the three primary colors, blue InP-like quantum dots have both small particle size scale and poor homogeneity, leading to energy loss due to FRET generation in the quantum dot film. The blue color requires having quantum dots with large bandgaps, which also leads to low carrier injection efficiency and transport processes, limiting the efficiency of blue InP quantum dots. In 2017, the Deng group first fabricated blue QDLEDs using InP/ZnS quantum dots as the active layer. They used (DMA)$_3$ P instead of P(SiMe$_3$)$_3$ as the phosphorus for InP quantum dot synthesis precursors [37]. The former has a moderate reaction rate, which facilitates the control of the size and homogeneity of InP nuclei in the experiment. In addition, the formation of halide-amine passivation layers by a colloidal method combined with ZnS halide can better eliminate surface defects. In addition, they further improved the PLQY by growing ZnS shells and successfully synthesized highly fluorescent blue InP/ZnS small nuclei/thick shell quantum dots with PLQY up to 76.1%. The corresponding device can obtain a maximum brightness of 90 cd m^{-1} at a bias voltage of 10 V, although they did not mention EQE in the literature. In 2020, the Du group prepared blue InP/GaP/ZnS//ZnS core/shell/shell quantum dots with high PLQY (∼81%) and high color purity (FWHM of 45 nm) [20]. The device energy level structure and test results are shown in Figure 9.6. In the experiment, the introduction of GaP intermediate shell layer successfully reduces the defects caused by the low crystal matching. The extended shell growth time and the timely replenishment of the body method before exfoliation both successfully increased the shell thickness, thus mitigating the efficiency loss of FRET for quantum dot films. They prepared devices with ITO/PEDOT:PSS/TFB/QD/ZnO/Al structure and were able to achieve an EQE of 1.01%, which is the only reported EQE value for InP-based blue devices.

In white QDLEDs, InP is often used as a component in a hybrid light source. It is usually used as a red or green-emitting light source, where InP and yellow phosphor powder or other layers produce emitted light under the blue excitation of GaN chip emission at the bottom of the device, and the final mixed color is white. As shown in Figure 9.7a–d, the Sang-Wook Kim group mixed YAG:Ce phosphor with red-emitting InP/GaP/ZnS quantum dots and applied it to WLED [38]. They mixed

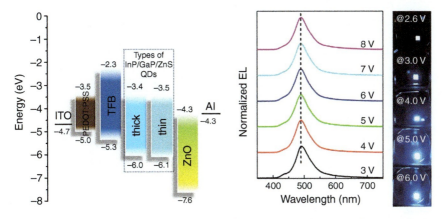

Figure 9.6 Blu-ray InP Class QDLED Device Energy Level Diagram and Performance Test.

the red quantum dots and phosphor with silicone resin in toluene solvent, removed the solvent, and then dropped the phosphor quantum dot silicone mixture onto the InGaN blue LED chip and cured it at 150 °C for 4 hours. The luminous efficiency of the white QDLED was 54.71 lm W^{-1}, Ra was 80.56, and the color temperature was 7864 K. The group immediately developed another InP-based WQDLED using a similar device structure, as shown in Figure 9.7e,f, with a CRI of 95 [33]. They improved the quality of white light by introducing more colors into the spectrum. In the new device, the InP-QD layer emits green light, the poly-TPD emits blue light, and the radical complex formed at the interface between the poly-TPD and TPBi emits red light.

9.2.2 ZnSe

As mentioned earlier, although the current InP material has achieved more than 21% EQE in red devices, it is difficult to make QDLEDs with high-efficiency blue InP due to its native material with small bandgap, while ZnSe semiconductor has been developed as an effective blue/violet emitting nanocrystal with a native bandgap energy of 2.8 eV and a quantum dot emission spectrum range between UV and blue [39].

In 2012, The So group prepared ZnSe/ZnS-based blue QDLEDs [3]. The structure of the device is as follows: PEDOT: PSS/poly-TPD or PVK/QD/ZnO/Al. In this device, PVK is used as a hole transport layer with its lower HOMO energy level to enhance the hole injection efficiency. They also optimized the thickness of the HTL and EL layers to further improve the device performance. The obtained devices exhibited an EQE of 0.65% [4]. They prepared inverted devices with an EL peak at 441 nm, a narrow FWHM of 15.2 nm, and a maximum luminance of 1170 cd m^{-2}, but again, no EQE was mentioned in the paper. They attributed the high performance of the devices to the thick shell of quantum dots (~8 nm) and the optimal charge balance between CBP and ZnO in the charge transport layer. In 2015, the Chen group adopted a novel integrated method of "low-temperature injection and high-temperature growth" [5]. The group used low-temperature

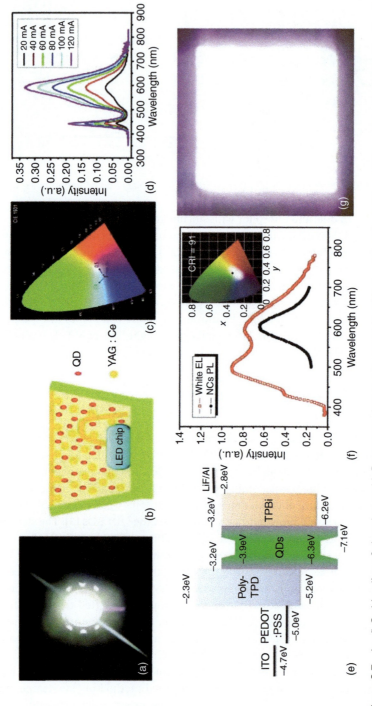

Figure 9.7 (a–d) Co-blending of phosphor and InP quantum dots to prepare white LED; (e–g) Quantum dots emitting composite light with other layers.

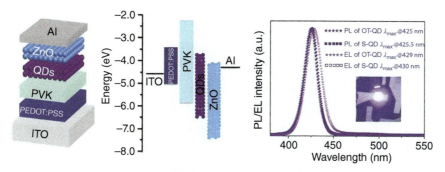

Figure 9.8 Schematic diagram of the structure, energy level diagram, and luminescence spectrum of the blue (violet) color cadmium-free QDLED device with the highest device efficiency at present.

injection and high-temperature growth in the synthesis of core/shell quantum dots such as ZnSe/ZnS. Unlike most conventional nucleation methods, which are grown at high temperatures with shell layers, the synthesized quantum dots exhibit monodispersity and high PLQY of 80%, with a FWHM of about 12–20 nm and a spectral range from 400 to 455 nm, with good color tunability, and the fabricated QDLEDs, as shown in Figure 9.8, can achieve a maximum brightness of 2632 cd m^{-2}, with a peak EQE of 7.83%, it achieves the highest EQE in the ZnSe blue QDLED series, but its EL wavelength (429 nm) is slightly beyond the desired blue range and glows purple.

To obtain pure blue QDLEDs, the Yang group recently reduced the bandgap width of ZnSe quantum dots by alloying ZnTe, which has a lower bandgap (2.25 eV), with ZnSe quantum dots, which presents a more feasible method for achieving blue light emission [6]. The wavelength can be tuned within 422–500 nm by adjusting the ratio of Te/Se. The final experiment of the double-shell ZnSeTe/ZnSe/ZnS quantum dots with optimal ZnSe inner shell thickness produced a suitable blue PL peak at 441 nm with 70% high QY and a half-peak width of 32 nm. As shown in Figure 9.9, the obtained QDs were used to fabricate the first ZnSeTe quantum dot-based devices prepared for blue QDLEDs, producing a peak brightness of 1195 cd m^{-2}, a current

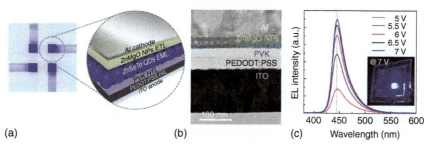

Figure 9.9 (a–c) Structure and luminescence spectra of blue ZnSeTe ternary quantum dot devices.

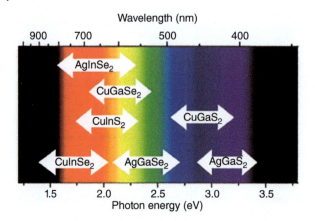

Figure 9.10 The luminescence spectral range of some I-III-VI quantum dots.

efficiency of 2.4 cd A^{-1}, and an EQE of 4.2%, corresponding to the recorded values reported for non-Cd blue QDLED.

9.2.3 I-III-VI

As another class of cadmium-free quantum dots, I-III-VI quantum dots have been studied by many researchers because of their low toxicity, high extinction coefficient, high defect tolerance, and attractive potential in optoelectronics and biomedical fields. Researchers have synthesized a series of such quantum dots, including CuGaSe$_2$, CuInS$_2$, AgInSe$_2$, CuInSe$_2$, CuGaS$_2$, and their derivatives [13, 40–45], the luminescence range is shown in Figure 9.10. From the beginning of the twenty-first century to the present, chemists have synthesized I-III-VI quantum dots by low-temperature pyrolysis of molecular single-source precursors, solvothermal methods, partial cation exchange, and water-based synthesis [9, 46–48].

More specifically, in the CuInS(CIS) class of I-III-VI quantum dots, the main contribution to luminescence comes not from its bandgap but from its intrinsic defects and the donor–acceptor pair (DAP) produced by the [Cu]/[In] composition ratio [49]. It is generally believed that the DAP major state is formed by the 2V$_{Cu^-}$ + In$_{Cu^{2+}}$ pair, where In$_{Cu^{2+}}$ is the donor produced by the substituted In ion at the Cu site and V$_{Cu^-}$ + is the acceptor produced by the Cu vacancy. The other donor (V$_{S^{2+}}$) produced by the sulfur vacancy can be systematically eliminated using an excess of alkanethiol. In addition, the defect state can be experimentally adjusted by post-synthetic heat treatments, such as annealing.

And although different types of I-III-VI quantum dots have been synthesized by researchers at the beginning of the twenty-first century, the application of these types of quantum dots in LEDs has only been developed in recent years. In 2011, the Xu group reported for the first time that quantum dot devices with CuInS$_2$-ZnS alloy (ZCIS) as the core and ZnSe/ZnS as the bilayer shell, synthesized by high-temperature organic solvent method, produced emission spectra at the near-band edge [50]. They can fine-tune the device's luminescence wavelength by adjusting the diameter of quantum dots. As shown in Figure 9.11, the maximum QDLED luminance reaches 1200, 1160, and 1600 cd m^{-2} in their prepared

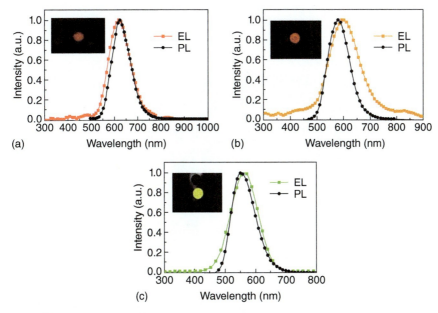

Figure 9.11 Fluorescence emission spectra and device electroluminescence spectra of red, yellow, and green colors achieved by $CuInS_2$-ZnS/ZnSe/ZnS.

red, yellow, and green devices, respectively, while the current efficiency of the corresponding devices is 0.58, 0.49 and 0.62 cd A^{-1} at an injection current density of 0.82 mA cm^{-2} and a luminance of 10 cd m^{-2}.

In 2012, the Zhong group discovered the effect of stoichiometry on the crystal structure and optical properties of $CuInS_2$, which explains the evolution of its PL emission in relation to the [Cu]/[In] molar ratio [51]. The synthesis method of quantum dot powder they developed can expand the amount of reaction by choosing a suitable [Cu]/[In] molar ratio and adjusting the alloy and core–shell structure, and more than ten grams can be prepared in one experiment with little effect on the quality of quantum dots after expanding the synthesis amount. The resulting quantum dots emitted light in the wavelength range from 500 to 800 nm, and the corresponding red (606 nm) and yellow (577 nm) QDLEDs had a maximum luminance of about 1700 and 2100 cd m^{-2}, respectively. In 2014, the Yang group fabricated the corresponding $CuInS_2$/ZnS quantum dots using highly luminescent colloidal QDLEDs [7]. They varied the thickness of the QD EML and optimized the device with a peak brightness of 1564 cd m^{-2} and a current efficiency of 2.52 cd A^{-1}. The EQE at 580 nm was 1.1%, the first reported EQE value for this class of quantum dot devices. In the same year, the group also designed and synthesized $CI_{1-x}G_xS$/ZnS quantum dots composed of $CI_xG_{1-x}S$/ZnS cores with different Ga elemental contents [52], with their PL peak wavelengths ranging from 479 to 578 nm and PLQY of 20%–85%. The QDLEDs reported in this article are in the stacking order of ITO/PEDOT: PSS/PVK/QDs/ZnO/Al. The luminance maximum of CGIS QDLED is 1673 cd m^{-2}, the CE value is 4.15 cd A^{-1} and the EQE decreases

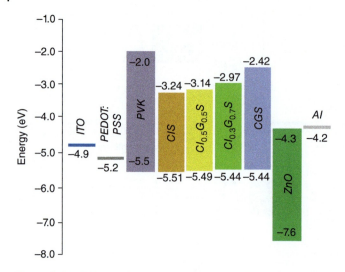

Figure 9.12 Effect of Ga doping on the energy level of quantum dots.

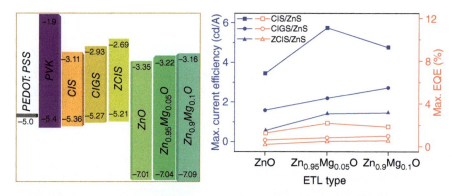

Figure 9.13 Energy level diagrams of different electron transport layers and the performance of the devices made.

from 1.54% to 0.007% as the Ga element content of quantum dots increases. They propose that the increase in Ga content makes the LUMO of the EML layer elevated, which makes electron injection more difficult and leads to a poor current and charge balance, resulting in a decrease in brightness (Figure 9.12).

The study of the ETL layer has also attracted the attention of scientists. In 2015, the group again studied the effect of the composition of the ETL on the performance of CIS QDLEDs [9]. They applied three $Zn_{1-x}Mg_x O$ ($x = 0, 0.05, 0.1$) NPs with different electron energy levels as ETLs in device fabrication and found that these ETL layers have a large degree of influence on the brightness and efficiency of QDLEDs, and compared to pure ZnO, as shown in Figure 9.13, by alloying ZnMgO can increase its HOMO energy level and reduce the electron injection barrier, which can significantly improve the devices performance.

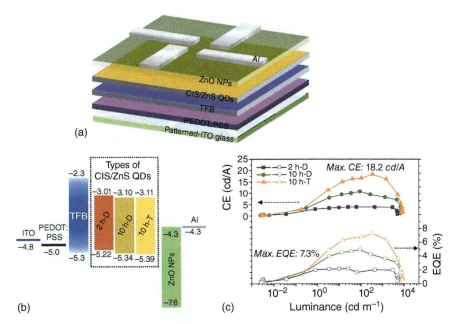

Figure 9.14 Energy level diagrams and fabricated device performance of quantum dots with different shell layer thicknesses.

In 2016, the Yang group investigated the effect of the shell thickness of quantum dots on their PLQY [53], the results of the study are shown in Figure 9.14, increasing the shell layer thickness can increase the PLQY from 79 to 89%. The corresponding devices of quantum dots with the thickest shell layer reached a CE of up to 18.2 cd A^{-1} and an EQE of 7.3%.

Due to the wide range of luminescence spectrum of CIS quantum dots, they are often used in making white QDLEDs for illumination [54]. In 2013, by controlling the synthesis steps, Zhong et al. assembled high-luminescence nanocrystals based on $CuInS_2$ [55] with a green quantum dot PLQY of 60% and a red quantum dot PLQY of 75%. By integrating selected $CuInS_2$-based quantum dots with blue emitting chips, WQDLEDs with tunable emission colors including white light can be realized. They also explored the application of $CuInS_2$-based NCs in high-power nc-WLEDs. The CRI of high-power devices in nc-WLEDs based on dual phosphors can reach 90 (Figure 9.15).

In 2014, the Liu group synthesized chalcopyrite CIS quantum dots with various [Cu]/[In] ratios by a simple solvothermal method [56–59]. The color-tunable CIS quantum dots were successfully synthesized by adjusting the [Cu]/[In] molar ratios. The radiation pathway was formed by DAP recombination, resulting in a broad PL band. WLEDs were fabricated using green phosphor Ba_2SiO_4:Eu^{2+} and CIS/ZnS quantum dots emitting orange and red colors. Their QD-WLEDs showed a luminous efficiency of 36.7 lm W^{-1} at a forward current of 20 mA and a high CRI value of about 90.

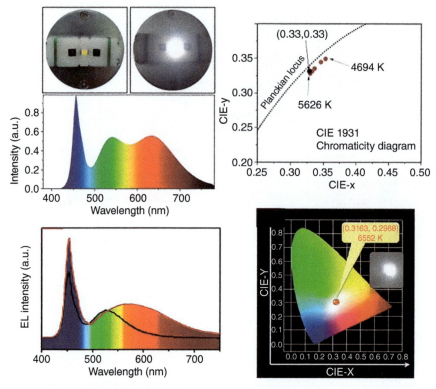

Figure 9.15 White QDLED based on CIS class.

There are few blue literatures reported on class I-III-VI quantum dots. In 2017, the Yang group investigated ternary Cu-Ga-S (CGS) quantum dots as well as quaternary Zn-Cu-Ga-S (ZCGS) obtained by alloying Zn after adding ZCGS. The prepared ZCGS/ZnS core–shell quantum dots can fluoresce blue and have a quantum yield of 78–83% [60]. In 2019, the group was able to achieve an EQE value of 7.1% for the blue QDLED devices prepared with this material [12].

CIS quantum dots generally have a wide luminescence peak, which is a shortcoming as a display material. Young-Shin Park's group found that thick-shell CIS/ZnS quantum dots showed higher photostability at the single-quantum dot level and greatly reduced PL half-peak width compared to conventional pure core or thin-shell samples, as shown in Figure 9.16 [61]. They point out that the large extension of the luminescence spectrum of quantum dots is not an intrinsic property but a result of the point-to-point variation of the emitted energy. This is related to the specificity of the emission mechanism DAP in CIS quantum dots.

9.3 Methods for Optimizing QDLED Performance

Methods to optimize QDLED performance can usually be classified into ligand engineering, core–shell structure, and QDLED device structure depending on the imposed object. For the device efficiency, EQE = $\eta_{out} \times \eta_{in}$, η_{out} is related to the

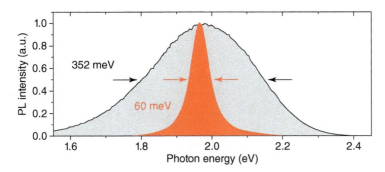

Figure 9.16 Study of narrow luminescence peaks of CIS class.

device optical coupling coefficient, while η_{in} is more correlated with the material properties such as PLQY. As for the display material, the color control is also an important influencing factor. Therefore, the devices are optimized by various methods to achieve the desired color and efficiency values.

9.3.1 Ligand Engineering

When synthesizing quantum dots, ligands are an integral part of the synthesis step. Because the surface defects of QDs can capture charges and inhibit their binding, which in turn reduces the performance of QDLEDs, ligands can reduce the surface defects of quantum dots and enhance device performance. However, the long alkyl chains of the ligands also inhibit charge injection into the quantum dots and reduce the exciton complexation. Therefore, the ideal solution is to synthesize a suitable ligand that facilitates the formation of crystals without severely hindering the charge injection. However, long alkyl lengths favor crystallization, and excessive reduction of alkyl chains is also detrimental to the surface quality of quantum dots. Therefore, researchers have developed an alternative method of ligand exchange. For example, in 2016, Huaibin Shen's group synthesized hydroxyl-capped CIS quantum dots using 6-mercaptohexanol (MCH) [10]. As shown in Figure 9.17, they first synthesized CIS quantum dots with OLA ligands, which have long alkyl chains that facilitate the growth of quantum dots with higher surface quality. They then performed ligand exchange and covered the quantum dots with MCH, and the finally obtained materials showed excellent performance in EL devices. The MCH ligand could regulate the barrier height between the electron transport layer ZnO and CIS quantum dots, thus improving the electron injection efficiency from the ZnO layer to the QD. After optimization, the maximum brightness of the device is 8 735 cd m^{-2}, and the EQE is 3.22%.

In 2019, Eunjoo Jang's group achieved the EQE record of cadmium-free QDLED [19]. In their experiments, reducing the harmful hindrance of ligands to electron or hole carriers and using ligand exchange by exchanging DDT with HA reduces the alkyl length and lowers the potential barrier for charge injection. The Xu group also used a novel exchange ligand 2-ethylhexanethiol (EHT) to improve the charge balance in QDLEDs [62]. As shown in Figure 9.18, using quantum dots exchanged with this ligand as the emitting layer, the optimized diode has a maximum luminance of 2354 cd m^{-2} and an EQE of 0.63%, as well as a lower turn-on voltage (2.7 V).

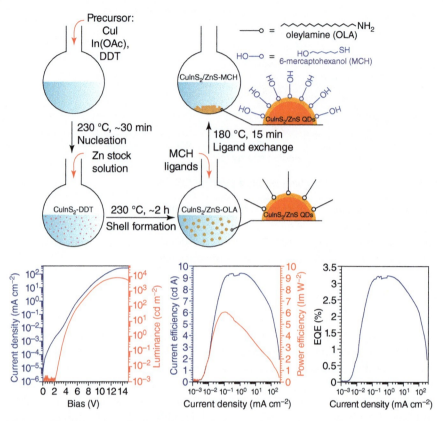

Figure 9.17 Schematic diagram of the experimental flow of ligand exchange and the corresponding device performance comparison.

9.3.2 Shell Engineering

The Osher composite process and FRET in quantum dot films largely affect the performance of QDLED devices. Therefore, the shell thickness of core/shell QDs is important to affect the performance of QDLEDs. For example, the Sun group used a thicker stress relieved ZnS shell to effectively passivate the surface defects of InP quantum dots, resulting in improved optical properties including quantum yield and reduced emission linewidth [33]. The PLQY of InP quantum dots rises from 12.3% to 38.1% after growing additional ZnS shells. In terms of the stability of the quantum dots, for the InP nanocrystals without additional ZnS shells, the PL intensity decreased rapidly (down to 70% of its initial value) within 10 hours of UV irradiation (365 nm, 3 mW cm^{-2}). For nanocrystals with additional ZnS shells, photodegradation was barely visible even after 25 hours of continuous irradiation.

In addition to the PLQY value, the shell thickness of quantum dots may affect the emission wavelength, and the color can be fine-tuned by adjusting the shell thickness. ZnSe quantum dots usually emit violet light due to their large bandgap. To obtain blue QDLEDs, the Yang group found that the degree of electron

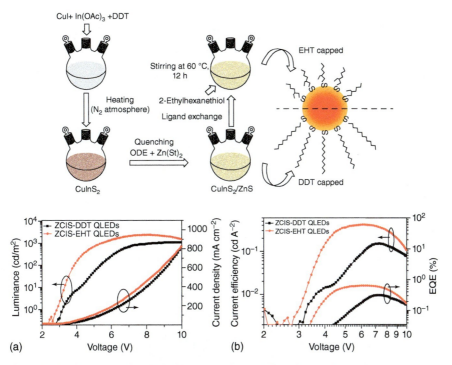

Figure 9.18 Schematic diagram of the experimental flow of ligand exchange and comparison of the performance of the corresponding devices.

delocalization depends on the presence and thickness of the ZnSe inner shell in the current QD system [6]. As shown in Figure 9.19, the pristine PL peaks of ZnSeTe/ZnS quantum dots are redshifted upon insertion into the 1 and 1.5 nm thick ZnSe inner shells. The tendency of PL redshift comes from rationalization (mainly electron) leakage into the adjacent shells. Considering that the conduction band shift at the ZnSeTe/ZnSe interface is smaller compared to ZnSeTe/ZnS, electron leakage may occur to some extent.

For quantum dots, the growth rate of the shell layer also has an impact on its performance. As shown in Figure 9.20, the Banin group controlled the growth rate of the shell layer by adjusting the activity of the precursor and found that the slower shell layer growth rate controlled thermodynamically for the same thickness was beneficial to enhance the fluorescence quantum yield and effectively suppressed the scintillation of quantum dots [63]. They concluded that the slower shell layer growth rate could reduce the defects at the interface between the core–shell of quantum dots and thus improve the performance of the material.

9.3.3 QDLED Device Structure Optimization

As an integral part of high-performance QDLEDs, HTL and ETL play an important role in charge transport and balancing carriers in the device. Figure 9.21a shows the energy level diagram of some charge transport materials for InP quantum dots [64].

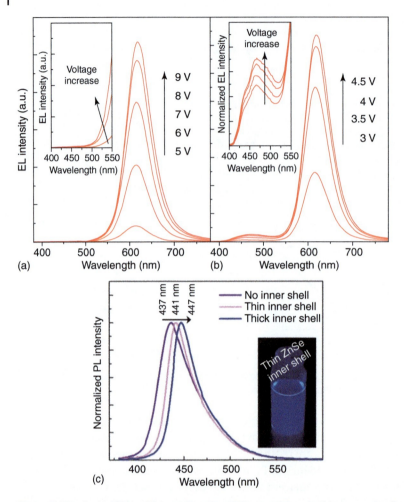

Figure 9.19 (a–b) Effect of increasing the thickness of the shell layer on the fluorescence of the material; (c) Effect of inserting different thicknesses of ZnSe layers on the quantum dot luminescence spectra.

The Kookheon Char group tried two different HTL materials in 2016: TFB and PVK. Unlike TFB, PVK-based devices do not have parasitic luminescence peaks, which means that such devices have better color purity [34]. They point out that PVK has a higher LUMO than TFB, which can inhibit electrons from crossing the light-emitting layer to bind to holes in the HTL layer and prevent radiative leap luminescence in that organic layer. However, experiments showed that the device with TFB as the HTL (EQE 2.5%) outperformed PVK (EQE 1.4%) because TFB had a higher mobility, which resulted in a higher CE. The Liu group also investigated the effect of ETL on device performance by comparing two different ETL layers, ZnMgO and ZnO, with the former showing better device performance because, in their devices, the LUMO with high ETL effectively enhanced electron injection and improved EQE. The Zhuo group investigated the effect of magnesium doping in the ZnO nanoparticles of the

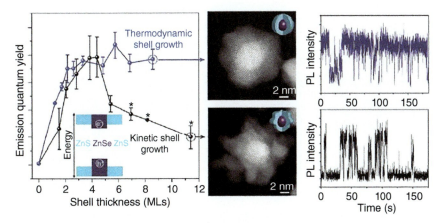

Figure 9.20 Effect of different shell growth rates on the fluorescence intensity and scintillation of the material.

n-type electron transport layer on the charge transfer balance [65]. As shown in Figure 9.21b, it was found that the increase in Mg doping could widen the bandgap of ZnO, change its energy level, increase its resistivity, reduce the electron current density, and improve the device efficiency. And the Deng group introduced chlorine atom doping in addition to Mg doping within ZnO nanoparticles and successfully increased the EQE to 4.05% [66]. Although many researchers believe that the energy level matching of ETL and HTL with the luminescent layer is important, the Kamat group suggests that the energy level matching cannot be considered as the only index for optimizing the device performance [67]. They designed using PVK and TPD as hole transport layer materials, as shown in Figure 9.21c and the results showed that higher conductivity and the presence of effective energy or charge transfer paths between the layers have a very important impact. They can reduce the on-state voltage by more than 50% and increase the relative efficiency of the device by more than five times.

9.4 Summary and Outlook

Although the overall performance of cadmium-free QDLEDs cannot match that of cadmium-containing QDLEDs. It has the greatest advantage of being nontoxic and still attracts the attention of scientists. In this chapter, we discuss the developments in quantum dot synthesis, ligand engineering, shell engineering, and QDLED structure optimization of cadmium-free QDLEDs and methods to improve the efficiency of cadmium-free QDLEDs. In QDLEDs, the Osher composite process and FRET severely impair the performance of QDLEDs. Therefore, a thick shell is grown on the quantum dots to alleviate the tight buildup of cores of quantum dots on each other after film formation. After the introduction of the core–shell structure, the core–shell structure generates defect traps on the core surface due to the mismatch degree of the lattice of two different layers, which reduces the luminescence efficiency of the

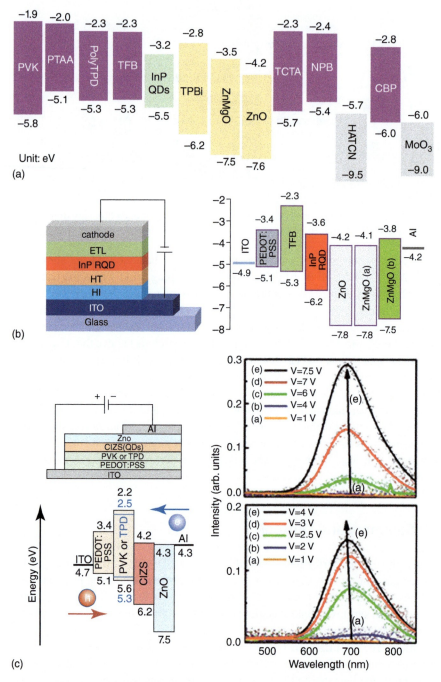

Figure 9.21 (a) Energy level diagram of common CTLs for InP quantum dots; (b) effect of doping Mg within the electron transport layer on the energy level structure of the device; (c) higher energy level matching for HTL and lower start-up voltage for QDLED.

material, so the intermediate shell and alloy core–shell structures are introduced, which largely improve the performance of QDLEDs. While in the crystal growth of quantum dots, long alkyl ligands are beneficial to the crystal growth, but they raise the charge potential barrier of the injected quantum dots in the preparation of devices, so the ligand exchange method is used to shorten the alkyl chain length of the ligands, while reducing the crystallization effect of quantum dots. In addition to the charge-emitting layer, the CTL plays an important role in carrier transport and equilibrium. In order to improve the performance of QDLEDs, attention should be paid to both CTL energy level matching and charge mobility.

The low EQE of cadmium-free QDLEDs remains an important challenge for putting QDLEDs into practical applications. Although the EQE of red InP-based QDLEDs reaches 21.4%, which is close to the efficiency of OLEDs, green and blue QDLEDs are still lagging behind. Therefore, new HTL materials with deeper HOMO energy levels and better hole mobility than conventional materials need to be introduced, especially in the case of blue QDLED light-emitting materials with deeper LUMO energy levels. In addition, to achieve charge balance, a buffer layer can be introduced by [68] that reduces excessive holes or electrons in appropriate amounts. In addition to the energy conversion efficiency, the lifetime of the device remains a challenge, and a thick case can prevent the QD from moisture and air entry outside the proper encapsulation. Finally, in the device structure, the optical coupling output efficiency can be improved by special design. The achievement of chalcogenide LEDs is the use of discontinuous chalcogenides and low-reflectivity polymers, which significantly improve the output coupling efficiency [69]. Similar structures with special designs may play an important role in cadmium-free QDLEDs.

Cadmium-based quantum dots have been widely used in light-emitting diodes and other optoelectronic devices due to their unique optical and electronic properties. However, the toxicity of cadmium and concerns about its potential environmental impact have prompted the exploration of non-cadmium-based quantum dot materials for optoelectronic applications.

One promising alternative to cadmium-based quantum dots is based on InP, which has been demonstrated to exhibit high efficiency and stability in light-emitting diodes. Other non-cadmium-based quantum dots that have been explored for light-emitting diodes and other applications include copper indium sulfide ($CuInS_2$) and copper zinc sulfide (Cu_2ZnSnS_4), among others.

In addition to these specific non-cadmium quantum dot materials, there are also several future research directions that could help to advance the development of non-cadmium quantum dot light-emitting materials and devices. Some potential research directions include:

Development of new materials: There is still a great deal of research needed to develop new non-cadmium quantum dot materials with improved performance and stability. This could involve exploring new compositions and structures, as well as optimizing synthesis methods to achieve high-quality materials.

Device engineering: In addition to material development, there is also a need for further research in the engineering of non-cadmium quantum dot devices.

This could involve optimizing device structures and fabrication processes to improve performance and stability.

Characterization and understanding of device properties: To develop high-performance non-cadmium quantum dot devices, it is also important to fully understand the properties of the materials and how they interact with other components in the device. This could involve characterizing the optical, electronic, and structural properties of the materials and using this information to optimize device performance.

Environmental impact studies: While non-cadmium quantum dot materials are generally considered to be less toxic than their cadmium-based counterparts, it is still important to conduct environmental impact studies to fully understand the potential risks and ensure that these materials are safe for widespread use.

In summary, the development of non-cadmium quantum dot materials and devices is an active area of research with significant potential for advancing optoelectronic technologies. Future research directions could include the development of new materials, optimization of device engineering, characterization of device properties, and environmental impact studies.

References

1 Dai, X., Zhang, Z., Jin, Y. et al. (2014). Solution-processed, high-performance light-emitting diodes based on quantum dots. *Nature* 515 (7525): 96–99.
2 Yang, X., Zhang, Z.-H., Ding, T. et al. (2018). High-efficiency all-inorganic full-colour quantum dot light-emitting diodes. *Nano Energy* 46: 229–233.
3 Xiang, C., Koo, W., Chen, S. et al. (2012). Solution processed multilayer cadmium-free blue/violet emitting quantum dots light emitting diodes. *Applied Physics Letters* 101 (5): 053303.
4 Ji, W., Jing, P., Xu, W. et al. (2013). High color purity ZnSe/ZnS core/shell quantum dot based blue light emitting diodes with an inverted device structure. *Applied Physics Letters* 103 (5): 053106.
5 Wang, A., Shen, H., Zang, S. et al. (2015). Bright, efficient, and color-stable violet ZnSe-based quantum dot light-emitting diodes. *Nanoscale* 7 (7): 2951–2959.
6 Jang, E.P., Han, C.Y., Lim, S.W. et al. (2019). Synthesis of alloyed ZnSeTe quantum dots as bright, color-pure blue emitters. *ACS Applied Materials & Interfaces* 11 (49): 46062–46069.
7 Kim, J.H. and Yang, H. (2014). All-solution-processed, multilayered CuInS(2)/ZnS colloidal quantum-dot-based electroluminescent device. *Optics Letters* 39 (17): 5002–5005.
8 Zhang, W., Lou, Q., Ji, W. et al. (2013). Color-tunable highly bright photoluminescence of cadmium-free Cu-doped Zn–In–S nanocrystals and electroluminescence. *Chemistry of Materials* 26 (2): 1204–1212.
9 van der Stam, W., Berends, A.C., Rabouw, F.T. et al. (2015). Luminescent CuInS2 quantum dots by partial cation exchange in Cu2–xS nanocrystals. *Chemistry of Materials* 27 (2): 621–628.

10 Bai, Z., Ji, W., Han, D. et al. (2016). Hydroxyl-terminated CuInS2 based quantum dots: toward efficient and bright light emitting diodes. *Chemistry of Materials* 28 (4): 1085–1091.

11 Gugula, K., Stegemann, L., Cywiński, P.J. et al. (2016). Facile surface engineering of CuInS2/ZnS quantum dots for LED down-converters. *RSC Advances* 6 (12): 10086–10093.

12 Yoon, S.-Y., Kim, J.-H., Kim, K.-H. et al. (2019). High-efficiency blue and white electroluminescent devices based on non-Cd I – III – VI quantum dots. *Nano Energy* 63: 103869.

13 Choi, D.B., Kim, S., Yoon, H.C. et al. (2017). Color-tunable Ag-In-Zn-S quantum-dot light-emitting devices realizing green, yellow and amber emissions. *Journal of Materials Chemistry C* 5 (4): 953–959.

14 Guan, Z., Tang, A., Lv, P. et al. (2018). New insights into the formation and color-tunable optical properties of multinary Cu-In-Zn-based chalcogenide semiconductor nanocrystals. *Advanced Optical Materials* 6 (10): 1701389.

15 Wang, T., Guan, X., Zhang, H., and Ji, W. (2019). Exploring electronic and excitonic processes toward efficient deep-red CuInS2/ZnS quantum-dot light-emitting diodes. *ACS Applied Materials & Interfaces* 11 (40): 36925–36930.

16 Lim, J., Bae, W.K., Lee, D. et al. (2011). InP@ZnSeS, core@composition gradient shell quantum dots with enhanced stability. *Chemistry of Materials* 23 (20): 4459–4463.

17 (2013). Highly efficient cadmium-free quantum dot light-emitting diodes enabled by the direct formation of excitons within InP@ZnSeS quantum dots. *ACS Nano* 7 (10): 9019–9026.

18 Zhang, H., Hu, N., Zeng, Z. et al. (2019). High-efficiency green InP quantum dot-based electroluminescent device comprising thick-shell quantum dots. *Advanced Optical Materials* 7 (7): 1801602.

19 Won, Y.H., Cho, O., Kim, T. et al. (2019). Highly efficient and stable InP/ZnSe/ZnS quantum dot light-emitting diodes. *Nature* 575 (7784): 634–638.

20 Zhang, H., Ma, X., Lin, Q. et al. (2020). High-brightness blue inp quantum dot-based electroluminescent devices: the role of shell thickness. *The Journal of Physical Chemistry Letters* 11 (3): 960–967.

21 Gao, P., Zhang, Y., Qi, P., and Chen, S. (2022). Efficient InP green quantum-dot light-emitting diodes based on organic electron transport layer. *Advanced Optical Materials* 2202066.

22 Yu, P., Cao, S., Shan, Y. et al. (2022). Highly efficient green InP-based quantum dot light-emitting diodes regulated by inner alloyed shell component. *Light: Science & Applications* 11 (1): 162.

23 Li, H., Bian, Y., Zhang, W. et al. (2022). High performance InP-based quantum dot light-emitting diodes via the suppression of field-enhanced electron delocalization. *Advanced Functional Materials* 32 (38): 2204529.

24 Zhao, X., Lim, L.J., Ang, S.S., and Tan, Z.K. (2022). Efficient short-wave infrared light-emitting diodes based on heavy-metal-free quantum dots. *Advanced Materials* 34 (45): 2206409.

25 Mićić, O.I., Sprague, J.R., Curtis, C.J. et al. (1995). Synthesis and characterization of GaP, InP, and GaInP2 quantum dots. *Journal of Physical Chemistry* 99 (19): 7754–7759.
26 Wolters, R.H., Arnold, C.C., and Heath, J.R. (1996). Synthesis of size-selected, surface-passivated InP nanocrystals. *Journal of Physical Chemistry* 100 (17): 7212–7219.
27 Mićić, O.I., Cheong, H.M., Fu, H. et al. (1997). Size-dependent spectroscopy of InP quantum dots. *Journal of Physical Chemistry* 101 (25): 4904–4912.
28 Adam, S., Talapin, D.V., Borchert, H. et al. (2005). The effect of nanocrystal surface structure on the luminescence properties: photoemission study of HF-etched InP nanocrystals. *Journal of Physical Chemistry* 123 (8): 084706.
29 Euidock Ryu, S.K., Jang, E., Jun, S. et al. (2009). Step-wise synthesis of InP/ZnS core-shell quantum dots and the role of zinc acetate. *Chemistry of Materials* 21 (4): 2621–2623.
30 Lim, K., Jang, H.S., and Woo, K. (2012). Synthesis of blue emitting InP/ZnS quantum dots through control of competition between etching and growth. *Nanotechnology* 23 (48): 485609.
31 Cao, F., Wang, S., Wang, F. et al. (2018). A layer-by-layer growth strategy for large-size InP/ZnSe/ZnS core–shell quantum dots enabling high-efficiency light-emitting diodes. *Chemistry of Materials* 30 (21): 8002–8007.
32 Li, Y., Hou, X., Dai, X. et al. (2019). Stoichiometry-controlled InP-based quantum dots: synthesis, photoluminescence, and electroluminescence. *Journal of the American Chemical Society* 141 (16): 6448–6452.
33 Yang, X., Zhao, D., Leck, K.S. et al. (2012). Full visible range covering InP/ZnS nanocrystals with high photometric performance and their application to white quantum dot light-emitting diodes. *Advanced Materials* 24 (30): 4180–4185.
34 Jo, J.H., Kim, J.H., Lee, K.H. et al. (2016). High-efficiency red electroluminescent device based on multishelled InP quantum dots. *Optics Letters* 41 (17): 3984–3987.
35 Kim, J.-H., Han, C.-Y., Lee, K.-H. et al. (2014). Performance improvement of quantum dot-light-emitting diodes enabled by an alloyed ZnMgO nanoparticle electron transport layer. *Chemistry of Materials* 27 (1): 197–204.
36 Wang, Y.H. and Herron, N. (1991). Nanometer-sized semiconductor clusters: materials synthesis, quantum size effects, and photophysical properties. *Journal of Physical Chemistry* 95 (2): 525–532.
37 Shen, W., Tang, H., Yang, X. et al. (2017). Synthesis of highly fluorescent InP/ZnS small-core/thick-shell tetrahedral-shaped quantum dots for blue light-emitting diodes. *Journal of Materials Chemistry C* 5 (32): 8243–8249.
38 Kim, S., Kim, T., Kang, M. et al. (2012). Highly luminescent InP/GaP/ZnS nanocrystals and their application to white light-emitting diodes. *Journal of the American Chemical Society* 134 (8): 3804–3809.
39 Liu, Y., Tang, Y., Ning, Y. et al. (2010). "One-pot" synthesis and shape control of ZnSe semiconductor nanocrystals in liquid paraffin. *Journal of Materials Chemistry* 20 (21): 4451–4458.
40 Nakamura, H., Kato, W., Uehara, M. et al. (2006). Tunable photoluminescence wavelength of chalcopyrite CuInS2-based semiconductor

nanocrystals synthesized in a colloidal system. *Chemistry of Materials* 18 (14): 3330–3335.

41 Castro, S.L., Bailey, S.G., Raffaelle, R.P. et al. (2004). Synthesis and characterization of colloidal CuInS2 nanoparticles from a molecular single-source precursor. *Journal of Physical Chemistry* 108 (33): 12429–12435.

42 Koo, B., Patel, R.N., and Korgel, B.A. (2009). Synthesis of CuInSe2 nanocrystals with trigonal pyramidal shape. *Journal of the American Chemical Society* 131 (9): 3134–3135.

43 Tang, J., Hinds, S., Kelley, S.O., and Sargent, E.H. (2008). Synthesis of colloidal CuGaSe2, CuInSe2, and Cu(InGa)Se2 nanoparticles. *Chemistry of Materials* 20 (22): 6906–6910.

44 Zhang, A., Dong, C., Li, L. et al. (2015). Non-blinking (Zn)CuInS/ZnS quantum dots prepared by in situ interfacial alloying approach. *Scientific Reports* 5: 15227.

45 Yao, D., Liu, H., Liu, Y. et al. (2015). Phosphine-free synthesis of Ag-In-Se alloy nanocrystals with visible emissions. *Nanoscale* 7 (44): 18570–18578.

46 Jiao, M., Huang, X., Ma, L. et al. (2019). Biocompatible off-stoichiometric copper indium sulfide quantum dots with tunable near-infrared emission via aqueous based synthesis. *Chemical Communications* 55 (100): 15053–15056.

47 Castro, S.L., Bailey, S.G., Raffaelle, R.P. et al. (2003). Nanocrystalline chalcopyrite materials (CuInS2 and CuInSe2) via low-temperature pyrolysis of molecular single-source precursors. *Chemistry of Materials* 15 (16): 3142–3147.

48 (2000). Synthesis of Nanocrystalline $CuMS_2$ (M = In or Ga) through a solvothermal process. *Inorganic Chemistry* 39 (7): 1606–1607.

49 Shin, S.J., Koo, J.J., Lee, J.K., and Chung, T.D. (2019). Unique luminescence of hexagonal dominant colloidal copper indium sulphide quantum dots in dispersed solutions. *Scientific Reports* 9 (1): 20144.

50 Tan, Z., Zhang, Y., Xie, C. et al. (2011). Near-band-edge electroluminescence from heavy-metal-free colloidal quantum dots. *Advanced Materials* 23 (31): 3553–3558.

51 Chen, B., Zhong, H., Zhang, W. et al. (2012). Highly emissive and color-tunable CuInS2-based colloidal semiconductor nanocrystals: off-stoichiometry effects and improved electroluminescence performance. *Advanced Functional Materials* 22 (10): 2081–2088.

52 Kim, J.-H., Lee, K.-H., Jo, D.-Y. et al. (2014). Cu – In–Ga – S quantum dot composition-dependent device performance of electrically driven light-emitting diodes. *Applied Physics Letters* 105 (13): 133104.

53 Kim, J.-H. and Yang, H. (2016). High-efficiency Cu–In–S quantum-dot-light-emitting device exceeding 7%. *Chemistry of Materials* 28 (17): 6329–6335.

54 Liu, Z., Guan, Z., Li, X. et al. (2020). Rational design and synthesis of highly luminescent multinary Cu-In-Zn-S semiconductor nanocrystals with tailored nanostructures. *Advanced Optical Materials* 8 (6): 1901555.

55 Chen, B., Zhong, H., Wang, M. et al. (2013). Integration of CuInS2-based nanocrystals for high efficiency and high colour rendering white light-emitting diodes. *Nanoscale* 5 (8): 3514–3519.

56 Chuang, P.H., Lin, C.C., and Liu, R.S. (2014). Emission-tunable CuInS2/ZnS quantum dots: structure, optical properties, and application in white light-emitting diodes with high color rendering index. *ACS Applied Materials & Interfaces* 6 (17): 15379–15387.

57 Kim, J.-H., Kim, B.-Y., Jang, E.-P. et al. (2017). A near-ideal color rendering white solid-state lighting device copackaged with two color-separated Cu–X–S (X = Ga, In) quantum dot emitters. *Journal of Materials Chemistry C* 5 (27): 6755–6761.

58 Song, W.-S., Kim, J.-H., Lee, J.-H. et al. (2012). Synthesis of color-tunable Cu–In–Ga–S solid solution quantum dots with high quantum yields for application to white light-emitting diodes. *Journal of Materials Chemistry* 22 (41): 21901–21908.

59 Song, W.S. and Yang, H. (2013). Solvothermal preparation of yellow-emitting CuInS2/ZnS quantum dots and their application to white light-emitting diodes. *Journal of Nanoscience and Nanotechnology* 13 (9): 6459–6462.

60 Kim, B.Y., Kim, J.H., Lee, K.H. et al. (2017). Synthesis of highly efficient azure-to-blue-emitting Zn-Cu-Ga-S quantum dots. *Chemical Communications* 53 (29): 4088–4091.

61 Zang, H., Li, H., Makarov, N.S. et al. (2017). Thick-shell CuInS2/ZnS quantum dots with suppressed "blinking" and narrow single-particle emission line widths. *Nano Letters* 17 (3): 1787–1795.

62 Li, J., Jin, H., Wang, K. et al. (2016). High luminance of CuInS2-based yellow quantum dot light emitting diodes fabricated by all-solution processing. *RSC Advances* 6 (76): 72462–72470.

63 Ji, B., Koley, S., Slobodkin, I. et al. (2020). ZnSe/ZnS Core/shell quantum dots with superior optical properties through thermodynamic shell growth. *Nano Letters* 20 (4): 2387–2395.

64 Wu, Z., Liu, P., Zhang, W. et al. (2020). Development of InP quantum dot-based light-emitting diodes. *ACS Energy Letters* 5 (4): 1095–1106.

65 Li, D., Kristal, B., Wang, Y. et al. (2019). Enhanced efficiency of InP-based red quantum dot light-emitting diodes. *ACS Applied Materials & Interfaces* 11 (37): 34067–34075.

66 Chen, F., Liu, Z., Guan, Z. et al. (2018). Chloride-passivated Mg-doped ZnO nanoparticles for improving performance of cadmium-free, quantum-dot light-emitting diodes. *ACS Photonics* 5 (9): 3704–3711.

67 Wepfer, S., Frohleiks, J., Hong, A.R. et al. (2017). Solution-processed CuInS2-based white QD-LEDs with mixed active layer architecture. *ACS Applied Materials & Interfaces* 9 (12): 11224–11230.

68 Lin, K., Xing, J., Quan, L.N. et al. (2018). Perovskite light-emitting diodes with external quantum efficiency exceeding 20%. *Nature* 562 (7726): 245–248.

69 Cao, Y., Wang, N., Tian, H. et al. (2018). Perovskite light-emitting diodes based on spontaneously formed submicrometre-scale structures. *Nature* 562 (7726): 249–253.

10

AC-Driven Quantum Dot Light-Emitting Diodes

Quantum dot light-emitting diodes (QDLEDs) offer a range of advantages, including narrow emission width, high quantum yield, tunable emission wavelength, and high stability, making them a promising technology for commercial applications. Two types of QDLEDs exist: direct current (DC) driven and alternating current (AC) driven. The DC is a simple driving mode for the realization of effective luminescence of EL devices. However, high current density and unidirectional continuous current injection are prone to be generated in the DC driving mode, leading to adverse charge accumulation at the luminescent layers and thereby poor operational stability. In addition, inevitable power loss is unavoidable when connecting the EL device with household 110/220 V and 50/60 Hz AC power lines, for which converters and rectifiers are required. Therefore the DC driving mode limits their practical application and performance. This has led to the emergence of AC-driven QDLEDs, which offer unique advantages and are seen as a promising alternative to DC-driven QDLEDs for various applications. As a result, scientists are exploring the use of AC-driven QDLEDs to develop next-generation full-color displays and solid-state lighting.

AC-driven QDLEDs have some potential advantages over DC-driven QDLEDs. One advantage is that they can provide a simpler and more efficient way to achieve full-color displays, as the color of the emitted light can be easily tuned by adjusting the frequency of the AC drive signal. In addition, AC-driven QDLEDs can potentially achieve higher efficiency and stability than DC-driven QDLEDs, as they can avoid the accumulation of charges and Joule heating that can occur in DC-driven devices. This is because the AC signal can periodically reverse the bias across the device, which allows charges to be more evenly distributed and dissipated, reducing the risk of charge accumulation and localized heating. Moreover, electrochemical reactions between the organic layer and the electrodes can be effectively prevented by the inert dielectric layer, which protects the device from degradation from external moisture and oxygen in the atmosphere. Furthermore, the AC-driven mode has the ability to make devices with light-emitting units of various colors operate alternately at positive and negative cyclic voltages. This results in the rapid change of light colors and the formation of a new color of light.

Colloidal Quantum Dot Light Emitting Diodes: Materials and Devices, First Edition. Hong Meng.
© 2024 WILEY-VCH GmbH. Published 2024 by WILEY-VCH GmbH.

However, there are also some potential drawbacks to AC-driven QDLEDs. One of the main challenges is the need to design and implement a suitable AC driving circuitry, which can be more complex than a DC power supply. In addition, the efficiency of AC-driven QDLEDs can be limited by the frequency of the AC signal, as higher frequencies may result in more charge loss through non-radiative decay processes. Additionally, the lifetime of AC-driven QDLEDs may be limited by the potential for electrical breakdown and damage from the high voltage required for AC driving. Nevertheless, AC-driven EL devices have obvious advantages in some important aspects, including stable operation and color-tunable integrated devices.

In this chapter, we will discuss the principle of AC-driven QDLEDs, performance metrics, materials required, and the research developments in recent years in QDLEDs.

10.1 Principle of Luminescence of DC and AC-Driven QDLEDs

The principle of QDLEDs can be understood through different theories, including photoexcitation, carrier injection, energy transfer, and ionization. In the photoexcitation theory, excitons are formed in quantum dots (QDs) by absorbing high-energy photons. In the carrier injection theory, electrons and holes are injected into the carrier transport layers (CTLs), which then enter the quantum dot layer to form excitons that recombine and release photons. The energy transfer theory proposes that excitons are first formed in the luminescent layer made of polymers, organic small molecules, or inorganic semiconductor materials and then transferred to the quantum dots through radiation-free coupling between dipoles. The ionization mechanism considers that a large electric field can ionize electrons from one quantum dot to another, resulting in the creation of holes. When these ionization events occur throughout the QD film, the resulting electrons and holes can meet on the same QD to form excitons [1].

The structure and operating mechanisms of DC-driven and AC-driven QDLED devices are different [2]. Figure 10.1a,b show the structure and operating mechanism of a typical DC-driven device, which consists of two functional layers above and below the light-emitting layer (EML). These layers transport externally injected carriers (electrons and holes) through the (hole and electron) transport layers, which are then injected from the outside via metal and indium tin oxide (ITO) transparent electrodes. The electrons and holes combine in the EML to form excitons, which emit light. Figure 10.1c,d show AC-driven QDLED devices with different structures and operating mechanisms. The first structure schematic shows an AC-QDLED device with a dual dielectric layer structure, where the dielectric layer prevents carrier injection from the outside, and the doped charge transport layer acts as a charge generation layer. Driven by an AC voltage, carriers are generated by the charge generation layer and delivered to the EML, where they recombine to form excitons that emit light. The second structure schematic shows a device with a single dielectric layer structure. In this case, a charge generation layer is adjacent to the EML and generates one type of carrier (electron or hole). Through a metal electrode, the other type of carrier is injected into the EML, where the carriers recombine to form excitons that emit light. The third structure schematic shows a device with a single dielectric layer structure without a charge generation layer. In this device, carrier charges are injected into the EML from one electrode, and the opposite charge is injected directly from the other electrode. The two carriers recombine in the EML to form excitons that emit light. The fourth structure schematic shows a device with a dual-injection structure, which is similar to the structure of a DC-driven EL device. The charge carriers are injected into the EML layer through two electrodes, and the two carriers recombine in the EML to form excitons that emit light.

When a DC power supply is connected to the two electrodes of the QDLED, the device is excited and emits light. This type of device is called a DC QDLED, and its device structure is shown in Figure 10.2a. When a positive bias is applied, electrons and holes are injected from the cathode and anode, respectively, into the electron transport layer and hole transport layer. The injected electrons and holes then recombine with each other in the emissive layer, which is made up of quantum dots, resulting in the emission of light.

AC-driven QDLEDs operate based on a process called AC electroluminescence (ACEL), which is a type of electroluminescence that occurs under the influence of an alternating electric field. The structure of the ACEFL can be modeled as a circuit consisting of three capacitors connected in series under the drive of an AC electric field (Figure 10.3). It shows that the thickness of the dielectric layer and the dielectric constant determine the electric field distribution around the phosphor particles within the emitting layer. The emission of light is based on the principle of hot electron impact excitation and consists of four sequential processes: injection of electrons at the interface into the phosphor layer under an AC electric field, acceleration of the injected electrons, excitation of the luminescent center by high-energy electron impact, and optical conversion of the excitation energy levels in the luminescent layer. In sandwich AC-driven EL devices, there is no demanding requirement on the work function of the electrodes, which is different from the electroluminescence mechanism of p–n junctions.

The operation of DC-driven electroluminescent devices is usually understood as direct carrier injection, energy transfer, or both. In contrast, AC-driven electroluminescent devices are generally explained in terms of ionization mechanisms.

Depending on the device structure and luminescence principle, there are three main device structures for AC-driven QDLEDs: symmetric, asymmetric, and hybrid structures. The symmetric structure has two identical layers, one of which is the QD layer, and the other is the transport layer. In contrast, the asymmetric structure has two different layers, where one layer is the QD layer, and the other layer is the electron or hole transport layer. The hybrid structure combines features of both symmetric and asymmetric structures and includes a QD layer sandwiched between two different transport layers.

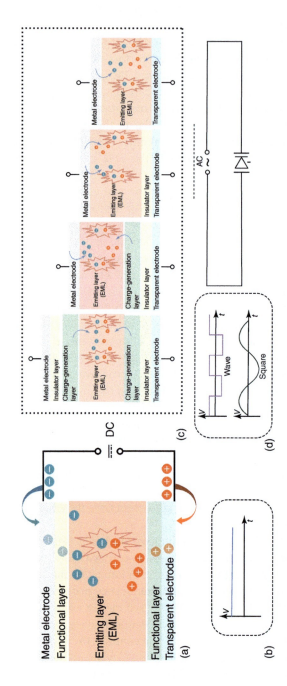

Figure 10.1 Schematic diagram of the structure and operating mechanism of (a) (b) DC-driven EL devices. (c) (d) AC-driven EL devices. Source: Yu et al. [2]/American Chemical society.

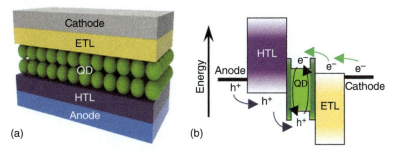

Figure 10.2 (a) Representative device structure of a DC-driven quantum dot light-emitting diode (QDLED). (b) Energy diagram of a typical QDLED showing the charge injected from the anode and cathode. Source: Choi et al. [3]/ CC BY 4.0 / Public domain / Springer Nature.

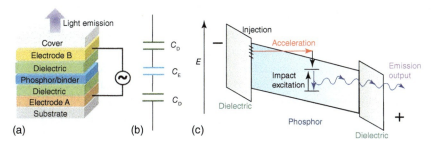

Figure 10.3 (a) Schematic illustration of a traditional sandwich AC-driven thin film electroluminescent (AC-TFEL) device. (b) The equivalent circuit of the sandwich AC-driven EL device. (c) Scheme of the hot-electron impact excitation principle.

10.2 Mechanism of Double-Emission Tandem Structure of AC QDLEDs

There are double-insulation, single-insulation, double-injection, and tandem structures of AC-driven QDLEDs as shown in Figure 10.1c. Same as the reported tandem structure of bi-emission layer as of AC-OLED, as shown in Figure 10.4, two different conductive particles, ITO and passivated gold nanoparticles, can be used to create bipolar charge-generation layers. The double-insulated AC-QDLED device is shown in Figure 10.4a. After a positive voltage is applied to the bottom electrode, charge carriers are generated in the charge generation region (Figure 10.4b(a)), and holes and electrons move toward the top and bottom electrodes, respectively. Under positive voltage, holes and electrons accumulate temporarily near the insulating layer (Figure 10.4b(b)). Changing the voltage polarity reverses the drift direction of the generated carriers. Electrons and holes are then supplied to the upper and lower EL layers, respectively. The newly generated carriers encounter returning, oppositely charged carriers in both the upper and lower layers (Figure 10.4b(c)). Thus, at each half-cycle of the voltage, charge recombination

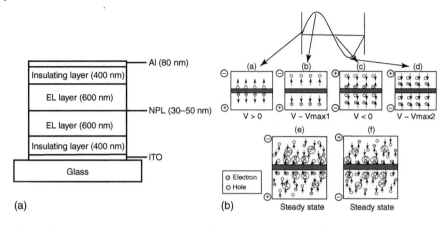

Figure 10.4 (a) Device structure with gold nanoparticle layer (NPL). (b) Schematic diagram of EL processes in double-insulation devices. Source: Adapted from Lee et al. [4].

occurs in both EL layers (Figure 10.4b(d)). At steady state, in the case of continuously applied AC voltage, charge recombination occurs at both positive and negative half-cycles (Figure 10.4b(e),(f)). Since no charge is injected from the electrodes, an electrode-independent characteristic is achieved in AC-driven QDLEDs, which is superior to conventional charge-injection type devices (e.g. DC-driven QDLEDs), where charge injection efficiency and electrode material stability are always incompatible [4].

10.2.1 Field-Generated AC QDLEDs

Field-generated AC QDLEDs, also known as electroluminescent quantum dot devices (ELQDs), are a type of AC-driven QDLED that utilizes the ionization mechanism for light emission. In this type of device, an alternating electric field is applied to the QD film, which generates electron and hole pairs through impact ionization. These electron and hole pairs then recombine radiatively in the QD film, resulting in light emission. ELQDs typically have a simple device structure consisting of a QD film sandwiched between two metal electrodes. The QD film is usually formed by spin-coating or drop-casting a solution of QDs onto a substrate, followed by annealing to remove any residual solvent. The thickness of the QD film is typically on the order of a few tens of nanometers. One advantage of ELQDs is their high luminous efficiency, which can be attributed to the efficient formation of electron and hole pairs through impact ionization. Another advantage is their simple device structure, which can lead to low fabrication costs. However, one drawback of ELQDs is their relatively low brightness, which can limit their practical applications.

A field-accelerated QDLED is a non-p–n junction device that simultaneously generates carriers and electric fields, which excite the electroluminescence of quantum dots in a new way. A typical structure of such a device is shown in Figure 10.5e. The QD light-emitting layer (QD layer) is sandwiched between two dielectric layers. The two electrodes are located at the top and bottom of the device. When a

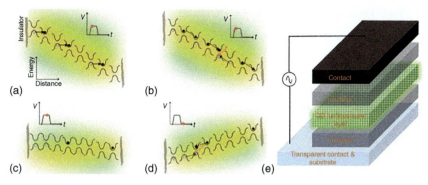

Figure 10.5 Typical AC field-driven QDLED electroluminescence mechanism. (a) In the presence of a sufficiently high applied electric field, electrons can be transferred from the valence band of one QD to the conduction band of an adjacent QD; (b) Electrons and holes generated by the field may undergo electric field-assisted transport to the excited states of adjacent quantum dots, during which electrons and holes form excitons and complex in a nonradiative or radiative manner; (c) Electrons and holes distributed within the QD film induce internal fields that shield the applied electric field and continuously redistributes the charge so that excitons and luminescence persist throughout the applied pulse; (d) if the applied electric field is removed (or reduced), the remaining charge creates an internal field in the opposite direction of the previously applied electric field, which causes the electrons and holes to move toward each other and generates a luminescence light. Source: Wood et al. [5]/American Chemical society (e) Schematic diagram of a typical field AC-driven QDLED structure.

voltage is applied to the device, the electrodes do not inject carriers into the QD layer, but electroluminescence generation can be observed. This phenomenon can be explained by the fact that when the voltage energy applied to the quantum dots exceeds the bandgap of the quantum dots, the quantum dots ionize in the electric field and produce free electrons. The ionized electrons are transferred from the valence band of one quantum dot to the conduction band (CB) of the adjacent quantum dot, resulting in spatially separated electron–hole pairs in the quantum dot layer. The electron–hole pairs recombine to emit light (see Figure 10.5a–d) or transfer energy to a luminescent impurity (e.g. Mn^{2+}), which excites the impurity to emit light. Applying a pulsed or sinusoidal voltage to the two electrodes of the device can cause the device to emit light.

When a pulsed voltage or sinusoidal voltage is applied, the ionized electrons and holes will move to the QD/dielectric layer interface under the electric field, respectively. This distribution of electrons and holes creates an internal electric field that attenuates the effect of the external electric field. When the external electric field is weakened or diminished, the internal electric field drives the electrons and holes to move toward the center of the QDs layer and then recombine to emit light. Thus, although the external electric field strength needs to reach $5\,MV\,cm^{-1}$ to excite free carriers, the field-driven QDLEDs can also emit light at lower electric fields ($\sim 1\,MV\,cm^{-1}$) due to the formation of the internal electric field [5].

10.2.2 Half-Field to Half-Injection AC QDLEDs

Half-field to half-injection (HFHI) AC QDLEDs are a type of AC-driven QDLEDs that operate based on a half-field ionization mechanism combined with a half-injection carrier transport process. This approach can effectively reduce the required driving voltage and prolong the lifetime of the device.

The device structure of HFHI AC QDLEDs consists of a QD layer sandwiched between an electron injection layer and a hole injection layer, with the two electrodes on either side of the layers. In contrast to field-accelerated QDLEDs, HFHI AC QDLEDs use a p–n junction-like structure and the electron and hole injection layers have different work functions.

The working mechanism of HFHI AC QDLEDs is as follows (Figure 10.6): When an AC voltage is applied to the device, the electron injection layer is positively biased and the hole injection layer is negatively biased during half of the AC cycle. During this time, electrons are injected from the electron injection layer into the QD layer, and holes are injected from the hole injection layer. At the same time, the applied electric field is not strong enough to cause significant ionization of the QDs. During the other half of the AC cycle, the polarity of the bias is reversed, causing electrons to be injected from the other side and holes to be injected from the opposite side. At this time, the applied electric field is strong enough to cause the ionization of QDs, and free electrons and holes are generated by impact ionization. These free carriers recombine with opposite carriers in the QD layer, leading to the emission of light. The electrons and holes in the QDLED device will migrate back and forth between functional layers and the emissive layer with the reversal of the applied voltage. As the frequency increases, the applied voltage changes faster, and the electrons and holes will be pulled back by the reverse electric field even before they reach the emissive layer.

The advantages of HFHI AC QDLEDs include the low driving voltage and the prolonged device lifetime due to the reduced ionization rate of the QDs. The major disadvantage of this device structure is its lower efficiency compared to other AC-driven QDLEDs.

Recently, another type of AC QDLED has been proposed. Unlike the above-mentioned field QDLED, this AC QDLED has only a dielectric layer, which is set below or above the light-emitting layer. A typical structure of such a device is shown in Figure 10.7a. In this structure, one type of carrier can be injected directly from the external electrode, while the other type of carrier is generated inside the device. In the positive drive cycle, one type of carrier can be injected and then combined with internally generated carriers to form excitons, which radiate decayed light. The carriers injected after binding accumulate on the surface of the dielectric layer. And when the negative drive cycle is performed, the accumulated carriers can be released from the external electrode and returned to their original state, as shown in Figure 10.7c,d [7].

During the first half-cycle of AC excitation (charging process), electrons can be injected directly from the Al electrode, but it is not possible to inject holes from the ITO side due to dielectric insulation. Therefore, it is then necessary to generate

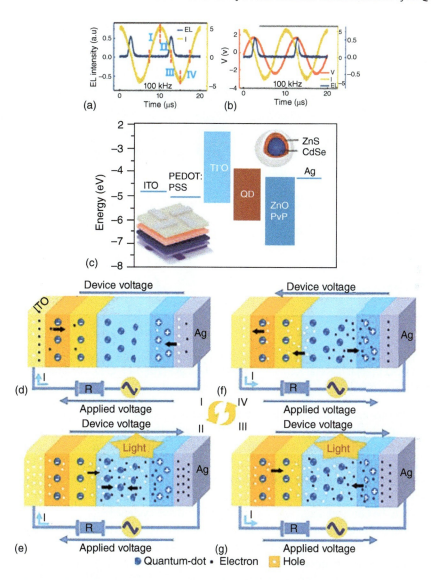

Figure 10.6 (a) The V-I-EL-t characteristics of QDLED at 100 kHz; (b) The EL-I-t characteristics at 100 kHz; (c) Energy band alignment diagram (the inset shows the structure of the QDLED device and quantum-dot); (d)–(f) Schematic diagram of carrier transfer model of QDLED under AC-drive at high frequencies. Source: Sun et al. [6]/American Chemical society.

holes inside the device. This is achieved through the MoO_3/TFB hole generation layer. Since MoO_3 has a low CB and is well matched to the highest molecular orbital (HOMO) of the TFB, electrons in the HOMO of the TFB can tunnel into the CB of MoO_3. Then, electrons and holes are generated in the CB of MoO_3 and the HOMO of TFB, respectively, and the injected electrons can then complex with the generated holes, during which luminescence can be observed. After recombination, the

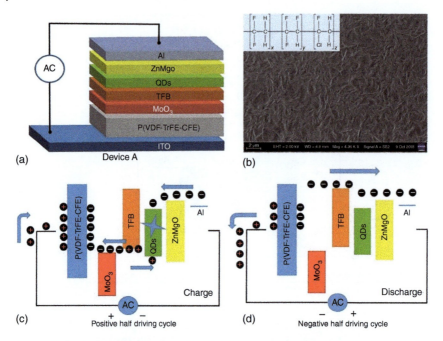

Figure 10.7 Example of a half-field-induced half-injection device. (a) Device structure of AC-driven QDLED; (b) molecular formula and SEM image of P(VDF-TrFE-CFE); operating process of AC-driven QDLED (c) charging; (d) discharging.

excess electrons accumulate on the dielectric surface, while an equal amount of positive charge rapidly accumulates on the other side of the dielectric layer to maintain charge neutrality. After the charging process is complete, no more electrons can be injected from the Al electrode, and the device is in an open circuit. When the reverse voltage is applied during the next half-cycle, the accumulated holes and electrons discharge and return the device to its original uncharged state, which is then ready to be charged again during the next half-drive cycle [8].

Compared to field-driven QDLEDs, such asymmetric AC QDLEDs can allow one carrier to be injected into the emitting layer, thus increasing the number of such carriers and thus the chance of carrier binding, thus showing higher luminance. However, improving the charge balance of such half-field-driven half-injected AC QDLEDs is the key and challenge to achieving low drive voltage, high brightness, and high-power efficiency.

10.2.3 AC/DC Dual Drive Mode QDLEDs

AC/DC dual-drive mode QDLEDs combine the AC and DC driving modes to achieve high efficiency and high brightness. The device structure includes a QD layer sandwiched between two transport layers, with an AC voltage superimposed on a DC bias voltage. The working mechanism of AC/DC dual-drive mode QDLEDs is that the DC bias voltage can inject electrons and holes into the QD layer, while the AC voltage can generate an electric field to accelerate the transport of carriers and improve the recombination efficiency of electron–hole pairs. The superposition of AC and DC voltages can also promote the migration and accumulation of charges in the QD layer, which further enhances the luminescence efficiency. The advantages of AC/DC dual-drive mode QDLEDs include high efficiency and high brightness due to the combined benefits of AC and DC driving modes. The device can also operate at a lower voltage than traditional DC QDLEDs, which can reduce power consumption and prolong the device's lifetime. The main disadvantage of AC/DC dual-drive mode QDLEDs is that the device structure is more complex than traditional DC QDLEDs, which can increase the manufacturing cost. Additionally, the performance of the device is highly dependent on the specific combination of AC and DC voltages, which can be challenging to optimize.

Another type of AC-driven QDLED structure is similar to the conventional DC-driven QDLED, using a wide bandgap semiconductor (most commonly ZnO) as an electron transport layer and a protective layer to isolate the vulnerable QD layer from the exposed air environment. The wide bandgap semiconductor in this structure can also be seen as an ideal replacement for the high dielectric constant dielectric and hole generation layers in the half-field-induced half-injection type AC QDLED described in the previous section. As shown in Figure 9.6a,b for a typical structure of this device and its energy band arrangement, holes can be injected from p-Si to $CsPbBr_3$ quantum dots using forward bias, thus breaking the energy band potential barrier in the valence band. Meanwhile, electrons can be injected through the ZnO layer and then recombine with the holes in the CsPbBr quantum dots.

Although this type of device produces light only under forward bias when driven with AC, the AC driving method has higher EL intensity and better stability than the DC driving method in this system. There are two reasons for the improved device performance with the AC drive method. One is that heat generation can be suppressed at high current densities because the device has a shorter runtime compared to the DC mode. The other reason is the reduced charge accumulation in the defect state due to the frequent reversal of the applied bias voltage. The bound charge can be extracted in each cycle, thus reducing defect generation and also contributing to the durability of the device.

10.3 Optimization Strategies for AC QDLEDs

Here are some optimization strategies that could potentially improve the performance of AC QDLEDs:

QD material optimization: The performance of QDLEDs is strongly influenced by the properties of the QD materials, including size, shape, composition, and surface ligands. Optimizing the QD synthesis and surface treatment processes can improve the quantum yield, stability, and charge transport properties of QDLEDs.

Device engineering: Optimizing the device structure, such as the layer thickness, doping concentration, and electrode materials, can improve the charge injection, transport, and recombination efficiency of QDLEDs. For example, using a thin and highly doped layer between the electrode and the QD layer can reduce the contact resistance and improve the carrier injection.

AC signal optimization: The performance of AC QDLEDs is strongly influenced by the frequency, amplitude, and waveform of the AC signal. Optimizing the AC signal parameters, such as the frequency, duty cycle, and amplitude, can improve the efficiency and stability of AC QDLEDs. For example, using a higher frequency or duty cycle can reduce the time for carrier accumulation and reduce the carrier recombination in the QD layer.

Surface modification: Surface modification of QDs and electrodes can improve the charge transfer and reduce the interfacial barrier of the device. For example, using a self-assembled monolayer on the electrode can improve the contact between the electrode and the QD layer.

Hybridization with other materials: Hybridizing QDs with other materials, such as polymers or small molecules, can improve the charge transport and light extraction efficiency of QDLEDs. For example, using a polymer matrix can improve the charge transport and prevent the aggregation of QDs.

Structural modification: Structural modification of the device, such as using a microcavity or distributed Bragg reflector, can improve the light extraction efficiency and reduce the optical loss of the device.

These strategies can be combined and optimized to achieve better performance of AC QDLEDs, and the specific optimization parameters may depend on the device structure and materials used.

Due to the difference in principle, AC QDLEDs should have the following advantages over DC QDLEDs. Under DC driving conditions, charge accumulation occurs at the electrode/active material interface state. The charge accumulation shields the applied electric field and reduces the effective injection. However, by applying an AC bias, the average amount of accumulated charge decreases due to frequent reversal of the applied electric field, which may reduce trilinear state exciton annihilation at high current densities. As a result, the device can be more effective and resistant to higher voltages. Under DC driving conditions, electrochemical reactions may occur between the active material and the electrodes. The thicker dielectric layer in AC QDLEDs can isolate the device from external moisture and oxygen, while the intermediate dielectric layer can effectively prevent such electrochemical reactions. Thus, the lifetime and stability of the device are improved. DC drivers connected to 110/220 V, 50/60 Hz residential AC power require complex back-end electronics such as power converters, and rectifiers, so losses are inevitable. However, AC QDLEDs can be easily integrated into AC power lines without any power loss. Band matching must be considered for DC carrier injection. However, field-induced AC QDLEDs avoid such troubles. Light-emitting materials with different chemical compositions and energy levels can be doped into the same device as needed so that the light-emitting peaks can be modulated from the visible region to the near-infrared region. As mentioned earlier, AC QDLEDs can be classified into three different types based on their principles and, therefore, different strategies to optimize their performance.

10.3.1 Optimization of the Field-Induced AC QDLED

There are several strategies that can be employed to optimize the performance of field-induced AC QDLEDs:

Optimization of quantum dot properties: The properties of the quantum dots, such as size, shape, and surface chemistry, can be optimized to achieve a higher luminescence efficiency and longer radiative recombination lifetime. For instance, larger QDs may lead to stronger Coulomb coupling and a higher degree of confinement, resulting in a larger exciton binding energy and more efficient radiative recombination.

Engineering the dielectric layers: The dielectric layers can be engineered to optimize the carrier injection and extraction efficiency. The dielectric layers can also be designed to facilitate charge transport, improve carrier confinement, and reduce electron leakage. The optimization of the thickness and the quality of the dielectric layer can significantly impact the devices performance.

Optimization of device structure: The device structure can be optimized to improve the charge carrier injection, transport, and extraction efficiency. For example, the use of a graded or interfacial layer between the QDs and the electrodes can facilitate charge injection and reduce the interface recombination.

Carrier balance control: The balance between the electron and hole injection and transport can be controlled to maximize the radiative recombination efficiency. Various methods, such as doping, use of buffer layers, and optimizing the energy levels of the transport layers, can be used to optimize the carrier balance.

AC voltage optimization: The AC voltage parameters, such as amplitude, frequency, duty cycle, and waveform, can be optimized to improve the devices performance. For instance, higher voltage amplitude and frequency can increase the probability of carrier generation and reduce the charging time of the quantum dots.

Surface passivation: The surface of the quantum dots can be passivated to reduce surface defects and trap states, which can lead to a higher luminescence efficiency and longer radiative recombination lifetime. Surface passivation can be achieved by using organic ligands or inorganic shells.

Device fabrication and packaging: The device fabrication and packaging process can significantly impact the devices performance. The use of advanced fabrication techniques and packaging materials can minimize device degradation and increase the devices stability and lifetime.

By combining these optimization strategies, the performance of field-induced AC QDLEDs can be significantly improved, leading to their widespread applications in the fields of optoelectronics and solid-state lighting.

Although field-driven QDLEDs can eliminate the design problems of traditional DC QDLEDs in terms of energy level matching and create a new type of QDLED, such devices still need further optimization. Since field-driven QDLEDs are relatively simple in structure and can use the same emitting layers as DC QDLEDs, some optimization strategies for DC QDLEDs can also be applied to field-driven QDLEDs. To achieve efficient AC field-driven QDLEDs, two main factors need to be considered. First, the distribution of carriers in the emitting layer must be controlled to increase the light intensity. Second, the dielectric constant of the insulating layer must be high in order to reduce the electric field strength required for ionization, reduce the driving voltage, and increase the breakdown voltage. The dielectric layer and the light-emitting layer are two important parts of the field-driven AC QDLED, so there are two main methods to improve the performance of the field-driven AC QDLED.

10.3.1.1 Dielectric Layer Optimization

To improve the performance of the device, including brightness and efficiency, it is important to increase the field strength at the location of the light-emitting layer, rather than increasing the field strength of the entire device. When an AC bias voltage is applied to the device, the dielectric is polarized, and the applied voltage can be divided between the devices based on the capacitance of each layer. All the capacitive elements (all the layers between the top and bottom electrodes) constitute the total capacitance of the device. The device will not emit light until the applied voltage is high enough. This threshold is called the AC bias threshold for light emission. When the applied bias is less than the threshold AC bias, the voltage drop between the emitting layers is not sufficient to ionize the QD and no luminescence can be observed. When the applied bias exceeds the threshold AC bias, the quantum dots ionize in a high electric field to form excitons, resulting in light emission.

The use of a dielectric layer with a high dielectric constant material allows the external field to be concentrated on the light-emitting layer and reduces the threshold AC bias.

The ionization of field-driven AC QDLEDs generates local carriers inside the emitting layer, eliminating the need for charge injection or remote carrier transport. As can be seen, the quantum dot layer can also be discontinuous, and even quantum dots can be mixed with dielectric materials to make field-driven QDLEDs, such as a quantum dot matrix in a transparent polymer film.

Wood et al. demonstrated that active layers consisting of QD clusters embedded in insulating polymers can be used in electrically excited emission devices as well, and are not limited to colloidal QDs for photoexcitation applications, to take advantage of the higher thin-film PL efficiency obtained when QDs are dispersed.

The QD ligand and the insulating polymer matrix, as well as the QD concentration, will determine the size and spatial segregation of the resulting QD clusters in the QD-polymer film. Figure 10.7a shows that increasing the mass fraction of QDs in the active layer corresponds to increasing the cluster size in the QD-polymer blend and decreasing the PL efficiency of the blend film. This result is expected because larger QD clusters are more likely to contain QDs with trap state defects that can quench the luminescence on any adjacent QD.

However, although the QD film has the highest PL efficiency $(70 \pm 5)\%$ at low QD concentrations, no significant EL signal could be observed under these conditions. This may be due to the fact that the ionization process of quantum dots requires at least multiple quantum dots in close proximity to each other, and only when this condition is satisfied can electrons be extracted from one quantum dot and transferred to the neighboring ones. Thus, more ionization processes can occur only in larger clusters of QDs because more QDs can transfer charges to their neighboring QDs.

Figure 10.7b shows the voltage drop across the sensing capacitor Vc using two different devices. It can be seen that more charge accumulates in the device with a pure QD film than in the device with a QD cluster. These preliminary observations of QD clusters inside insulators confirm that electric field-driven luminescence in QD films is a highly localized process that does not require long-range transport in the QD film.

To further verify this, the researchers attempted to construct similar LED structures using light-emitting organic molecular films prepared by co-evaporation of dopant/wide-bandgap matrix systems, which are similar to QD/polymer blends. Molecular films of *fac* tris(2-phenylpyridine)iridium (Irppy3) embedded in the main body of bis(triphenylsilyl)benzene (UGH2) were prepared. For the films containing 25,50,75, and 100 w% Irppy3, the photoluminescence (PL) spectra showed progressive red-shift spectral data, indicating more aggregation with increasing mass fraction of Irppy3. The PL and EL responses of molecular organic films with the main body/dopant system show the same trend as QD films with insulating substrates, which confirms that larger clusters can provide more adjacent molecules and ionized electrons can tunnel through organic molecules and QD films.

The variation of field strength with the dielectric constant of the substrate at a single QD grain location can be approximated by Eq. (10.1) as:

$$E_Q = E_{\text{matrix}} \left[\frac{3\varepsilon_1}{2\varepsilon_1 + \varepsilon_1 - \varphi(\varepsilon_2 - \varepsilon_1)} \right] \tag{10.1}$$

where the quantum dots are approximated as regular spherical particles with relative permittivity ε_2 surrounded by a dielectric matrix with permittivity ε_1. Ematrix = U/d is the average electric field on the matrix (the QD matrix in the dielectric film), while φ is the volume fraction of the QD.

From Eq. (10.1), it is clear that we need a dielectric material with a high dielectric constant to reduce the emission threshold voltage and improve the performance of field-driven AC QDLEDs for both laminar dielectric-QDs systems and matrix dielectric-QDs systems. Wood et al. have shown that the threshold voltage of devices using silica as an insulating layer is greater than that of devices using alumina as an insulating layer due to the fact that the dielectric constant of silica ($\varepsilon \sim 3.9$) is smaller than that of alumina ($\varepsilon \sim 9$) [5]. The commonly used dielectric layer materials in field-driven AC QDLEDs are Al_2O_3 [5, 8], SiO_2 [5], HfO_2 [8], TaO_x [9], etc.

It has also been reported that by adding an electron-blocking layer to the device, the electric field required for electroluminescence can be obtained with reduced current density through the quantum dot layer, and there is little burst of PL on the QD film with voltage change [9, 10]. Figure 10.8 shows a typical approach to optimize the performance of field-driven QDLEDs by introducing a thin film of ceramic material (ZnS) to create a potential barrier for most carriers, limiting the current through the quantum dot layer, and achieving stable quantum dot electroluminescence for long periods of time in air without encapsulation.

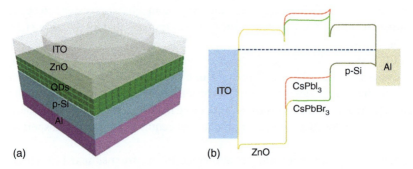

Figure 10.8 (a) Typical structure of AC/DC dual-drive mode QDLED. (b) Energy band diagram of AC/DC dual-drive mode QDLED.

A typical current density versus voltage (J–V) is given in Figure 10.8b. Although device structures 2A, 2B, and 2C (with ZnS barrier films) require higher voltages than structure 1 (without ZnS barrier films), the current densities in structures 2A, 2B, and 2C are significantly smaller than those in structure 1. This data suggests that the ZnS film limits the QD charge and prevents the luminescence burst of the QD layer at high current densities.

10.3.1.2 Quantum Dot Layer Optimization

AC-driven QDLEDs use a similar layer of QDs as conventional DC-driven QDLEDs, so the optimization strategy is also similar to the QDs in DC QDLEDs.

Kobayashi et al. used a low-temperature preparation of inorganic AC QDLEDs in which a CdSe/ZnS quantum dot layer was used as the light-emitting layer, formed by a new ion beam deposition process [11]. An ion-beam direct deposition technique called liquid dispersed quantum dots ion-beam deposition (LIQUID) was used to form an approximately 50 nm thick QD-structured luminescent layer on the ZnS buffer layer. An important feature of this technique is the formation of surfactant-free polycrystalline films from pre-synthesized quantum dots at relatively low temperatures, maintaining significant luminescence properties. They chose a field-driven AC quantum diode structure to achieve precisely matched electroluminescence properties, photon energies, and linewidths for PL from source solutions. Compared to typical luminescence centers consisting of a single ion (e.g. manganese), high-field accelerated electrons are expected to have larger collision cross-sections in this structure due to the larger geometry of the wave function of electrons in the QD.

Similar work has been done by Omata et al. [9]. They fabricated inorganic multilayer thin-film QDLEDs using the same LIQUID method as Kobayashi et al. See Figure 10.9a and achieved ultraviolet (UV) electroluminescence (EL) at 3.30 eV using colloidal ZnO QDs as the light-emitting layer. The EL spectrum is the same as the PL spectrum of the source solution of ZnO QDs, as shown in Figure 10.9b, and can be explained such an electroluminescence phenomenon in terms of quantum-limited electron–hole pair complexes. The thin MgO layer on both sides of the ZnO QD layer is the key to obtain the UV-EL emission, while

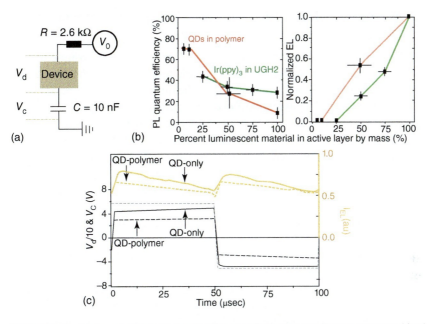

Figure 10.9 (a) A measurement circuit. This circuit studies the charge generated in the active layer of the device by measuring the voltage across the device and the voltage across the sense capacitor (10 nF) connected in series with the device. The resistor (2.6 kΩ) is used to minimize damage to the device electrodes during sudden voltage pulses. (b) Plot of photoluminescence (PL) efficiency and normalized EL intensity with the proportion of luminescent material present in the active layer. Two different semiconductor nanoscale systems, i.e. quantum dots embedded in an insulating polymer matrix (red line) and small organic molecules embedded in a molecular organic body matrix with a wide bandgap (green line), show the same trend, i.e. a decreasing trend in photoluminescence efficiency and an increasing trend in photoluminescence response with increasing emitter mass fraction (which is related to the increase in quantum dot or luminescent molecular cluster size). (c) Time-resolved data plots for two different devices: one with a luminescent layer consisting of QDs only (solid line) and the other with a QD-polymer co-blended active layer (dashed line). The applied voltage was chosen so that an electric field intensity of about 3.3×10^5 kV m^{-1} was dropped on both the QD and QD-polymer layers. The applied waveform for the QD-only device is shown as a gray dashed line. The electroluminescence intensity is shown in orange, and the voltage across the sense capacitor is shown in black.

only the visible emission associated with defects appears without the MgO layer. This is due to the fact that the MgO layer can help achieve surface passivation and form quantum well structures. A series of low-toxicity, cadmium-free colloidal tripartite chalcopyrite I-III-VI2 QDs, and QD-activated field-driven AC QDLEDs using the LIQUID method to reduce residual organic matter in the QD film were also developed by Omata et al. [12]. A schematic diagram of the LIQUID system used in the above work is shown in Figure 10.9c.

10.3.2 Optimization of Half-Field-Driven Half-Injected AC QDLEDs

Half-field-driven half-injected AC QDLEDs are a type of QDLED that has been shown to have promising properties such as high efficiency and color purity. Here are some optimization strategies that could potentially improve the performance of these devices:

Optimization of quantum dot size and composition: The size and composition of the quantum dots can have a significant impact on the performance of the device. By carefully selecting the size and composition of the quantum dots, it is possible to tune the emission wavelength and improve the efficiency of the device.

Optimization of the injection layer: The performance of the device can be improved by optimizing the injection layer, which is responsible for injecting electrons and holes into the quantum dot layer. The injection layer can be optimized by selecting materials with appropriate energy levels to facilitate efficient charge injection.

Optimization of the electrode materials: The electrode materials used in the device can also affect the performance. By selecting appropriate materials with high conductivity and low work function, it is possible to reduce the series resistance and improve the overall efficiency.

Optimization of the device structure: The device structure can also be optimized to improve performance. For example, using a graded heterojunction structure can improve the injection efficiency and reduce the recombination of charge carriers.

Optimization of the driving voltage and frequency: The performance of the device can be improved by optimizing the driving voltage and frequency. By selecting appropriate driving parameters, it is possible to achieve higher efficiency and faster response time.

Overall, optimizing the quantum dot size and composition, injection layer, electrode materials, device structure, and driving parameters can all contribute to improving the performance of half-field-driven half-injected AC QDLEDs.

Xia et al. introduced P(VDF-TrFE-CFE) as a dielectric layer for a half-field-driven half-injected AC QDLED because of its higher dielectric constant than commonly used polymer dielectrics such as polyvinyl phenol (PVP) and some inorganic oxide dielectrics including SiO_2 and Al_2O_3 [7]. The device structure is shown in Figure 10.10a. The importance of selecting dielectric materials with high dielectric constants was introduced in the previous section, and the principle is similar for the dielectric materials in the half-field-induced half-injection AC QDLED. They chose P(VDF-TrFE-CFE) as the dielectric layer because P(VDF-TrFE-CFE) can be prepared by the solution method, and therefore AC-driven electroluminescent devices prepared by the full solution method can be realized. They demonstrated that the brightness of the device is strongly influenced by the thickness of the dielectric layer because it affects the overall capacitance of the device and thus determines the charge storage capacity of the device.

$$C = \frac{\varepsilon_0 s}{4\pi k d} \tag{10.2}$$

where ε_0 is the vacuum dielectric constant, and s, k, and d represent the surface area, the relative dielectric constant, and the thickness of the dielectric layer, respectively. From Eq. (10.2), it can be seen that the thinner the dielectric layer, the higher the capacitance, which means that more charge can be accumulated and thus more electrons can be injected, and thus more excitons can be generated. However, if the

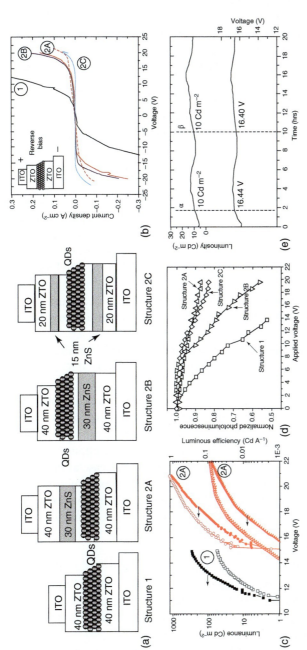

Figure 10.10 (a) Schematic cross-sectional view of the device structure described in this part. Structure 1 consists of a colloidal QD sandwiched between two layers of ZnO:SnO$_2$ (ZTO). Structures 2A, 2B, and 2C have at least one layer of ZnS added within the ZTO layer 15 nm from the QD film. ZnS can be located above the QD in the ZTO (structure 2A), below the QD (structure 2B), or on either side of the QD (structure 2C). (b) Current density versus voltage characteristics of device structures 1 (black line), 2A (red line), 2B (purple line), and 2C (blue line) under forward and reverse DC bias conditions. The inset indicates the voltage direction convention. (c) Luminance (solid line symbols) and luminous efficiency (hollow symbols) for device structure 1 (black) and structure 2A (red). The data for optimized structure 2A are shown in red with circle symbols. (d) Normalized photoluminescence (PL) as a function of voltage applied in forward bias for device structures 1 (square), 2A (upward-pointing triangle), 2B (downward-pointing triangle), and 2C (diamond). In the case of forward-biased device structures 2A and 2C, the electrons injected into the ZTO encounter the ZnS layer before the QD film. In contrast, the electrons in the ZTO in structures 1 and 2B do not encounter the ZnS layer and do not accumulate on the QD film. Thus, device structures 2A and 2C show fewer PL bursts at increasing voltages compared to structures 1 and 2B, suggesting that the accumulation of electrons at the ZnS layer rather than at the QD film reduces the charging of the QD layer. Device structures 1 and 2B have the same trend of PL quenching, but the voltage is shifted due to the presence of an insulating ZnS layer in structure 2B. This figure illustrates that the degree of electron accumulation on the QD film, rather than the electric field at the ends of the QD, determines the amount of QD charging and luminescence burst. (e) Plot of luminance (upper axis, left axis) and voltage (lower axis, right axis) versus time for device structure 2A operated at a constant current of 30 mA cm^2 under forward bias. The device has been unpacked and operated in air.

dielectric layer is too thin, a DC current may be injected due to the poor insulating ability of the thin dielectric layer, leading to the breakdown of the device at high voltage. By comparing the brightness-frequency characteristics of devices with different thicknesses of P(VDF-TrFE-CFE), they found that the optimal thickness of P(VDF-TrFE-CFE) is 680 nm, as shown in Figure 10.9b. Further increasing the thickness of P(VDF-TrFE-CFE) leads to a decrease in capacitance, which reduces the amount of stored charge and eventually reduces the brightness of the device.

10.3.2.1 Charge Generation Layer Optimization

As we mentioned above, due to the operating principle of such AC-QDLED devices, it is critical and challenging to enhance the charge concentration balance in order to achieve low drive voltage, high brightness, and high-power efficiency. Therefore, the ability of the charge generation layer to generate charges is another key factor affecting the devices performance.

Xia et al. chose MoO_3/TFB hole-generating layer to generate holes inside the device [7]. As shown in Figure 10.10c, due to the perfect alignment of the MoO_3 CB with the highest occupied molecular orbital (HOMO) energy level of the TFB, electrons in the HOMO of the TFB can be transferred to the CB of MoO_3, which leads to the generation of holes in the poly(9,9-dioctylfluorene-alt-N-(4-sec-butylphenyl)-diphenylamine) (TFB). Then, the injected electrons can complex with the generated holes to produce excitons, leading to the generation of photoemission. Figure 10.10c shows the luminescence frequency characteristics of the device at different MoO_3 thicknesses.

The device without MoO_3 shows a very low brightness due to the inability to generate holes efficiently. By introducing MoO_3, the electrons in the HOMO from the TFB can be efficiently transferred to the CB of MoO_3, thus generating holes in the TFB. The device with 9 nm MoO_3 shows the best performance. When driven with a 60 V square pulse voltage at 250 kHz, it has the highest brightness of 65 000 cdm^{-2}. When the thickness of MoO_3 is further increased, the performance decreases. This may be because the hole generation capacity is saturated and the number of holes generated in the hole generation layer (HGL) will no longer increase as the MoO_3 layer is increased. Or, in this case, the number of injected electrons cannot catch up with the number of holes generated in the HGL, and the insulating property of the thick MoO_3 layer becomes the main factor affecting the devices performance.

10.3.2.2 Tandem Structure

Due to the unique principle of the half-field-induced semi-injected AC QDLED, the performance can be improved by connecting in series an orthogonal and an inverted half-field-induced semi-injected AC QDLED to produce two regions of light [7]. As shown in Figure 10.8d, the symmetric HGLs on both sides of the dielectric layer generate cavities under forward and reverse bias.

Half-field-driven semi-injected AC QDLEDs with a series structure can generate light during both the positive and negative half-drive cycles. In the positive half-drive cycle, the positive QDLED exciton recombines to emit light, when the inverting QDLED discharges. In the upcoming negative half-drive cycle, the

10.3 Optimization Strategies for AC QDLEDs

Table 10.1 Key Performance Data for Forward, Inverted, and Series AC QDLEDs.

Device	Structure	Light-emitting half cycle	Luminance (cd m^{-2}) Peak	@1000 mA cm^{-2}	Current efficiency (cd A^{-1}) Peak	@1000 mA cm^{-2}
A	Regular	Positive	65 760	39 200	4.4	3.9
B	Inverted	Negative	30 570	19 250	2.1	1.9

operating processes of the two devices are exchanged, i.e. the inverted QDLED emits light while the positive QDLED discharges, thus ensuring that emitting occurs throughout the drive cycle.

This is equivalent to using the discharge current to emit light. The efficiency of the series device is increased by 30% compared to the conventional device due to the increase in the luminescence time-share. Figures 10.8e,f compare the luminance and current efficiency of the forward device, inverted device, and series device. The key performance data of the devices is shown in Table 10.1.

10.3.2.3 AC/DC Dual Drive Mode QDLED Optimization

There are still relatively few reports on the optimization of this type of device, but this type of device has achieved relatively low start-up voltage and power consumption and can achieve good operating stability in AC mode. Further improvements in brightness and luminescence efficiency can be expected in the future by optimizing the device structure, such as passivation of quantum dots, film thickness control, and the introduction of hole transport layers.

Here are some optimization strategies for improving the performance of AC/DC dual-drive mode QDLED:

(1) Optimization of device structure: The device structure plays a critical role in determining the performance of QDLEDs. In AC/DC dual-drive mode QDLEDs, the design, and optimization of the device structure can improve the efficiency of charge transport, reduce charge trapping, and improve the injection and extraction of carriers. For example, optimizing the thickness and composition of the transport layers and using interfacial modification techniques can improve the devices performance.

(2) Optimization of QD materials: The properties of QD materials, such as size, shape, and composition, can significantly affect the performance of QDLEDs. The optimization of QD materials can improve the charge transport and exciton formation, resulting in a higher quantum yield and brightness. For example, using highly luminescent QDs, such as CsPbBr3, or modifying the surface chemistry of QDs can improve the devices performance.

(3) Optimization of charge injection: Efficient charge injection is essential for high-performance QDLEDs. In AC/DC dual-drive mode QDLEDs, the optimization of charge injection can improve the overall device performance. For

example, using a graded doping profile or using a charge-blocking layer can improve the charge injection efficiency.

(4) Optimization of carrier balance: In QDLEDs, the balance between electrons and holes plays a crucial role in determining the devices performance. The optimization of carrier balance can improve the device performance, such as reducing the turn-on voltage and increasing the efficiency. For example, using a dual charge injection layer or a hole transport layer with a high mobility can improve the carrier balance and device performance.

(5) Optimization of electrical and optical properties: The optimization of the electrical and optical properties of QDLEDs can improve the devices performance. For example, optimizing the voltage waveform and frequency in AC/DC dual-drive mode QDLEDs can improve the light output and power efficiency. Additionally, using optical structures, such as distributed Bragg reflectors or microcavities, can improve the light extraction efficiency and directionality.

10.3.3 Conclusion and Future Direction of AC-QDLED

AC-driven QDLEDs are a relatively new area of research, and there is ongoing work aimed at improving their performance and understanding their fundamental properties. Future research directions for AC-driven QDLEDs will be:

1) Improving device efficiency: One of the main goals of QDLED research is to improve the efficiency of the devices, which is currently lower than that of traditional OLEDs. Researchers are exploring new materials and device architectures to improve the efficiency, such as using different types of quantum dots, optimizing the thickness of the device layers, and engineering the device interfaces.

2) Understanding the fundamental properties of QDLEDs: There is ongoing research aimed at understanding the fundamental properties of QDLEDs, such as the dynamics of charge and energy transfer processes within the device. This knowledge can be used to design new device architectures and improve device performance.

3) Developing new fabrication methods: Researchers are exploring new methods for fabricating QDLEDs, such as solution-based methods that are more scalable and cost-effective than traditional vacuum deposition techniques.

4) Exploring new device applications: QDLEDs have potential applications in a range of fields, such as displays, lighting, and sensing. Researchers are exploring new device architectures and materials to enable these applications.

5) Developing AC-driven QDLEDs for neuromorphic computing: Recently, researchers have demonstrated that QDLEDs can be used for neuromorphic computing, which is a type of computing that mimics the way the brain processes information. Future research in this area will focus on developing AC-driven QDLEDs that can be integrated into neuromorphic computing systems.

6) Developing flexible and stretchable QDLEDs: Flexible and stretchable QDLEDs have potential applications in wearable devices and flexible displays. Researchers are exploring new materials and device architectures to develop QDLEDs that can withstand mechanical stress and deformation.

7) Investigating the impact of AC driving on quantum dot properties: AC driving can affect the optical and electronic properties of quantum dots, such as their emission spectra and charge transfer dynamics. Researchers are investigating the impact of AC driving on quantum dot properties and exploring how to optimize device performance by tuning these properties.
8) Developing quantum dot-based lighting technologies: QDLEDs have potential applications in lighting, as they can be engineered to emit light in a specific range of wavelengths. Researchers are exploring the use of QDLEDs in lighting applications and investigating new device architectures to enhance the performance of these devices.

Overall, the field of AC-driven quantum dot light-emitting devices is still in its early stages, and there is much to be learned about these devices. However, the potential applications and benefits of these devices make them an exciting area of research for the future.

References

1 Supran, G.J., Shirasaki, Y., Song, K.W. et al. (2013). QLEDs for displays and solid-state lighting. *MRS Bulletin* 38 (9): 703–711.
2 Yu, S., Hu, J., Zhang, H. et al. (2022). Recent progress in AC-driven organic and perovskite electroluminescent devices. *ACS Photonics* 9 (6): 1852–1874.
3 Choi, M.K., Yang, J., Hyeon, T., and Kim, D.-H. (2018). Flexible quantum dot light-emitting diodes for next-generation displays. *npj Flexible Electronics* 2 (1): 10.
4 Lee, S.-B., Fujita, K., and Tsutsui, T. (2005). Emission mechanism of double-insulating organic electroluminescence device driven at AC voltage. *Japanese Journal of Applied Physics* 44 (9R): 6607.
5 Wood, V., Panzer, M.J., Bozyigit, D. et al. (2011). Electroluminescence from nanoscale materials via field-driven ionization. *Nano Letters* 11 (7): 2927–2932.
6 Sun, W., Xie, L., Guo, X. et al. (2020). Photocross-linkable hole transport materials for inkjet-printed high-efficient quantum dot light-emitting diodes. *ACS Applied Materials & Interfaces* 12 (52): 58369–58377.
7 Xia, F., Sun, X.W., and Chen, S. (2019). Alternating-current driven quantum-dot light-emitting diodes with high brightness. *Nanoscale* 11 (12): 5231–5239.
8 Wood, V., Halpert, J.E., Panzer, M.J. et al. (2009). Alternating current driven electroluminescence from ZnSe/ZnS:Mn/ZnS nanocrystals. *Nano Letters* 9 (6): 2367–2371.
9 Omata, T., Tani, Y., Kobayashi, S. et al. (2012). Ultraviolet electroluminescence from colloidal ZnO quantum dots in an all-inorganic multilayer light-emitting device. *Applied Physics Letters* 100 (6): 061104.
10 Wood, V., Panzer, M.J., Caruge, J.-M. et al. (2009). Air-stable operation of transparent, colloidal quantum dot based LEDs with a unipolar device architecture. *Nano Letters* 10 (1): 24–29.

11 Kobayashi, S., Tani, Y., and Kawazoe, H. (2007). Quantum dot activated all-inorganic electroluminescent device fabricated using solution-synthesized CdSe/ZnS nanocrystals. *Japanese Journal of Applied Physics* 46 (40): L966–L969.

12 Omata, T., Tani, Y., Kobayashi, S., and Otsuka-Yao-Matsuo, S. (2012). Quantum dot phosphors and their application to inorganic electroluminescence device. *Thin Solid Films* 520 (10): 3829–3834.

11

Stability Study and Decay Mechanism of Quantum Dot Light-Emitting Diodes

11.1 Quantum Dot Light-Emitting Diode Stability Research Status

Quantum dot (QD) materials are widely recognized for their excellent properties, such as high photoluminescence quantum yield (PLQY), good monochromaticity, controllable particle size, continuously tunable emission spectra, good thermal and photochemical stability, and solution processability. Since their first introduction in 1994, quantum dot light-emitting diodes (QDLEDs) have received significant research attention due to their promising potential in display and lighting applications [1–3]. Conventional white LED backlights used in LCDs typically employ yellow phosphors as down-converters, which are limited to displaying a color gamut covering only 70% of the National Television Systems Committee (NTSC) standard. However, quantum dot materials offer a wider color gamut and excellent monochromaticity, making QDLEDs the most promising candidates for next-generation full-color displays and solid-state lighting [1–6].

At the outset of the development of QDLEDs, there was little optimism about their potential for use in display applications, mainly due to the extremely low external quantum efficiency (EQE, less than 1%) and a maximum brightness of only $100\,cd\,m^{-2}$ displayed by early QDLEDs, which paled in comparison to the performance of OLEDs (short for organic light-emitting diodes) developed by C.W. Tang. However, with the maturation of OLED technology after 2000, and the experience gained from the optimization of OLED structures and understanding of their working mechanisms, QDLED technology has rapidly advanced, with increasing efficiency and improved performance. As a result, QDLEDs are now considered a promising technology for next-generation displays and solid-state lighting, thanks to their wider color gamut, good monochromaticity, tunable emission spectra, and other excellent properties [7, 8]. To date, research has shown that the EQE of the three primary colors of red, green, and blue QDLEDs can all exceed 20%, meeting the requirements for high luminous efficiency. Additionally, QDLEDs can be produced in a cost-effective and efficient manner, making them an ideal choice for large-screen displays. However, one of the biggest challenges for large-scale industrial production of QDLEDs is their stability, which remains

Colloidal Quantum Dot Light Emitting Diodes: Materials and Devices, First Edition. Hong Meng.
© 2024 WILEY-VCH GmbH. Published 2024 by WILEY-VCH GmbH.

Table 11.1 Summary of stability studies of some red-, green-, and blue-QDLEDs [10–17].

Device structure	QD	Emission peak (nm)	Turn-on voltage (V)	Peak EQE (%)	Life time (h) (T_{50} at 100 cd m^{-2})
ITO/ZnO/QDs/NPB/LG101/Al	CdSe/CdS	620	1.7	18.5	4000
ITO/PEDOT: PSS/p-TPD/PVK/QDs/PMMA/ZnO/Ag	CdSe/CdS	640	1.7	20.5	100 000
ITO/PEDOT: PSS/TFB/QDs/ZnO/Ag	CdSe/CdS	536	2.8	15.45	3513
ITO/PEDOT: PSS/TFB/QDs/ZnO/Al	$Zn_xCd_{1-x}S_ySe_{1-y}$	526	2.1	21.0	90 000
ITO/PEDOT: PSS/PVK/QDs/ZnO/Al	CdSe/ZnS	468	5.1	19.8	47.4
ITO/PEDOT: PSS/PVK/QDs/ZnO/Al	$ZnCdS/Cd_xZn_{1-x}S/ZnS$	445	6.5	18.0	47.4
ITO/PEDOT: PSS/p-TPD/PVK/QDs/ZnMgO/Ag	CdSe/CdZnS	624	1.7	18.2	190 000
	CdZnSeS/ZnS	526	2.2	18.1	—

Where lifetime is the time required for the device to decay by half at a brightness of 100 cdm^{-2}.

an area of active research and development [9, 10]. Table 11.1 presents the stability results of selected red, green, and blue QDLEDs. The lifetime of red and green QDLEDs can surpass 10 000 hours, with a maximum T50 of 190 000 hours, while the T50 of blue QDLEDs is only 47.4 hours. In the literature, the longest T50 of OLEDs can reach 1 000 000 hours (with an initial brightness of 1000 cd m^{-2}). Thus, it is apparent that the stability issue of QDLEDs is still a significant challenge to overcome.

11.2 Factors Affecting the Stability of Quantum Dot Light-Emitting Diodes

To gain a better understanding of the stability of QDLEDs, it is important to have a clear understanding of their structure and working mechanism. The typical structure of a QDLED consists of an ITO cathode, an electron injection layer, an electron transport layer (ETL), a QD light-emitting layer, a hole transport layer (HTL), a hole injection layer, and a metal anode [18]. Figure 11.1a shows the schematic of a QDLED structure under the applied bias, where electrons are injected from the cathode, pass through the ETL, and reach the quantum dot light-emitting layer, while holes are injected from the anode, pass through the HTL, and reach the quantum dot light-emitting layer, where they combine with electrons and emit light, as shown in Figure 11.1b. Figure 11.2 summarizes the materials commonly used in each functional layer of a QDLED, and the selection of appropriate materials for each functional layer is crucial to improve device efficiency and stability. To study the factors that affect the stability of QDLEDs, it is necessary to investigate the impact of each functional layer on the device. In the following sections, we will discuss the factors affecting the stability of QDLEDs from three aspects.

11.2.1 Quantum Dot Light-Emitting Layer

Thanks to their high luminescence color purity, tunable emission wavelength, high PLQY, and excellent intrinsic stability, QD materials are well-suited for use

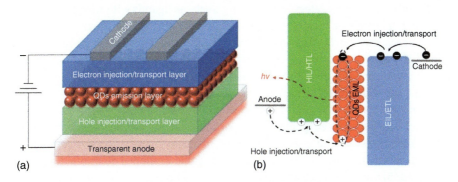

Figure 11.1 (a) Structure of quantum dot light-emitting diode; working mechanism (b).

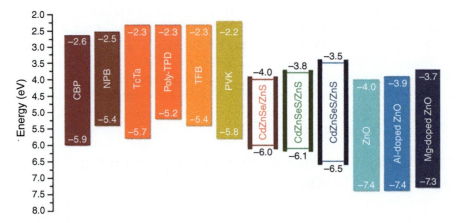

Figure 11.2 Common materials for the functional layer of quantum dot light-emitting diodes.

as the light-emitting layer in high-brightness LEDs. However, in the early stages of quantum dot synthesis, only core CdSe quantum dots were available with a limited emission range, low PLQY, and poor optical properties. Subsequently, researchers developed a method to encapsulate wide-bandgap shells such as CdS, ZnS, and ZnSe around the CdSe core. The inorganic shell layer can effectively passivate the surface defects caused by dangling bonds, significantly improving the PLQY and stability of the quantum dots.

To gain more insight into the effect of the core–shell structure on quantum dot performance, Lim et al. investigated the relationship between photoluminescence (PL)/electroluminescence (EL) performance and shell thickness of CdSe/$Zn_{1-x}Cd_xS$ quantum dots. They found that the lifetime of quantum dot materials improved with the increase in the shell layer thickness, and the EQE of the films showed a clear trend of gradual improvement. This indicates that the energy loss caused by interdot fluorescence diffraction can be significantly suppressed with the increase in the shell layer thickness [19].

To overcome the challenge of stability in quantum dot materials, ligand engineering is another effective strategy. By replacing unstable ligands with strongly binding ones, such as thiols or oleates, the quantum dots can maintain their initial state even after repeated purification cycles. As illustrated in Figure 11.3, high-brightness CdSe/CdS/ZnS core–multishell QDs without protective atmosphere were prepared in paraffin and oleate media using a phosphine-free precursor injection method. The CdSe/CdS/ZnS core–multishell quantum dots were coated with polymers (PEG, PVA, PVP, and PAA) to improve their stability. Furthermore, the core–multishell-structured quantum dots exhibited enhanced emission in the range of 355 410 nm by suppressing defect-sensitive nuclei and non-radioactive complexes in the PL spectrum. XRD analysis revealed that cubic Zn-doped quantum dots with a grain size in the range of 2244 nm were obtained, and the absorption spectra of all samples showed good UV absorption in the range of 302–380 nm [20].

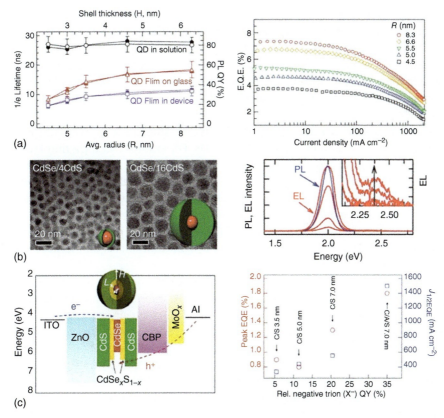

Figure 11.3 PL decay lifetime (filled) and PLQY (empty) and EL efficiency of QDLEDs (right) based on QDs with different total radii for CdSe/Zn$_{1-x}$Cd$_x$S quantum dots with different shell layer thicknesses in different states (left). (b) Comparison of TEM images of CdSe-core QDs with Cd shell layer thickness of 4 and 16 layers, respectively (left), and PL spectrum of active layer with EL spectrum of QDLED at elevated driving voltage (right). The inset shows the EL spectra of the high-energy partial amplification.

11.2.2 Hole Transport Layer

The HTL is an important component in QDLEDs, as it facilitates the injection of holes from the anode into the QD light-emitting layer. The HTL also needs to provide good hole transport and block electrons to prevent quenching of the QD emission. The choice of HTL material and structure can have a significant impact on the stability and performance of QDLEDs.

Several materials have been investigated for use as HTLs in QDLEDs, including poly(3,4-ethylenedioxythiophene) polystyrene sulfonate (PEDOT:PSS), molybdenum oxide (MoOx), and copper phthalocyanine (CuPc). Among these, PEDOT:PSS is commonly used due to its high hole mobility, good conductivity, and excellent film-forming properties. However, PEDOT:PSS can also cause oxidation of the QD surface and lead to degradation of the devices performance and stability.

To address this issue, alternative HTL materials have been developed, such as organic small molecules and polymers. For example, Zhang et al. reported the use of a new HTL material, 2-(4-(diphenylamino)phenyl)-5-(4-(diphenylamino)phenyl) thiophene (TPD-PAB), which showed good hole transport and blocking properties without causing degradation of the QD surface. The resulting QDLEDs exhibited improved stability and higher performance compared to those using PEDOT:PSS.

In addition to material selection, the thickness and structure of the HTL can also impact the performance and stability of QDLEDs. For example, an ultra-thin HTL can reduce the resistance of the device and improve the current injection, but it can also lead to poor hole transport and charge balance. On the other hand, a thicker HTL can provide better hole transport and charge balance, but it may also lead to increased resistance and lower device efficiency.

Overall, the development of HTL materials and structures is an active area of research in QDLED technology, as it can have a significant impact on the stability and performance of these devices.

The HTL in QDLEDs typically exhibits lower charge mobility than the ETL, leading to carrier buildup and exciton complexation at the interface with the quantum dot layer. The accumulation of carriers at this interface results in space charge formation, quenching of excitons, and device performance degradation. To address this issue, an ultra-thin 1,3,5-tris(1-phenyl-1*H*-benzimidazol-2-yl)benzene (TPBi) interface layer can be inserted between the quantum dots and the cavity transport layer in QDLEDs. This layer separates the carrier buildup interface from the exciton formation zone, thereby suppressing exciton quenching and improving device performance [21].

In 2015, Peng et al. prepared QDLEDs with an inverted structure using a ZnO ETL, CdSe/ZnS quantum dot light-emitting layer, N,N'-bis(naphthalen-1-yl)-*N*,*N*'-bis(phenyl)benzidine (NPB) HTL, and HATCN hole injection layer. In this device structure, the same issue exists where electron injection is easier than hole injection, leading to performance degradation [22]. Therefore, they incorporated a TCTA layer as an interface layer between the CdSe/ZnS quantum dots and the NPB HTL to regulate the charge balance. Although TCTA and NPB have similar LUMO energy levels, the electron mobility of TCTA is much lower than that of NPB, which effectively suppresses excessive electron tunneling through NPB to directly reach the anode and helps to confine electrons in the quantum dot emitting layer. On the other hand, since the HOMO energy level of TCTA is precisely between that of CdSe/ZnS quantum dots and NPB, it forms a step potential barrier, which effectively promotes hole injection. Based on these two reasons, the EQE of QDLEDs after modification by the TCTA interface layer is increased by 2.7 times, and the luminescence intensity is increased by 2 times compared to the original device, as shown in Figure 11.4. Currently, our basic understanding of the relationship between hole-injection layer (HIL)/HTL interface design and the operating stability of QDLEDs is limited. Dai et al. demonstrated that during the preparation of red QDLEDs, electron leakage at the device interface induces in situ electrochemical reduction reactions of polyfluorene (HIL), which generate

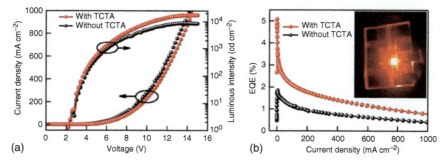

Figure 11.4 (a) Current density-voltage and luminous intensity-voltage curves and (b) EQE-current density curves for QDLED with and without TCTA interface layer.

trap states and deteriorate the charge transport properties [23]. In their study, they employed oxygen plasma-treated PEDOT:PSS as the HIL to create an optimized HIL/HTL interface with improved injection properties. This straightforward method enhanced exciton generation in the quantum dot layer and mitigated HTL degradation caused by electron leakage, resulting in superior performance of red-emitting QDLEDs. Specifically, these QDLEDs exhibited high EQE of over 20.0% in the luminance range of 1000 to 10 000 cd m^{-2}, and an operating lifetime of T95 at 1000 cd m^{-2} was extended to 4200 hours.

11.2.3 Electronic Transport Layer

The choice of electron transport material (ETM) in QDLEDs can significantly affect the devices performance and stability. One of the major challenges with ETMs is achieving balanced electron and hole injection into the QD layer, as excessive electron injection can cause charge accumulation and exciton quenching, leading to device instability and degradation.

One commonly used ETM is ZnO, which has good electron transport properties and can be easily deposited through low-cost solution-based processes. However, ZnO can also act as an electron trap and cause charge accumulation, leading to device instability and degradation. To address this issue, various strategies have been developed, such as modifying the surface of ZnO with ligands to reduce trap states and optimizing the thickness of ZnO to balance electron and hole injection. Other ETMs, such as TiO_2, have also been explored for use in QDLEDs. TiO_2 has high electron mobility and good stability, but suffers from poor compatibility with QDs and can cause exciton quenching. To mitigate these issues, surface modification of TiO_2 with organic ligands has been employed to improve the compatibility and reduce trap states.

In recent years, organic ETMs, such as fullerene derivatives and small molecules, have gained attention due to their high electron mobility, good compatibility with QDs, and tunable energy levels. However, organic ETMs can also suffer from instability and degradation under operating conditions, which can limit the devices lifetime.

Overall, the choice of ETM in QDLEDs requires a careful balance between electron transport properties, compatibility with QDs, and stability under operating conditions, and continued research and development in this area is necessary to further improve the performance and stability of QDLEDs.

The ETL is responsible for transporting the electrons injected from the cathode to the light-emitting layer. The challenge is to reduce the loss of electrons during the transport process and improve the light-emitting performance of the device. To address this, Lim et al. used ZnO as the ETL, InP/ZnSeS QDs as the light-emitting layer, TCTA as the HTL, and MoO_3 as the hole injection layer in their QDLEDs. However, in this structure, the potential barrier of electron injection from ZnO into InP/ZnSeS quantum dots (about 0.5 eV) is higher than that of hole injection from TCTA (about 0.2 eV), leading to charge injection imbalance [24]. The ETL plays a crucial role in transporting electrons from the cathode to the light-emitting layer in QDLEDs. However, loss of electrons during the transport process can significantly affect the light-emitting performance of the device. Lim et al. used ZnO as the ETL, InP/ZnSeS quantum dots as the light-emitting layer, TCTA as the HTL, and MoO_3 as the hole injection layer to prepare QDLEDs. The high potential barrier for electron injection from ZnO into InP/ZnSeS quantum dots (about 0.5 eV) compared to hole injection from TCTA (about 0.2 eV) led to an unbalanced charge injection and non-radiative compound formation of excitons.

To address this issue, they embedded a lead niobate ferrate (PFN) interface layer between the ZnO ETL and the InP/ZnSeS quantum dot layer, optimizing its thickness. They found that the insertion of PFN formed a dipole layer, causing the vacuum energy level to shift downward and reducing the potential barrier height for electron injection. However, increasing the thickness of PFN widened the potential barrier for electron tunneling, hindering electron tunneling from the ZnO layer to the InP/ZnSeS quantum dot layer. After optimization, they prepared the best QDLED performance with 0.5 mg ml^{-1} PFN solution, showing the lowest turn-on voltage (2.2 eV), highest luminescence intensity (3900 cd m^{-2}), and highest EQE (3.46%).

Lee et al. explored the synthesis of cadmium-free blue QDs with high-quality PL properties and the preparation of efficient QDLEDs, as shown in Figure 11.5 [25]. Lee et al. investigated the synthesis of cadmium-free blue QDs with high-quality PL properties, as well as the preparation of efficient QDLEDs, as depicted in Figure 11.5. They successfully prepared true blue multishell ZnSeTe quantum dots with an emission spectrum of 445 nm, a PLQY of 84%, and a forbidden bandwidth of 27 nm.

To improve the ETL materials, they modified the surface of ZnMgO nanoparticles (NPs) by reacting them with Mg to form a $Mg(OH)_2$ overlay on the surface. The presence of the $Mg(OH)_2$ overlay led to a decrease in electron mobility and an improvement in the charge balance of the QD emitting layer. Furthermore, the $Mg(OH)_2$ layer was found to alleviate the emission burst of the QD-EML. By combining blue ZnSeTe quantum dots with m-ZnMgO as the ETM, they obtained a high-brightness and high-efficiency blue QDLED with a luminance of 2904 cd m^{-2} and an EQE of 9.5%.

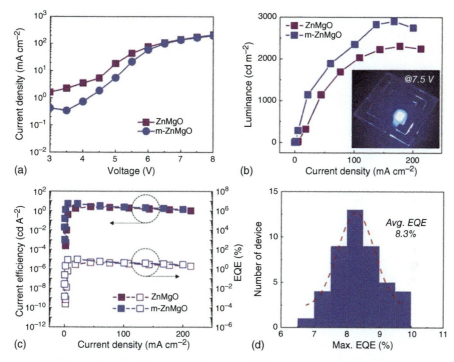

Figure 11.5 (a) Variation of current density with applied voltage; (b) brightness (inset: EL image acquired at 7.5 V); (c) current efficiency EQE with current density of pristine ZnMgO and m-ZnMgO-based QDLEDs; and (d) m-ZnMgO-based QDLEDs obtained from 45 devices histograms of peak EQEs.

11.2.4 Other Functional Layers

The cathode-interface regulation layer and anode-interface regulation layer are important functional layers in QDLEDs, as they can improve the devices stability and efficiency. The cathode-interface regulation layer is responsible for improving electron injection and reducing the energy barrier between the cathode and the ETL, while the anode-interface regulation layer can enhance hole injection and reduce the energy barrier between the anode and the HTL.

For example, Chen et al. reported a QDLED with a Cs_2CO_3/Al cathode and a MoO_3/Alq_3 anode, where a thin $CsPbBr_3$ interface layer was introduced between the ETL and the cathode to improve electron injection and reduce energy barriers. The QDLED exhibited a high maximum EQE of 14.4% and an operational lifetime of 250 hours at a brightness of 1000 cd m^{-2}.

Similarly, Kim et al. used a solution-processed ZnO nanoparticles anode and a solution-processed MoO_3 hole injection layer and introduced a thin LiF layer between the anode and the HTL. This LiF layer acted as an anode-interface regulation layer, and effectively reduced the energy barrier for hole injection, resulting in a higher EQE of 5.5% and improved operational lifetime of 2000 hours at a brightness of 1000 cd m^{-2}.

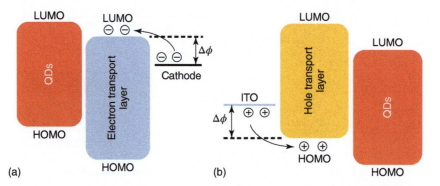

Figure 11.6 (a) Electron injection barrier diagram and (b) hole injection barrier diagram.

In cathode interface regulation, the electron injection barrier is mainly determined by the difference between the Fermi energy level of the cathode material and the lowest unoccupied orbital (LUMO) energy level of the ETL material. Generally, a larger energy level difference results in a higher electron injection barrier and a lower injection efficiency, as shown in Figure 11.6a. To promote electron injection in QDLEDs, an interface layer can be embedded between the cathode and the ETL.

For the anode interface, ITO has a high work function energy (around 4.7 eV), but when used as an anode material, the energy corresponding to the work function is lower than the highest occupied orbital (HOMO) energy level of most HTL materials, resulting in an interfacial barrier for hole injection, as shown in Figure 11.6b, where $\Delta\Phi$ is the difference between the energy corresponding to the work function of ITO and the HOMO energy level of the HTL. To weaken the hole injection barrier and promote hole injection and transport, scholars have proposed several feasible methods to modulate the ITO anode interface. Among them, the most widely applied ones are ozone plasma treatment and the introduction of a poly-3,4-ethylene-dioxothiophene (PEDOT:PSS) interfacial layer. Ozone plasma treatment can remove residual organic matter on the ITO surface and improve its wetting properties, facilitating film formation, and it can also improve the work function of the ITO, reducing the hole injection potential barrier. PEDOT:PSS, whose energy level is between the ITO and the HOMO energy levels of the HTL material, can form a step barrier and improve the hole injection efficiency.

Table 11.2 shows the performance comparison of QDLEDs with WO_3 nanoparticles and PEDOT:PSS as the anode interface layer, demonstrating the significant impact of anode interface modulation on device performance [26–28].

11.3 Quantum Dot Light-Emitting Diode Efficiency Decay Mechanism

One major efficiency decay mechanism in QDLEDs is related to the stability of the quantum dots themselves. Quantum dots can suffer from degradation over time

Table 11.2 Comparison of QDLED performance with WO_3 nanoparticles and PEDOT:PSS as the anode interface layer.

Anode interface layer	Turn-on voltage (V)	Maximum luminance (cd m^{-2})	Maximum EQE (%)	Maximum current efficiency (cd A^{-1})	Maximum luminous efficacy (lm W^{-1})	Lifetime (s)
WO_3 nanoparticle	3.8	30 006 (@50 mA)	3.32	9.75	6.8	6530
PEDOT:PSS	4.2	25 202 (@40 mA)	3.02	8.74	6.0	3130

due to the effects of high temperatures, humidity, and oxygen exposure, leading to a reduction in their PLQY and hence the overall device efficiency. Another factor that can contribute to efficiency decay is related to the stability of the interface layers in QDLEDs. The organic and inorganic interface layers in QDLEDs can also be affected by environmental factors, leading to changes in their electronic properties, and hence a reduction in device performance. Additionally, charge imbalance and recombination losses can also lead to efficiency decay in QDLEDs. Charge imbalance can occur when one type of charge (either electrons or holes) is favored over the other, leading to unbalanced carrier injection and a reduction in overall device performance. Recombination losses can occur due to non-radiative recombination processes, where the energy is lost as heat rather than emitted as light, resulting in a decrease in device efficiency. To address these challenges, ongoing research is focused on developing stable and efficient materials for use in QDLEDs, including improving the stability of the quantum dots and interface layers, and reducing charge imbalance and recombination losses through material and device engineering.

Up to now, numerous promising studies have focused on improving the performance of QDLEDs, particularly their device efficiency. EQE is widely acknowledged as the most crucial parameter for measuring the efficiency of QDLEDs, and it is determined by the following equation:

$$EQE = \gamma \cdot \eta_{rad} \cdot \eta_{out} \qquad (11.1)$$

where EQE is defined as the ratio of the total number of photons emitted by the QDLED to the number of electrons injected, γ is the percentage of injected charge used to form excitons, η_{rad} is the ratio of excitons recombined by radiative decay, and η_{out} is the efficiency of outward radiation, which determines how many of the generated photons can escape from the device. Therefore, selecting an appropriate energy band level and functional layer with high mobility can achieve uniform charge injection, meaning the number of electrons injected into the emitting layer is equal to the number of holes, resulting in a high EQE value. The η_{rad} value can be improved by adjusting the composition and structure of the QDs. From a device point of view, η_{rad} can be effectively improved by suppressing exciton bursts due to excess charge, interface defects, or adjacent CTLs. For the η_{out} efficiency, materials with better light transmission can be chosen to reduce the absorption of light by materials inside the device piece, or fully transparent devices can be prepared using some transparent electrodes [29].

Efficiency roll-off is a phenomenon in which the EQE of a device declines at high current density (J). It is often used as a measure of the stability of light-emitting devices, with smaller roll-off magnitudes indicating better stability. Efficiency roll-off can also be used to determine the critical current density, which represents the current density at which the device's EQE drops to half its maximum value. The relative change in the critical current density is an indicator of the device's efficiency roll-off effect, with smaller changes corresponding to less severe roll-off [30].

Attention has been focused on factors that affect the efficiency roll-off of QDLEDs at high current densities. Efficiency roll-off is a concern that must be addressed for most types of LEDs, and although its origin remains somewhat controversial, it is generally accepted that competition between radiative recombination and trap-mediated non-radiative decay and the intermittent recombination mechanism are significant factors that cause devices to undergo efficiency roll-off at high current densities, with the latter often playing a decisive role [31–33].

Auger recombination is a non-radiative process that can affect the stability and efficiency of QDLEDs. It occurs when a charge carrier (an electron or a hole) loses energy by transferring it to another charge carrier in the same region. This process competes with radiative recombination, which is the desirable process that produces light emission. At low current densities, radiative recombination is the dominant process in QDLEDs. However, at higher current densities, Auger recombination can become more significant, leading to reduced device efficiency and stability. This is because the energy lost through Auger recombination is converted into heat, which can increase the temperature of the device and affect its performance. To mitigate the impact of Auger recombination, researchers have explored various strategies such as using core–shell QDs with thicker shells or optimizing the size and shape of the QDs to reduce the probability of Auger recombination. In addition, reducing the density of QDs in the emitting layer and improving the charge injection balance can also help to suppress Auger recombination and improve device efficiency and stability.

To investigate the effect of intermittent complexation on the efficiency roll-off, Zou et al. adjusted the width of quantum wells in 2D/3D chalcogenides using a multi-quantum well structure to suppress intermittent complexation. Some studies suggest that significant EQE roll-off occurs at higher current densities ($J = 100 \, \text{mA cm}^{-2}$), and the peak EQE is higher at current densities $J \approx 10\text{–}30 \, \text{mA cm}^{-2}$. This indicates a significant competition between trap-mediated nonradiative decay and radiative recombination in these devices. Achieving high EQE at low current density and low-efficiency roll-off at high current density remains a challenge in developing devices with high efficiency [4].

Several factors have been suggested to contribute to the efficiency roll-off of QDLEDs at high current densities. In addition to trap-mediated non-radiative decay and intermittent Auger recombination mechanisms, which can cause luminescence bursts and reduce the luminescence efficiency, Joule heating and charge imbalance within the device have also been identified as important factors. Kim et al. investigated the efficiency roll-off of QDLEDs using pulsed drives and observed that the devices could sustain up to 150 A cm^{-2} without signs of intermittent compounding. Instead, they attributed the efficiency roll-off to Joule heating and unbalanced charge injection within the device. Joule heating can increase the local temperature, leading to exciton dissociation and alteration of charge transport properties. Unbalanced charging can cause excess carriers to leak out, thereby reducing the EQE. In addition, for some QD thin-film materials, degradation and disruption of the film morphology at high current densities may also contribute to the efficiency roll-off. Achieving a combination of high EQE at low current density and low-efficiency roll-off at high current density remains a challenge for QDLEDs [34–37].

11.4 Aging Mechanisms of QDLEDs

One major aging mechanism in QDLEDs is the degradation of the organic materials that make up the device. Exposure to air and moisture can cause the organic materials to degrade over time, leading to a decrease in device performance. This can be mitigated by encapsulating the device in a protective layer. Another aging mechanism is the formation of defects and traps within the device over time, which can lead to non-radiative recombination and a decrease in efficiency. This can be caused by factors such as thermal stress, exposure to UV radiation, and electrochemical reactions. Strategies such as using stable materials and device architectures, and optimizing device fabrication processes, can help to mitigate this issue. In addition, as mentioned earlier, the Auger recombination process can also contribute to device aging by causing efficiency roll-off at high current densities. Strategies to reduce Auger recombination, such as improving the materials and device design, can help to extend the lifetime of the device. Overall, understanding and mitigating these aging mechanisms is critical for the development of stable and long-lasting QDLEDs.

Stability remains the biggest challenge for QDLEDs to enter the commercialization market. To accelerate progress in this area, it is urgent to comprehensively understand the degradation mechanisms of QDLEDs. Different mechanisms have been proposed to describe QDLED degradation, including Auger-induced degradation, creation of quenching sites due to thermal effects and leakage current, and electrochemical reactions, which can lead to the degradation of functional layers (HTL, ETL, and QDs) and their interfaces. QDLEDs are typically constructed with multiple functional layers, which are often amorphous organic and inorganic films. Under long-term electric stress, the film topography and/or chemical properties can change, significantly influencing device performance. Moreover, the multiple and stacked functional layers in QDLEDs create various interfaces between two adjacent layers. The electric field and Joule heating effects inevitably cause interface reconstruction. Thus, suitable aging treatments can dramatically enhance QDLED performance. Device degradation is closely related to the aging of the functional layers and their interfaces. Research on device aging, both positive and negative, is essential to gain a deeper understanding of the working mechanism of QDLEDs and develop more efficient and stable devices. Positive aging is related to the enhancement of the devices performance during QDLED operation, while negative aging is characterized by performance degradation with driving times for the device, and is used to measure device stability.

11.4.1 Positive Aging

Positive aging effect, also known as aging-induced enhancement, is a phenomenon where the performance of a QDLED improves over time under electrical stress. This effect has been observed in some QDLED devices and has attracted much attention as it provides a potential method for enhancing device performance. One of the mechanisms that may contribute to the positive aging effect is the gradual formation of more favorable energy levels in the QD layers. Under electrical stress, the QDs experience a gradual modification of their electronic structure, which can lead to an increased overlap between the electron and hole wave functions, resulting in enhanced radiative recombination efficiency. The aging process may also lead to an increase in the number of trap states, which can assist in electron injection and transport. Several studies have reported the positive aging effect in QDLEDs. For example, Zhang et al. found that the aging effect could increase the peak EQE of green QDLEDs by up to 36%. They attributed this improvement to the reduction of non-radiative recombination channels, the enhanced electron injection, and the improved energy transfer efficiency from the charge transport layer to the QD layer. Other studies have reported similar positive aging effects in blue and red QDLEDs.

However, it is worth noting that the positive aging effect is not universal and is not observed in all QDLED devices. The effect is also highly dependent on device architecture, materials, and operation conditions. The aging process may also cause degradation and instability in some devices, making it a double-edged sword for QDLEDs.

Overall, the positive aging effect is an interesting phenomenon that has been observed in QDLEDs and may provide a new method for enhancing device performance. However, more studies are needed to fully understand the underlying mechanisms and optimize device design for this effect.

The positive aging phenomenon observed in QDLEDs, especially in hybrid devices containing metal oxide ETL, is an intriguing and fascinating effect. Various interpretations have been proposed to explain this phenomenon, but all of them are related to the inorganic ETMs ZnO and ZnMgO, which are believed to be the main contributors to the positive aging effect. The positive aging effect was first reported by the Holloway group, who observed it only when the devices were encapsulated with a UV-curable acidic resin. The green device showed the highest positive aging rate, followed by the red device, and the blue device had the lowest rate. Typically, the enhancement of device efficiency approached a saturation point in about a week. Chen et al. declared that the aging process is mainly due to the chemical reaction that occurs at the Al/ZnMgO (or ZnO) interface in the device. After eight days of storage, the performance of the blue, green, and red devices increased by 2.6, 1.3, and 1.25 times, respectively. This shelf-aging phenomenon was due to the slow chemical reaction between the metal electrode and the ZnMgO ETL, which formed an AlZnMgO alloy interface, improving the conductivity of ZnMgO. The formation of AlOx at the interface effectively blocked the electron trapping pathway and prohibited the exciton quenching induced by the metal electrode. Similarly, thermal annealing of the prepared devices could achieve positive aging by accelerating the chemical reaction at the interface. Ding et al. attributed the aging effects of the QDLEDs to the resistive switching effect of metal oxides due to the migration of ions, proposing that in the initial aging time period, the improvement of device efficiency was due to the migration of off-lattice oxygen ions, which accumulated at the QD/ZnMgO interface, forming conductive filaments within the ZnMgO layer. This process greatly improved the conductivity of the ZnMgO layer and suppressed exciton quenching. After this short positive aging process, the device experienced a long-term efficiency decrease due to the oxidation of the QDs by the accumulated active oxygen at the QD/ZnMgO interface, resulting in an increased barrier for electron injection from ETL to QDs and leading to device degradation. Recently, the effect of UV-curable glue on the positive aging of QDLEDs was investigated in detail by Chen et al. For the commonly employed ZnO or ZnMgO, the dominant channels for hole transport/sites are surface defects originating from $Zn(OH)_2$ and/or oxygen vacancies. The formation of a hydroxy group-terminated surface, i.e. Zn-OH component outside the ZnO or ZnMgO nanoparticles, also provides a hole transport pathway due to its suitable energy level of 5.5–6.5 eV. During encapsulation with UV-resin, these defects in the ZnMgO (or ZnO) were gradually passivated through the chemical reaction between ZnMgO (or ZnO) and various kinds of fatty acids contained in the resin, which served as weak acids. Chen et al. proposed that ZnO could react with these acids, leading to the formation of zinc carbonate, which helped to reduce the defect density of ZnO nanoparticles, optimizing the ZnO/QDs and ZnO/Al interfaces. Positive aging is an amazing and desirable effect because it significantly improves the efficiency and operation lifetime of QDLED devices, without requiring QD optimization or specific device design.

11.4.2 Negative Aging

The negative aging effect in QDLEDs refers to a decrease in their device performance over time due to various factors, such as material degradation, chemical reactions, and trap-assisted recombination. One factor contributing to the negative aging effect is the oxidation of the QDs, which can result from exposure to environmental factors such as moisture and oxygen. This process causes the QD surfaces to become less passivated, leading to an increase in trap states and a decrease in charge carrier mobility. As a result, the devices efficiency can decline significantly over time. Another factor that can contribute to negative aging is the formation of defects at the QD/ETL interface. These defects can arise from chemical reactions between the QDs and the ETL material, which can result in the formation of trap states that can capture and recombine charge carriers. This effect can cause a reduction in the devices efficiency over time.

Furthermore, the negative aging effect can also result from trap-assisted recombination, which can occur due to the presence of trap states within the QD or ETL material. These traps can capture charge carriers and release them through nonradiative pathways, leading to a decrease in the devices efficiency over time. Several methods have been proposed to mitigate the negative aging effect in QDLEDs, such as passivating the QD surface with organic or inorganic materials, encapsulating the devices with protective materials, and optimizing the device structure to reduce trap states and improve charge carrier transport. Nonetheless, further research is still needed to address these challenges and achieve long-term stable performance of QDLEDs.

Negative aging, also known as degradation, is a widely discussed topic and is a crucial factor in evaluating the practicability of a device. State-of-the-art QDLEDs have made great progress in terms of efficiency and stability. However, the operation and shelf lifetime still pose a formidable challenge to their widespread commercial application. The device's performance naturally and continuously deteriorates under high electric stress due to the degradation of the functional layers, including the QD emissive layer, charge transport/injection layers, as well as the electrodes. Physical and chemical reactions occur in these layers during device degradation, induced by intrinsic factors such as interfacial delamination, electro- or photo-chemical reactions, and Joule heating effects, as well as extrinsic factors such as penetration of water and oxygen, dust introduction during fabrication, and impurities. Although some groups have studied the failure mechanisms, it is still not well understood. HTL degradation, QD–ETL junction, and QDs themselves have been discussed in these reports. The close relation of different layers and interfaces to the photo-electrical properties of QDLEDs suggests that the apparent appearances of degradation may be caused by several mechanisms simultaneously. In other words, various factors leading to the degradation interfere with each other, and the evolution of photo-electrical properties of the device might originate from their combined effect. Therefore, analyzing the individual origin of the degradation remains challenging. Based on the typical sandwich structure of QDLEDs, recent progress has been summarized to better understand the degradation mechanism.

11.4.3 Electron Transport Layer

The ETL plays an important role in the performance and stability of QDLEDs. The degradation of QDLEDs can occur due to various reasons, such as the diffusion of moisture, oxygen, or other impurities, and the accumulation of charges or defects in the device layers. The ETL can impact the aging and degradation of QDLEDs in several ways:

Stability: The ETL is responsible for transporting electrons from the cathode to the quantum dots. A stable and efficient ETL can minimize electron accumulation, which can reduce the formation of traps and other defects in the quantum dot emissive layer, leading to better device stability.

Interface issues: The interface between the ETL and the quantum dot layer is crucial for electron injection and transport. Any defects or impurities in this interface can lead to charge accumulation, localized heating, or other degradation issues.

Energy level alignment: The energy level alignment between the ETL and the quantum dot layer can affect the electron injection efficiency and the resulting device performance. If the energy levels are not aligned properly, it can lead to inefficient electron injection, which can result in device degradation.

Therefore, the ETL can have a significant impact on the aging and degradation of QDLEDs, and it is important to carefully select and optimize the materials and properties of the ETL to ensure device stability and long-term performance.

There are very few reports on ETL degradation in metal oxide-based hybrid QDLEDs. Generally, metal oxide ETLs are more stable than their organic counterparts and are less likely to deteriorate under moderate external electric fields. This seems to be the case for red and green hybrid QDLEDs. However, a significant difference was found for blue QDLEDs. By measuring the electro-absorption (EA) spectroscopy and capacitance of the blue device, Qian found that the aging of red QDLEDs is mainly due to the degradation of the HTLs. Nevertheless, the performance evolution of blue QDLEDs with test times is closely related to the ETL-QDs junction. Due to the conduction band barrier between the ZnO ETL and QDs, electrons are easily transferred to the ETL, which leads to space-charge accumulation, operating-voltage rise, and ETL degradation. Given this, the development of ETL and QD materials with smaller electron injection barriers can effectively improve the operation lifetime of blue light devices. Recently, Ding et al. suggested that the resistive switching effect could be responsible for the low operation lifetime of ZnMgO-based QDLEDs. The off-lattice oxygen ions in the ETL migrated to the ETL/QDs interface, oxidizing the QDs and forming a high injection barrier at the QD/ETL interface, which induced the device degradation based on ZnMgO.

11.4.4 Hole Transport Layer

The HTL is a critical component in QDLEDs that can significantly impact their aging and degradation. The HTL is responsible for facilitating the transport of holes from the anode to the QDs in the emissive layer. During device operation, the HTL can degrade due to various factors such as the penetration of water and oxygen, interfacial delamination, and electro- or photo-chemical reactions.

Several studies have reported that the aging of red QDLEDs is mainly due to the degradation of the HTLs. The degradation of the HTL can lead to a decrease in hole injection efficiency, which causes an imbalance in charge transport and leads to a reduction in the quantum yield of the device. This can result in a lower luminance and shorter device lifetime.

In addition, some studies have found that the performance evolution of blue QDLEDs is also closely related to the HTL-QDs junction. The degradation of the HTL can affect the energy level alignment at the HTL-QDs interface, resulting in a reduction in hole injection efficiency and a shift in the emission color. Therefore, developing more stable HTL materials and optimizing the HTL-QDs interface are critical for improving the performance and longevity of QDLEDs.

In traditional QDLEDs, HTLs are typically made of organic small molecules such as CBP and TCTA, as well as organic polymers like TFB and PVK. These organic materials can be sensitive to water and oxygen and can deteriorate or become structurally deformed under external electrical stress and charge carrier impact. This can create nonradiative recombination centers. The hole mobility of commonly used HTLs is lower than that of ETLs, and the injection barrier for holes is much larger than that for electrons, resulting in an imbalanced charge injection into the QDs. As a result, holes accumulate at the HTL/QD interface and excessive electrons can leak into the HTL, leading to operational instability and degradation of the HTL. Non-radiative recombination sites in the HTL have been identified as a primary cause of device degradation in some research studies.

One study demonstrated that the degradation of the HTL can create non-radiative defects that induce static quenching of QD emission. Another study proposed a degradation mechanism involving exciton-induced degradation of the HTL, which creates quenched states that reduce the luminous efficiency of the QDs and the stability of QDLEDs. By improving hole injection and reducing electron leakage, the stability of the device can be greatly improved. Reversible degradation in QDLEDs has also been observed after turning off the driving voltage, which is due to the QD charging effect.

Insulating layers can be introduced as a strategy to improve device stability. One optimized QDLED was achieved by introducing double insulating modifying layers into the device, which dramatically suppressed charge leakage and led to an operation lifetime of over 110 000 hours at 100 cd m^{-2}. This indicates that the suppression of leakage current is a feasible way to improve device stability. The injection barrier and its resulting charge imbalance are the primary degradation mechanisms, and thus carrier injection and distribution in the device must be carefully considered to improve device performance.

11.4.5 QDs Layer

The QD layer can also have an impact on QDLED aging and degradation. One potential issue is the stability of the QDs themselves. QDs are typically composed of inorganic semiconductors, and these materials can undergo degradation over time, which can result in a decrease in the devices performance. In addition, QDs can suffer from structural deformation under external electrical stress and charge carrier impact, which can create nonradiative recombination centers that reduce the luminous efficiency of the device.

Moreover, the QD layer can also affect the stability of the charge transport layers in the device. For example, it has been reported that the injection of excessive electrons can lead to the degradation of the HTL in the device, which is a primary cause of operational instability. As mentioned earlier, the injection barrier for holes is often much larger than that for electrons, which leads to an imbalance in charge injection into the QDs. This results in the accumulation of holes at the HTL/QD interface and the leakage of electrons into the HTL, leading to degradation of the HTL.

To mitigate these issues, various strategies have been proposed to improve the stability of QD layers in QDLEDs. For example, passivation of the QD surface can improve the stability of the QDs and reduce the occurrence of nonradiative recombination. In addition, the use of encapsulation layers can help to protect the QDs from exposure to air and moisture, which can degrade the QDs over time. Furthermore, improving the charge balance between the electron and HTLs can reduce the accumulation of charges at the interface, which can help to mitigate degradation of the HTL.

As the emissive layer, the stability of QDs is critical for the operational lifetime of QDLEDs. Imbalanced carrier injection into the QDs in QDLEDs is inevitable, particularly for ZnO-based hybrid devices. Thus, QD charging should be taken into account in a working QDLED device. Negative charging of the QDs during QDLED operation has been observed due to excessive electrons injected into the device, as reported by Bae [38]. The fate of negatively charged QDs was systematically studied, and it was found that they can be neutralized by applying reverse bias to extract the extra electron or by injecting a hole. Additionally, a trion state can be formed by injecting another two carriers, with at least one hole. However, the generation of multi-carrier states for QDs leads to the generation of Joule heat, which induces permanent damage to the QDs. The significant Joule heat generated during device operation, particularly at high current density, significantly deteriorates the devices stability due to thermal-induced emission quenching. The extra electrons also cause electrochemical reactions of the QDs, leading to QD destruction. Therefore, the QD degradation induced by the heating effect and electrochemical reaction during the operation of QDLEDs is discussed. Elevated temperature can cause debonding of ligands, leading to defects in the QDs, and excessive heat can increase heat-assisted recombination and cause irreversible quenching. Zhang et al. reported that the thermal effect can be mitigated by adopting 1-dodecanethiol (DDT) ligands that have stronger binding to the surface of QDs compared with conventional oleic acid ligands. With DDT ligands, effective thermally assisted radiative recombination was maintained, and the QDLED with DDT-modified QDs as the emissive layer greatly improved the thermal stability and inhibited thermally induced emission quenching. Recently, electrochemical reactions of QDs with injected electrons have received significant attention. Carboxylate is commonly used as ligand anchoring on the outside of the QDs. Some of them are bonded onto surface cadmium atoms on the polar facets, and the others are cadmium-carboxylate ligands weakly adsorbed onto the nonpolar facets. Jin et al. [39] found that the surface cadmium-carboxylate ligands are highly sensitive to the injected electrons during QDLED device operation, leading to reduction reactions of surface ligands with injected charge carriers that produce CdO, which is the dominant origin leading to the extremely low efficiency of QDLEDs. Electrochemical reactions occurring on the QD surface seriously affect the surface chemical features of QDs, which reduce QD performance and are partly responsible for QDLED degradation. Therefore, introducing an electrochemically inert ligand to suppress the reduction reaction on the ligand significantly improves the working life of the device.

11.5 Characterization Technologies for QDLEDs

There are several methods for characterizing QDLEDs, including:

Current–voltage (IV) measurements: This measures the electrical behavior of the device under different voltage and current conditions. It can be used to extract the device performance parameters such as turn-on voltage, current efficiency, and EQE.

Electroluminescence (EL) spectroscopy: This measures the light output of the device as a function of wavelength. It provides information on the peak wavelength, full width at half maximum (FWHM), and emission intensity.

Time-resolved photoluminescence (TRPL): This measures the decay time of the excited carriers in the QDs, which can provide information about the efficiency of the radiative and non-radiative recombination processes.

Photoconductivity measurements: This measures the change in the electrical conductivity of the device under illumination. It can provide information on the mobility of the charge carriers in the device.

Atomic force microscopy (AFM): This is a technique used to image the surface of the device at high resolution. It can be used to determine the surface morphology of the QD layer and the quality of the device interface.

Transmission electron microscopy (TEM): This is a technique used to image the internal structure of the device at high resolution. It can be used to determine the size and distribution of the QDs and the thickness of the device layers.

X-ray diffraction (XRD): This is a technique used to determine the crystal structure and orientation of the device layers. It can be used to verify the crystal quality of the QD layer and the device interface.

These are just a few examples of the many techniques available for characterizing QDLEDs, and the specific techniques used may depend on the research goals and the specific properties of the device being studied. So far, much progress has been made in understanding the EL process in QDLEDs. However, due to the complex optoelectronic properties under applied electrical fields, many fundamental physics or scientific issues in QDLEDs are still not clear or addressed. These include charge mobility properties, charge transport/distribution, exciton formation processes, sub-voltage turn-on phenomena, charge transport materials/QDs interface issues, charge recombination/trapping/decay dynamics, device efficiency "roll-off" at high current densities, etc. The classical and reliable characterization of individual QDs is no longer applicable to the properties of QD assembly films. We need more direct and efficient tools to characterize the assembly behavior of QDs in thin-film form. The combination of in-situ electrical and optical measurements is expected to provide more valuable and reliable information. Therefore, methods such as transient electroluminescence (TrEL) spectroscopy have been developed to study the charge transport, distribution, and accumulation in QDLEDs. Electrical absorption spectroscopy has also been introduced to characterize the changes of functional layers in devices under an electric field. Differential absorption spectroscopy, which allows direct measurement of charge accumulation and electric fields in working devices, provides an alternative technique to study EL processes in situ. There is also a widely used electrical measurement for organic LEDs, namely displacement current measurement (DCM), which is also worth developing to interpret carrier trapping processes in QDLEDs. We believe that more effective in-situ measurements will reveal the physical mechanisms and accelerate the development of QDLED technology. In the following, we will briefly introduce these characterization techniques mentioned above.

11.5.1 Transient Electroluminescence

TrEL is a technique used to study the mechanism of device performance in QDLEDs. It involves measuring the temporal evolution of the EL signal under pulsed electrical excitation. This technique provides valuable insights into the charge transport and recombination processes occurring in the device.

In a typical transient EL measurement, a short voltage pulse is applied to the QDLED, which leads to the injection of charge carriers into the device. These carriers recombine in the emissive layer, leading to the emission of photons. The emitted photons are detected using a photodetector, such as a photomultiplier tube (PMT) or a silicon photodiode, and the resulting EL signal is recorded as a function of time.

By analyzing the temporal evolution of the EL signal, researchers can extract information about the charge carrier dynamics in the device. For example, the decay time of the EL signal can provide insights into the radiative and nonradiative recombination processes in the device, as well as the mobility of the charge carriers. Additionally, the shape of the EL signal can be used to distinguish between different recombination mechanisms, such as direct recombination, trap-assisted recombination, and Auger recombination. Transient EL is a powerful technique for characterizing the performance of QDLEDs and can be used to optimize device design and improve device efficiency.

TrEL spectroscopy is often used to measure the transport, distribution, and accumulation characteristics of charges in devices. TrEL measurements of a device's TrEL profile and time-resolved spectra are the basis for characterizing and studying the device principle. In TrEL measurements, a periodic pulsed voltage provided by a signal generator is applied to the device, and the corresponding EL signal is collected by a PMT, as shown in Figure 11.7. The electrical signal of the PMT is recorded by a sampling oscilloscope. A typical TrEL measurement system can be roughly divided into the following four parts according to the measurement process sequence: (i) the delay time between the turn-on of the pulse voltage and the onset of EL, which may be attributed to the charge mobility of the charge transport and emission layers; (ii) the initial EL rise, consisting of a fast process and a slow process, which is usually determined by the exciton formation process and the charge distribution; (iii) the saturation or flat EL process, which occurs when the drive pulse voltage is turned off and the charge injection, transport, and distribution will show an initial fast delay process and a long-lasting tail; and (iv) the decay period (also referred to as the falling edge).

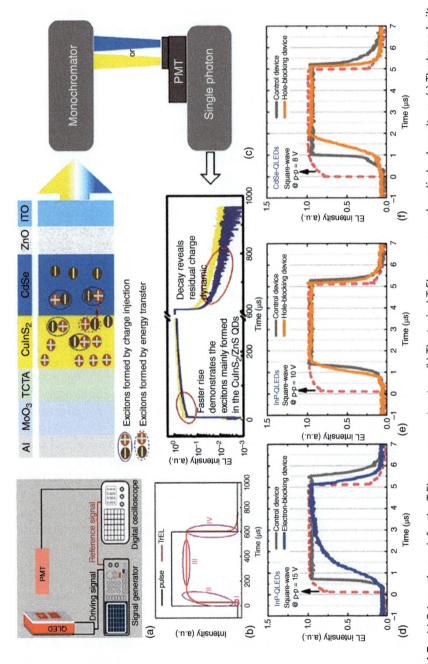

Figure 11.7 (a) Schematic graph for the TrEL measurement system. (b) The typical TrEL response and applied pulse voltage. (c) The home-built wavelength-resolved TrEL system. (d)–(f) The TrEL of InP- and CdSe-QDLED with different structures.

The resistor–capacitor (RC) discharge characteristics involve fast processes, and the slow and persistent tail is closely related to the recombination characteristics of the excitons formed in the emitting layer and the charge distribution within the device, as well as the influence of the charge transport layer on the excitons. An overfast process is usually observed at the rising or/and falling edges, which remains unclear and may be due to Auger recombination or/and residual charge. Moreover, the carrier dynamics (including charge transport, accumulation, trapping, and recombination) can also be partially obtained by varying the mode of the driving pulse voltage. Energy transfer in QDLEDs, especially the formation of phosphorescent molecules to QDs, is usually indirectly described by means of PL measurements. Recently, the energy transfer from the phosphorescent molecules of Firpic to the QD during QDLED operation has been demonstrated by TrEL measurements. Compared to PL characterization, TrEL provides direct evidence and a new method to study the energy transfer properties of QDLEDs during luminescence. In addition, the effect of energy level and mobility of HTL on device performance, especially the EL mechanism, can also be investigated by TrEL measurements. Recently, a wavelength-resolved TrEL system has been built, through which the EL mechanism of dichroic white QDLEDs was explored, as shown in Figure 11.7c. This measurement system is able to distinguish the EL characteristics of different emitters and their interactions in the color mixing process and electrically driven devices. It also explained the charge balance characteristics of InP-based QDLEDs by TrEL measurements. In conclusion, TrEL is a powerful tool for characterizing charge dynamics.

11.5.2 Electro-Absorption (EA) Spectroscopy

EA spectroscopy is a technique that can be used to study the electronic structure and dynamics of materials, including QDLEDs. In this technique, an electric field is applied to the material, and the resulting changes in absorption spectra are measured. By studying the changes in absorption spectra, information about the energy levels and exciton dynamics in the material can be obtained.

In the context of QDLEDs, EA spectroscopy can be used to study the properties of the QD layer, including the bandgap, the energy levels of the excitons, and the exciton binding energy. By measuring the changes in absorption spectra under applied electric fields, the electronic properties of the QD layer can be characterized. Additionally, the EA spectra can be used to study the exciton dissociation process, which is critical for understanding the charge carrier dynamics in the QDLEDs. Overall, EA spectroscopy is a powerful tool for understanding the electronic properties and dynamics of materials, including QDLEDs. It can provide insights into the mechanism of the devices performance and help in the design of new and improved QDLEDs.

EA spectroscopy is a powerful tool used to measure the change in the absorption spectrum induced by an externally applied electric field, thereby obtaining the electronic state and optical properties of materials (see Figure 11.8). By monitoring the Stark shift of excitonic levels, typical quantities such as the change in permanent

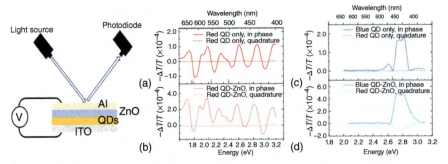

Figure 11.8 (Left): The schematic diagram of the electro-absorption spectrum measurement setup. (Right): Electro-absorption (EA) spectra of the quantum dot–ZnO junction. (a) (in-phase and quadrature) of the sample with a structure of ITO/red ($Cd_{1-x}Zn_xSe_{1-y}S_y$) quantum dots (QDs)/Al. (b) EA spectra (in-phase and quadrature) of the sample with a structure of ITO/red ($Cd_{1-x}Zn_xSe_{1-y}S_y$) QDs/ZnO/Al. (c) EA spectra (in-phase and quadrature) of the sample with a structure of ITO/blue ($Cd_{1-x}Zn_xS$) QDs/Al. (d) EA spectra (in-phase and quadrature) of the sample with a structure of ITO/blue ($Cd_{1-x}Zn_xS$) QDs/ZnO/Al.

dipole moment, μ, and the change in polarizability, α, on electronic and vibrational transitions can be obtained from the EA spectrum of a solid film containing the absorbing molecules of interest. This technique can also detect molecular aggregation or molecular complex formation that cannot be identified in typical absorption spectra.

The degradation of materials under electrical stress can be intuitively reflected in their characteristic absorption. Unlike the very stable red devices, whose lifetime is primarily limited by the slow degradation of the hole-transporting layer, the poor lifetime of blue QDLEDs originates from the fast degradation at the QD–ETL junction. It has been suggested that material engineering for efficient electron injection is a prerequisite for boosting the operating lifetime [40].

11.5.3 In-Situ EL–PL Measurement

In-situ EL–PL measurement technique is a nondestructive method to study the performance of QDLEDs by simultaneously measuring their EL and PL spectra under operational conditions. This technique allows researchers to obtain real-time information about the exciton and charge carrier dynamics in the device, such as carrier recombination and energy transfer processes, as well as the occurrence of degradation mechanisms such as quenching and surface oxidation. By varying the applied voltage, temperature, or other experimental conditions, it is possible to study the effects of these factors on the devices performance and to optimize the device design for improved efficiency and stability. In-situ EL–PL measurements can be performed using a variety of setups, including microscope-based systems and customized electro-optical characterization tools, and can provide valuable insights into the underlying physical processes in QDLEDs.

The in-situ and simultaneous EL–PL measurements can be classified into steady-state and transient-state processes. In 2013, Yasuhiro et al. investigated

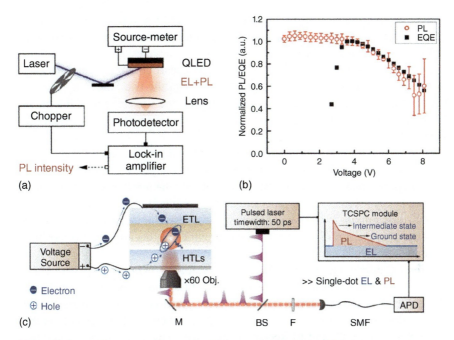

Figure 11.9 (a) The schematic graph of the steady-state EL and PL measurement system. (b) The similar trace of the EQE roll-off and the decreased PL efficiency. (c) The experimental system to detect the steady EL and transient PL simultaneously [41, 42].

the origin of efficiency roll-off in QDLEDs by simultaneously measuring the steady-state EL and PL intensities with the test system shown in Figure 11.9a. They recorded pulsed or periodic PL signals with a photodetector combined with a Lock-in amplifier and showed that strong electric fields decreased the emission efficiency of QDLEDs as the driving voltages increased.

More recently, Jin et al. developed an in-situ steady-state EL and transient PL measurement to elucidate the exciton formation mechanisms, as shown in Figure 11.9c. During QDLED operation, they introduced a short laser pulse to excite the QD and then collected the transient PL signals of the QD using an avalanche photodiode to estimate the charge state of the QD. The in-situ transient PL measurement provided reliable support for carrier dynamic processes. With these measurements, they clarified the microscopic images of electrically excited single QDs and the macroscopic EL processes of QDLEDs. They found that usually one electron is injected into the QD first, followed by injection of a hole, which forms an exciton. These results provide constructive guidance for material synthesis and structural design. In addition to the applications discussed above, we believe that these EL–PL testing technologies can also be used to assess the degradation behavior of QDLEDs during aging tests.

11.5.4 Differential Absorption Spectroscopy

Differential absorption spectroscopy is a technique used to investigate the electronic transitions in materials by measuring the difference in the absorption of light before and after a perturbation is applied. In the context of QDLEDs, differential absorption spectroscopy can be used to study the dynamics of charge carrier recombination and the effects of trap states on the devices performance. In a typical experiment, a short optical pulse is used to excite the QDLED, and the absorption of a second, weaker pulse is measured as a function of time delay between the two pulses. By varying the energy and duration of the excitation pulse and the wavelength of the probe pulse, information about the density of states and relaxation times of the QDs can be obtained.

Differential absorption spectroscopy can provide information about the charge carrier dynamics and trap states in QDLEDs, which are important factors that affect the devices performance. It can also be used to identify the mechanisms responsible for non-radiative recombination processes and the origin of PL quenching in QDs.

Overall, differential absorption spectroscopy is a powerful tool for characterizing the electronic properties of QDLEDs and can provide insights into the mechanisms of their operation and degradation (Figure 11.10).

Differential absorption spectroscopy was developed by Kim et al. [43] to characterize the internal electrical properties of QDLEDs under operation, as shown in Figure 11.11. A tunable supercontinuum laser was used as the light source for the absorption measurement, with a voltage-tuned balanced photodetector as the detector. Steady-state PL and EL spectra were collected by a spectrometer, and the

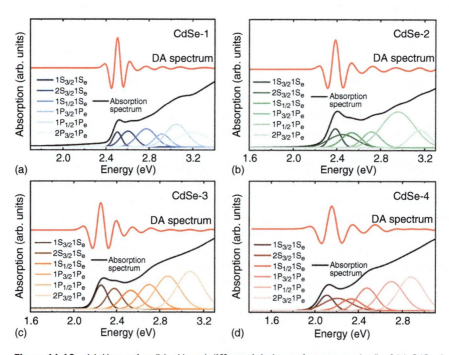

Figure 11.10 (a) Absorption (black) and differential absorption spectra (red) of (a) CdSe-1, (b) CdSe-2, (c) CdSe-3, and (d) CdSe-4 with first through sixth excitonic transitions extracted from differential absorption (DA) spectroscopy analysis. Source: Kim et al. [43]/MDPI/CC BY 4.0.

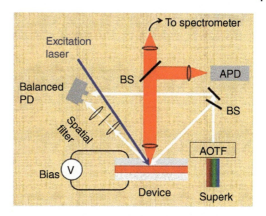

Figure 11.11 The schematic illustration of the differential absorption spectrum measurement system.

transient (time-resolved) PL spectrum was recorded by an avalanche photodiode. The effect of charging on the QD emission was isolated by building a capacitor-type device with the structure of $ITO/ZnO/QD/TCTA/SiO_2/Al$. The differential absorption spectrum reflects the population states of electron orbitals in the QDs, while the quantum-confined Stark effect induced by the electric field can also be obtained by measuring the differential absorption spectrum with the QDLED under a reverse driving voltage. The transient PL spectra can be used to estimate the charging of the QDs. Unlike EA technology, differential absorption spectroscopy is carried out without electrical modulation to measure the absorption spectrum. A significant advantage of differential absorption spectroscopy is that it successfully distinguishes the influence of charging and field on the QD emission in QDLEDs. Therefore, this technology should become a powerful tool to characterize the exciton dynamics of QDs in QDLED devices.

11.5.5 Displacement Current Measurement DCM Technology

DCM is a technique used to investigate the charge dynamics and carrier transport in QDLEDs. It is a noninvasive electrical characterization technique that allows for the measurement of the transient current and voltage response of the device under different operating conditions. In DCM, a small voltage step is applied to the device, and the resulting displacement current is measured as a function of time. The displacement current arises from the charging and discharging of the capacitances in the device, including the parasitic capacitances. By analyzing the transient response of the displacement current, information about the mobility and recombination dynamics of the charge carriers in the device can be obtained. DCM can be used to investigate a wide range of device parameters, including the mobility of charge carriers, the density of trap states, and the recombination kinetics. By varying the applied voltage, it is possible to probe the charge transport in different regions of the device, from the injection layer to the QD layer and the ETL. Overall, DCM is a powerful technique for investigating the electronic properties of QDLEDs, and it provides valuable insights into the fundamental mechanisms that govern device performance.

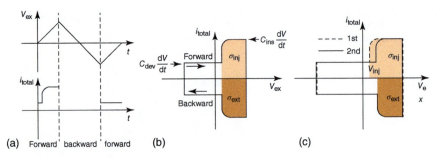

Figure 11.12 The sweep triangular voltage, current (a), and DCM curve (b). The representative DCM curve of the device with charge trapped (c).

In DCM measurement, a triangular wave voltage is applied to the device to measure the quasi-static current response, which contains information about charge injection, extraction, accumulation, and trapping behavior. Generally, the total current (i_{total}) in the device is composed of the conduction current (i_c) and displacement current (i_{dis}). In the DCM measurement, the device is in the quasi-static system, which means that a steady state is immediately reached in the device at a given voltage, and the conduction current can be ignored. The time derivative of the induced charge on the electrode can be used to calculate the i_{dis} by [222].

$$i_{dis} = \partial \sigma_{elect}/\partial t \tag{11.2}$$

$$i_{dis} = C_{app}\, dV/dt \tag{11.3}$$

$$C_{app} = d\sigma_{elec}/dV \tag{11.4}$$

where σ_{elec} is the charge density on the electrode, C_{app} represents the apparent capacitance of the device, and dV/dt is the sweep rate of the triangular wave voltage. The sweep voltage mode, current response, and typical DCM curves are shown in Figure 11.12a–c. The $dV/dt > 0$ and $dV/dt < 0$ are defined as a forward and backward sweep, respectively. During the forward sweep, the carriers start to be injected into the device. The total amount of charge (σ_{inj}) injected into the organic layer per unit area can be obtained by $\sigma_{inj} = \int t\, t_{inj}\, i_{dis} dt' = \int V V_{inj} C_{app}\, dV'$ (4), with the t_{inj} (V_{inj}) corresponding to the time (voltage) while the carrier injection begins. Similarly, the extracted charge, σ_{ext}, can also be calculated from i_{dis}. Based on this, the trapped charge in the device can be calculated from the first and second sweeps. Additionally, the schematic illustration of the Differential Absorption Spectrum measurement system is shown in Figure 11.11.

11.6 Outlook

Despite the challenges facing QDLEDs, such as device stability, research and application of this technology continue to advance, with new mechanisms being proposed to solve stability problems. Future research should focus on promoting carrier injection balance in QDLEDs, suppressing exciton quenching in the emitting layer, and reducing non-radiative compound formation in the device. These efforts will significantly improve luminous intensity, efficiency, and stability of QDLEDs. Although QDLEDs face strong competition in the display field, they offer excellent performance and a wide range of applications, with significant potential for future development. QDLEDs are poised to become a strong competitor to OLED and PeLED, and we look forward to their successful commercialization and large-scale use in our daily lives. Understanding the mechanism of degradation and aging in QDLEDs is crucial to improving their stability and commercial viability. Here are some potential research directions that could be pursued in the future:

(1) Investigation of long-term device performance under various operating conditions, including high temperature, high humidity, and high current density. This will provide insights into the degradation mechanism of QDLEDs and help identify the critical factors that lead to device failure.
(2) Characterization of the surface chemistry and morphology of the QD films and interfaces and their evolution over time. This can be done using various analytical techniques such as X-ray photoelectron spectroscopy (XPS), atomic force microscopy (AFM), and scanning electron microscopy (SEM).
(3) Exploration of new materials and approaches to improve the stability and reliability of QDLEDs. For example, new encapsulation strategies, device architectures, and surface passivation techniques can be investigated to reduce the exposure of QDs to external factors that lead to their degradation.
(4) Development of advanced in situ and operando characterization techniques that can provide real-time information on the dynamics of QDLED degradation. This can include techniques such as time-resolved photoluminescence (TRPL) and TrEL measurements, which can provide insights into the behavior of charge carriers and excitons under various operating conditions.

By pursuing these research directions and others, we can gain a better understanding of the mechanism of QDLED degradation and aging, and develop strategies to improve their stability and longevity.

References

1 Chang, Y., Li, K., Feng, Y.L. et al. (2016). Crystallographic facet-dependent stress responses by polyhedral lead sulfide nanocrystals and the potential "safe-by-design" approach. *Nano Research* 9 (12): 3812–3827.
2 Liu, S.H., Liu, X.Y., and Han, M.Y. (2016). Controlled modulation of surface coating and surface charging on quantum dots with negatively charged gelatin for substantial enhancement and reversible switching in photoluminescence. *Advanced Functional Materials* 26 (48): 8991–8998.
3 Li, X., Hu, B., Zhang, M. et al. (2019). Continuous and controllable liquid transfer guided by a fibrous liquid bridge: toward high-performance QLEDs. *Advanced Materials* 31 (51): e1904610.

4 Zou, Y., Ban, M., Yang, Y. et al. (2018). Boosting perovskite light-emitting diode performance via tailoring interfacial contact. *ACS Applied Materials & Interfaces* 10 (28): 24320–24326.

5 Makarov, N.S., Lin, Q.L., Pietryga, J.M. et al. (2016). Auger up-conversion of low-intensity infrared light in engineered quantum dots. *ACS Nano* 10 (12): 10829–10841.

6 Chiba, T., Hayashi, Y., Ebe, H. et al. (2018). Anion-exchange red perovskite quantum dots with ammonium iodine salts for highly efficient light-emitting devices. *Nature Photonics* 12 (11): 681–687.

7 Jiang, X.B., Li, B., Qu, X.L. et al. (2017). Thermal sensing with CdTe/CdS/ZnS quantum dots in human umbilical vein endothelial cells. *Journal of Materials Chemistry B* 5 (45): 8983–8990.

8 Xie, Y.Y., Geng, C., Liu, X.Y. et al. (2017). Synthesis of highly stable quantum-dot silicone nanocomposites via in situ zinc-terminated polysiloxane passivation. *Nanoscale* 9 (43): 16836–16842.

9 Balan, A.D., Olshansky, J.H., Horowitz, Y. et al. (2019). Unsaturated ligands seed an order to disorder transition in mixed ligand shells of CdSe/CdS quantum dots. *ACS Nano* 13 (12): 13784–13796.

10 Singh, S., Samanta, P., Srivastava, R. et al. (2017). Ligand displacement induced morphologies in block copolymer/quantum dot hybrids and formation of core-shell hybrid nanoobjects. *Physical Chemistry Chemical Physics* 19 (40): 27651–27663.

11 Yin, H., Truskewycz, A., and Cole, I.S. (2020). Quantum dot (QD)-based probes for multiplexed determination of heavy metal ions. *Microchimica Acta* 187 (6): 336.

12 Liu, Z.J., Lin, C.H., Hyun, B.R. et al. (2020). Micro-light-emitting diodes with quantum dots in display technology. *Light: Science & Applications* 9 (1): 83.

13 Dong, H., Xu, F., Sun, Z.Q. et al. (2019). In situ interface engineering for probing the limit of quantum dot photovoltaic devices. *Nature Nanotechnology* 14 (10): 950.

14 Basu, K., Zhang, H., Zhao, H.G. et al. (2018). Highly stable photoelectrochemical cells for hydrogen production using a SnO_2-TiO_2/quantum dot heterostructured photoanode. *Nanoscale* 10 (32): 15273–15284.

15 Ashree, J., Wang, Q., and Chao, Y.M. (2020). Glyco-functionalised quantum dots and their progress in cancer diagnosis and treatment. *Frontiers of Chemical Science and Engineering* 14 (3): 365–377.

16 Pyo, K., Thanthirige, V.D., Yoon, S.Y. et al. (2016). Enhanced luminescence of au-22(SG)(18) nanoclusters via rational surface engineering. *Nanoscale* 8 (48): 20008–20016.

17 Righetto, M., Privitera, A., Carraro, F. et al. (2018). Engineering interactions in QDs-PCBM blends: a surface chemistry approach. *Nanoscale* 10 (25): 11913–11922.

18 Cho, Y.J. and Lee, J.Y. (2011). Low driving voltage, high quantum efficiency, high power efficiency, and little efficiency roll-off in red, green, and deep-blue phosphorescent organic light-emitting diodes using a high-triplet-energy hole transport material. *Advanced Materials* 23 (39): 4568–4572.

19 Lim, M.H., Jeung, I.C., Jeong, J. et al. (2016). Graphene oxide induces apoptotic cell death in endothelial cells by activating autophagy via calcium-dependent phosphorylation of c-Jun N-terminal kinases. *Acta Biomaterialia* 46: 191–203.

20 Yan, F., Xing, J., Xing, G. et al. (2018). Highly efficient visible colloidal lead-halide perovskite nanocrystal light-emitting diodes. *Nano Letters* 18 (5): 3157–3164.

21 Yang, D., Zou, Y., Li, P. et al. (2018). Large-scale synthesis of ultrathin cesium lead bromide perovskite nanoplates with precisely tunable dimensions and their application in blue light-emitting diodes. *Nano Energy* 47: 235–242.

22 Peng, Y.D., Yang, A.H., Xu, Y. et al. (2016). Tunneling induced absorption with competing nonlinearities. *Scientific Reports* 6.

23 Dong, Y., Chien, L.-C., Lee, S.-D., Yoon, T.-H. (2017). Photoluminescent (PL) or electroluminescent (EL) quantum dots for display, lighting, and photomedicine (Conference Presentation). Advances in Display Technologies VII.

24 Lin, J., Hu, D.D., Zhang, Q. et al. (2016). Improving photoluminescence emission efficiency of nanocluster-based materials by in situ doping synthetic strategy. *Journal of Physical Chemistry C* 120 (51): 29390–29396.

25 Li, G.P., Wang, H., Zhang, T. et al. (2016). Solvent-polarity-engineered controllable synthesis of highly fluorescent cesium lead halide perovskite quantum dots and their use in white light-emitting diodes. *Advanced Functional Materials* 26 (46): 8478–8486.

26 Hussein, R., Jaurigue, L., Governale, M., and Braggio, A. (2016). Double quantum dot cooper-pair splitter at finite couplings. *Physical Review B* 94 (23): 235134.

27 Iotti, R.C., Dolcini, F., Montorsi, A., and Rossi, F. (2016). Electron-phonon dissipation in quantum nanodevices. *Journal of Computational Electronics* 15 (4): 1170–1178.

28 Jang, D.M., Kim, D.H., Park, K. et al. (2016). Ultrasound synthesis of lead halide perovskite nanocrystals. *Journal of Materials Chemistry C* 4 (45): 10625–10629.

29 Wang, H., Zhang, X., Wu, Q. et al. (2019). Trifluoroacetate induced small-grained CsPbBr3 perovskite films result in efficient and stable light-emitting devices. *Nature Communications* 10 (1): 665.

30 Shi, Z., Li, Y., Zhang, Y. et al. (2017). High-efficiency and air-stable perovskite quantum dots light-emitting diodes with an all-inorganic heterostructure. *Nano Letters* 17 (1): 313–321.

31 Yuan, F., Xi, J., Dong, H. et al. (2018). All-inorganic hetero-structured cesium tin halide perovskite light-emitting diodes with current density over 900 A cm^{-2} and its amplified spontaneous emission behaviors. Physica status solidi (RRL). *Rapid Research Letters* 12 (5): 1800090.

32 Yuan, S., Hao, Y., Miao, Y. et al. (2017). Enhanced light out-coupling efficiency and reduced efficiency roll-off in phosphorescent OLEDs with a spontaneously distributed embossed structure formed by a spin-coating method. *RSC Advances* 7 (69): 43987–43993.

33 Yusoff, A.R.B.M., Gavim, A.E.X., Macedo, A.G. et al. (2018). High-efficiency, solution-processable, multilayer triple cation perovskite light-emitting diodes with copper sulfide–gallium–tin oxide hole transport layer and aluminum-zinc oxide–doped cesium electron injection layer. *Materials Today Chemistry* 10: 104–111.

34 Zhang, X.M., Wang, B., and Liu, Z.Q. (2016). Tuning PbS QDs deposited onto TiO_2 nanotube arrays to improve photoelectrochemical performances. *Journal of Colloid and Interface Science* 484: 213–219.

35 Zhao, Z.Y., Liu, J.Q., Li, A.W., and Xu, Y. (2016). Strong coupling between J-aggregates and surface plasmon polaritons in gold nanodisks arrays. *Acta Physica Sinica* 65 (23): 231101.

36 Zheng, K., Lu, M., Rutkowski, B. et al. (2016). ZnO quantum dots modified bioactive glass nanoparticles with pH-sensitive release of Zn ions, fluorescence, antibacterial and osteogenic properties. *Journal of Materials Chemistry B* 4 (48): 7936–7949.

37 Zheng, K.B., Zidek, K., Abdellah, M. et al. (2016). High excitation intensity opens a new trapping channel in organic - inorganic hybrid perovskite nanoparticles. *ACS Energy Letters* 1 (6): 1154–1161.

38 Bae, W.K., Park, Y.S., Lim, J. et al. (2013). Controlling the influence of Auger recombination on the performance of quantum-dot light-emitting diodes. *Nature Communications* 4 (1): 2661.

39 Pu, C., Dai, X., Shu, Y. et al. (2020). Electrochemically-stable ligands bridge the photoluminescence-electroluminescence gap of quantum dots. *Nature Communications* 11 (1): 937.

40 Song, C., Weiran, C., Taili, L. et al. (2019). On the degradation mechanisms of quantum-dot light-emitting diodes. *Nature Communications* 10 (1): 765.

41 Yasuhiro, S., Geoffrey, J.S., and William, A.T. (2013). Origin of efficiency roll-off in colloidal quantum-dot light-emitting diodes. *Physical Review Letters* 110 (21): 217403.

42 Yunzhou, D., Xing, L., Wei, F. et al. (2020). Deciphering exciton-generation processes in quantum-dot electroluminescence. *Nature Communications* 11 (1): 2309.

43 Kim, S.H., Man, M.T., Lee, J.W. et al. (2020). Influence of size and shape anisotropy on optical properties of CdSe quantum dots. *Nanomaterials* 10 (8): 1589.

12

Electron/Hole Injection and Transport Materials in Quantum Dot Light-Emitting Diodes

12.1 Introduction

Quantum dot light-emitting diodes (QDLEDs) are an exciting technology with the potential for high-performance displays and lighting. The efficiency and stability of QDLEDs are critical factors in their widespread adoption, and both the hole transport layer (HTM) and electron transport layer (ETM) play significant roles in these properties. The HTM facilitates hole injection and transport from the anode to the quantum dot (QD) layer, while the ETM facilitates electron injection and transport from the cathode to the QD layer. As a result, the development of high-performance HTMs and ETMs is critical for improving the performance of QDLEDs. In this book chapter, we will review the current state of research and development in HTMs and ETMs for QDLEDs. We will discuss the various materials and processing techniques that have been explored to develop high-performance HTMs and ETMs, including materials selection, optimization of thickness, doping and surface modification, multifunctional HTMs and ETMs, and scalability and processing techniques.

The chapter will be organized into several sections, each focused on a different aspect of HTMs and ETMs in QDLEDs. The first section will provide an overview of the key requirements and challenges associated with HTMs and ETMs in QDLEDs, including their impact on device performance and stability. The next section will review the different types of materials that have been explored as HTMs and ETMs in QDLEDs, including traditional small-molecule organic materials, polymers, metal oxides, and perovskites. This section will also discuss the advantages and disadvantages of each material type, as well as their impact on device performance. The following sections will focus on specific aspects of HTM and ETM development, such as the optimization of layer thickness, doping and surface modification, and the development of multifunctional HTMs and ETMs. We will also discuss the role of device architecture in achieving efficient and stable QDLEDs, including the use of interfacial layers to improve charge injection and transport. Finally, we will discuss recent advancements in processing techniques for HTMs and ETMs, including solution processing, printing techniques, and vacuum deposition.

Colloidal Quantum Dot Light Emitting Diodes: Materials and Devices, First Edition. Hong Meng.
© 2024 WILEY-VCH GmbH. Published 2024 by WILEY-VCH GmbH.

Throughout the chapter, we will provide examples of recent research in the field of HTMs and ETMs for QDLEDs, including the development of high-performance materials and devices, as well as the identification of key challenges and opportunities in this field. We will also highlight the potential impact of QDLEDs on emerging technologies, such as flexible displays and lighting, and discuss the future directions for research in this field.

By highlighting the key advancements and remaining challenges in this area, we hope to stimulate further research and innovation in the field of QDLEDs and related technologies. Furthermore, this chapter will serve as a platform for researchers and industry professionals to exchange ideas and collaborate on new research projects.

Overall, this book chapter aims to provide a comprehensive overview of the current state of research and development in HTMs and ETMs for QDLEDs, as well as the challenges and opportunities in this field. It will be of interest to researchers, engineers, and industry professionals working in the field of QDLEDs and related technologies. By providing a roadmap for future research in this field, we hope to inspire new ideas and collaborations that will ultimately lead to the development of high-performance and commercially viable QDLEDs.

12.2 Charge-Transport Mechanisms

In QDLEDs, the charge transport mechanism involves the movement of electrons and holes through the QD layer, which leads to the recombination of electron–hole pairs and the emission of light. The QD layer in QDLEDs is typically made up of semiconductor materials, such as CdSe or CdTe, which have a bandgap that can be tuned by adjusting the size and composition of the QDs. The QDs are sandwiched between an electron transport layer (ETL) and a hole transport layer (HTL), which are typically made up of organic or inorganic materials that have appropriate energy levels to facilitate charge injection and transport. When a voltage is applied to the QDLED, electrons are injected from the ETL into the QD layer, while holes are injected from the HTL into the QD layer. These injected charges then move through the QD layer and recombine with each other, releasing energy in the form of light. The emitted light has a wavelength that depends on the size and composition of the QDs. The efficiency of charge transport in QDLEDs is an important factor that affects the devices performance. One of the challenges in QDLEDs is to ensure efficient charge injection and transport across the QD layer, which requires careful selection of materials and device design.

In addition to charge injection and transport, other factors that can affect the efficiency of QDLEDs include nonradiative recombination processes, surface traps, and device degradation over time. One approach to improving the charge transport efficiency in QDLEDs is to use hybrid structures that combine inorganic QDs with organic materials. Organic materials can provide better charge transport properties and help to reduce nonradiative recombination, while inorganic QDs can provide higher brightness and color purity. Another approach is to use graded heterojunctions, where the bandgap of the QD layer is gradually increased from the ETL to the HTL. This can help to reduce energy barriers for charge transport and improve charge injection and extraction efficiency.

In summary, the charge transport mechanism in QDLEDs involves the injection and transport of electrons and holes through a QD layer, which leads to the recombination of electron–hole pairs and the emission of light. The efficiency of charge transport in QDLEDs is a key factor that affects device performance and can be improved through careful selection of materials and device design.

12.3 Electron Transport Materials (ETMs) for QDLED

The ETM plays a critical role in facilitating electron injection and transport from the cathode to the QD layer, which is responsible for emitting light. The development of high-performance ETMs is thus essential for improving the performance of QDLEDs. In this section, we will review the current state of research and development in ETMs for QDLEDs. We will discuss the various materials and processing techniques that have been explored to develop high-performance ETMs, including materials selection, optimization of thickness, doping and surface modification, multifunctional ETMs, and scalability and processing techniques. The section will provide insights into the challenges and opportunities in the development of ETMs for QDLEDs and the potential of this technology for future display and lighting applications.

12.3.1 Metal-Doped ETMs

ZnO has become a commonly used material for electron transport and has been proved to be the most favorable ETL material due to its high transparency, low work function, and high electron mobility. A solution-process crystalline ZnO nanoparticle (NP) layer as the ETL demonstrated maximum brightness levels of 4200 cd m^{-2}, 68 000 cd m^2, and 31 000 cd m^{-2} for emission of blue, green, and orange–red light, respectively (Figure 12.1) [1].

However, QDLED devices with ZnO as the ETM lead to a significant decrease in exciton dissociation and EL energy efficiency due to electron transfer from QD to ZnO caused by the energy difference between conduction band minima (CBM). This mechanism is believed to be the cause of QD luminescence quenching. It is shown that the CBM energy level of ZnO-based ETLs can be controlled by doping, which is a promising approach to optimize the electronic structure and device performance. To address this issue, doping has been found to be an effective way to modify the electronic structure of ZnO and improve its performance in devices. Inspired by this and based on their observation of spontaneous charge transfer at the QD/ZnO interface, Cao et al. sought to improve the performance of QDLEDs by modifying Ga-doped ZnO NPs to achieve controlled band structure at room temperature (RT) and solubility without the involvement of bulk organic ligands [2]. Their modification led to a significant reduction in transfer between the interface of CdSe/ZnO QD NPs and ZnO NPs, which served as the ETM, resulting

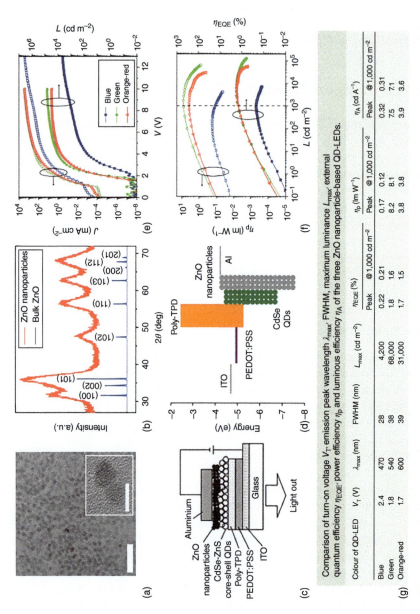

Figure 12.1 (a) TEM image of ZnO nanoparticles (scale bar, 20 nm). Inset: High-resolution TEM images of same ZnO nanoparticles showing lattice fringes (scale bar, 3 nm). (b) XRD pattern from ZnO nanoparticles (red) together with that from bulk wurtzite ZnO (blue). (c) Schematic of layers in the device structure. (d) Energy level diagram for the various layers. (e) Current–density (J) and luminance (L) versus driving voltage (V). (f) Luminance power efficiency (η_P) and EQE η_{EQE} versus luminance.

in an outstanding luminous brightness of up to 44 000 cd m^{-2}. When doped with 8% Ga, the efficiency of the QDLEDs reached as high as 15 cd A^{-1}, which was the most efficient red QDLED reported at the time.

Meanwhile, to further promote the balance of charge transport, Wang et al. proposed the idea of using $Zn_{0.95}Mg_{0.05}O$ as an electron transport layer material [3]. This doping strategy results in a higher conduction band minimum (CBM) of 0.07 eV compared to pure ZnO, which can effectively confine excess electrons and suppress exciton bursts. Moreover, doping ZnO with Mg elements can effectively reduce the generation of oxygen vacancies, resulting in lower defect density and conductivity, which further contributes to the inhibition of exciton bursts. The resulting device shows a maximum luminous current efficiency of 24.6 cd A^{-1} and a power efficiency of 25.8 lm W^{-1}, which are 19% and 38% higher, respectively, than the undoped version. This strategy holds potential for developing high-efficiency white QDLEDs and other optoelectronic devices.

Sun et al. used Al as a ZnO doping element for the electron transport layer in green QDLED devices. They found that a device with a 10% doped ZnO electron transport layer showed a maximum current efficiency of 59.7 cd A^{-1} and an external quantum efficiency (EQE) of 14.1%, which was 1.8 times higher than that of the undoped device. This indicates that Al doping is an effective strategy for improving the performance of QDLED devices [4].

Group-III elements, such as aluminum (Al) and gallium (Ga), were widely used as n-type dopants for fabrication of ZnO ETLs (Figure 12.2) [5]. The use of doped ZnO NPs in the ETL is a powerful approach for tuning QDLED performance that allows for achieving record-setting device characteristics. A comparative study of the most common doping strategies for ZnO NPs used as an ETL in the direct QDLED structure has been investigated by Samokhvalov's group. Their findings indicated that AZO NPs with a doping level of 10% were the best candidates to be used in the ETL of future QDLEDs.

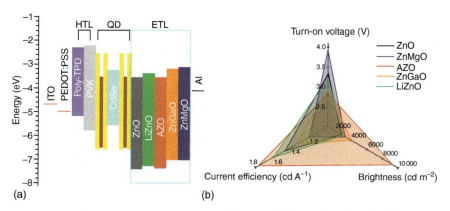

Figure 12.2 (a) Structure and the corresponding fat-band energy levels of the QDLED devices. (b) Brightness, current efficiency, and turn-on voltage of the fabricated QDLEDs with different electron transport layer materials: ZnO (black line), ZnMgO (blue), AZO (red), ZnGaO (orange), and LiZnO (green).

12.3.2 Metal Salt-Doped ETMs

In a study conducted by Lee et al., it was found that applying Rb_2CO_3 to Mg-doped ZnO ETMs improved the efficiency roll-off and operational lifetime of red QDLEDs. However, the results showed that the improvements were reversed. The optimal doping concentration of Rb_2CO_3 was found to be 4% and a thickness of 90 nm, which led to a long lifetime (T90) of 14 672 hours at a starting luminosity of 100 cd m^{-2}, a maximum luminosity of 129 100 cd m^{-2}, and a maximum current efficiency of 13.6 cd A at a luminosity of 60 000 cd m^{-2}. The performance enhancements are attributed to the strong n-type doping effect of Rb_2CO_3. Additionally, the analysis of the effect of different films shows that the thermal stability of the Mg-doped ZnO is enhanced by this method [6].

12.3.3 Design of Composite Materials ETMs

Qin Zhang achieved excellent efficiency in QDLED devices by combining a metal-doped layer with a ZnO layer. Specifically, they developed a composite ETM comprising both ZnO and Mg-doped ZnO NPs, which increased the EQE up to 13.5%. This value is 1.29 and 1.33 times higher than that of single ZnO and Mg-doped ZnO NPs, respectively. Additionally, the luminous intensity was enhanced to 22 100 cd m^{-1} at a voltage of 8.8 V. When used in electron-transporting devices for current–voltage measurements, the composite electron-transporting layer produced a higher current than the Mg-doped ZnO NP layer. Furthermore, the devices with composite ZnMgO:ZnO electron transport layers demonstrated lower leakage current densities compared to undoped ZnO materials at the same turn-on voltage. Transient measurements also showed that the composite electron transport layer ensured a more efficient charge transport balance than conventional ZnO NP layers.

12.3.4 Polymer-Modified ETMs

Yuan et al. worked on improving the electron transport properties by applying polyethoxy polyethyleneimine (PEIE) modified conventional ZnO layers as electron transport layers. The device enhancement was investigated using time-resolved photoluminescence and transient electroluminescence methods, which demonstrated that the PEIE-modified ZnO layer can restrict electron injection into the QD luminescent layer and reduce electron accumulation at the ZnO/QD interface. This provides a practical approach to suppressing the QD mass transfer process and thus improving device efficiency. The results show that the PEIE-modified layer has a synergistic effect on $CuInS_2$/ZnS-based QDLEDs, achieving a record-breaking increase in current efficiency up to 2.75 cd A^{-1} under red light excitation at 650 nm.

12.3.5 Inorganic Organic Hybrid ETMs

The use of polymers and metals in the design of electron transport layers to achieve organic–inorganic hybridization is a valuable approach. Rashed Alsharafi et al. observed that the large difference in electron and hole mobility can lead to exciton bursts and unbalanced charge injection at the interface. To address this issue, they replaced the ZnO layer with a combination of Mg and polyvinylpyrrolidone (PVP) to form an inorganic–organic hybrid layer and co-doped the ZnO NPs to form an electron transport layer. By adjusting the doping concentration of Mg, they could modify the energy band structure of the ZnO NPs, effectively suppressing spontaneous charge transport at the interface between the QD layer and the electron transport layer. Additionally, doping with PVP enabled adjustment of both the surface exciton burst site and the electron mobility of the electron transport layer. The resulting solution led to a significant increase in current efficiency and EQE, up to 16.16 cd A^{-1} and 15.45%, respectively, representing a 2.5 fold increase compared to conventional undoped ZnO NPs.

12.3.6 Double-Stacked ETMs

One way to address the carrier imbalance issue in QDLEDs is to employ a double layer of electron transport layers. For example, Myeongjin Park et al. stacked two metal oxide electron transport layers to create a double-layer stack consisting of ZnO and SnO_2, with SnO_2 serving as the second electron transport layer [7]. This approach improves the charge balance in the QDLED by a factor of 1.6, by preventing the spontaneous injection of electrons into the ZnO electron transport layer. These results indicate that the bimetallic oxide electron transport layer can significantly enhance the performance of QD-based light-emitting devices (Table 12.1).

Here are some of the key breakthroughs in ETMs development in colloidal QD QDLEDs during the years 2000 and 2023:

1. Introduction of inorganic ETMs: In 2007, researchers introduced inorganic materials, such as ZnO and ZnS, as electron transport layers in QDLEDs, which resulted in improved device performance. The efficiency of QDLEDs with inorganic electron transport layers increased from 0.1% to 1.5%.
2. Organic/inorganic hybrid ETMs: In 2012, researchers reported using organic/inorganic hybrid electron transport layers, such as poly(3,4-ethylenedioxythiophene): poly(styrenesulfonate) (PEDOT:PSS)/ZnO or PTAA/ZnO, in QDLEDs. These hybrid layers resulted in improved electron injection and transport, leading to higher efficiency and longer lifetime of the devices.
3. Surface ligand engineering: Surface ligand engineering has been an important breakthrough in improving electron injection and transport in QDLEDs. In 2016, researchers reported using pyridine-terminated ligands on the surface of QDs, which significantly improved electron injection and transport.
4. Introduction of metal chalcogenides: In 2020, researchers reported using metal chalcogenides, such as SnSe and SnS, as electron transport layers in QDLEDs. These materials exhibited high electron mobility and allowed for efficient electron injection, resulting in high-efficiency devices.

Here are some device performance data for QDLEDs with improved ETMs:

- In 2007, QDLEDs with inorganic electron transport layers achieved an efficiency of 1.5%.

Table 12.1 Some representative electronic transport layers and their device parameters.

Electronic transport layer	Device structure	External quantum efficiency (EQE) (%)	Current efficiency (cd A^{-1})	References
ZnO	Al / ZnO nanoparticles / CdSe-ZnS core-shell QDs / poly-TPD / PEDOT / ITO / glass	0.09		[8]
ZnO:SnO$_2$	Ag / ZnO:SnO$_2$ / QDs / NiO / ITO			[9, 10]
TiO$_2$	Al / TiO$_2$ / QDs / TFB / PEDOT:PSS / ITO			[11]
Ga-doped ZnO	Ag / Doped ZnO NPs / CdSe/ZnS QDs / PVK / PEDOT:PSS / ITO / Glass		15.0	[2]
Mg-doped ZnO	Al / Zn$_{0.95}$Mg$_{0.05}$O / B-QDs / ZnO / Y-QDs / TFB / PEDOT:PSS / ITO		24.6	[3]
Al-doped ZnO	Al / AZO / ZnO / QDs / PVK / PEDOT:PSS / ITO	14.1	59.7	[4]
Rb$_2$CO$_3$-doped MZO	Al / Organic / R-QDs / Rb$_2$CO$_3$:MZO / LZO / ITO	13.5	13.6	[6]

Table 12.1 (Continued)

Electronic transport layer	Device structure	External quantum efficiency (EQE) (%)	Current efficiency (cd A^{-1})	References
Composite materials	Al / ZnMgO/ZnO / QDs / poly-TPD / PEDOT:PSS / ITO/Glass	13.5		[12]
PEIE-doped ZnO	Al / MoO$_3$ / CBP / QDs / PEIE / ZnO / ITO		2.75	[13]
Mg and PVP co-doped with ZnO	Ag / Mg, PVP doped / QDs / TFB / PEDOT:PSS / ITO	15.45	16.16	[14]
ZnO and SnO$_2$ double stacking	Al / MoO$_3$ / TCTA / QDs / SnO$_2$ NPs / ZnO NPs / ITO / Glass	7.16	5.31	[7]

- In 2012, QDLEDs with organic/inorganic hybrid electron transport layers achieved a peak EQE of 6.4% and a half-lifetime of 35 hours.
- In 2016, QDLEDs with pyridine-terminated ligands achieved a peak EQE of 12.3% and a half-lifetime of 70 hours.
- In 2020, QDLEDs with SnSe electron transport layers achieved a peak EQE of 16.3% and a half-lifetime of 216 hours.

12.4 Electron Injection Materials for QDLED

Electron injection materials play a crucial role in determining the performance of QDLEDs, including their efficiency, brightness, and lifetime. There are several materials used for electron injection in QDLEDs, including:

Metal oxides: Metal oxides such as molybdenum oxide (MoO$_x$), tungsten oxide (WO$_x$), and zinc oxide (ZnO) are commonly used as electron injection materials in QDLEDs. These materials have a high work function, which facilitates electron injection from the electrode into the QD layer.

Organic materials: Organic materials such as PEDOT:PSS and polyethyleneimine (PEI) are also used as electron injection materials in QDLEDs. These materials can be easily processed, and their work function can be tuned by modifying their chemical structure.

Nanoparticles: NPs such as gold nanoparticles (AuNPs) and silver nanoparticles (AgNPs) have also been used as electron injection materials in QDLEDs. These materials have a high surface area, which facilitates electron injection into the QD layer.

Graphene: Graphene, a two-dimensional material made of carbon atoms, has also been investigated as an electron injection material in QDLEDs. Graphene has a high electrical conductivity and a low work function, which makes it a promising candidate for electron injection.

Conjugated Polyelectrolytes (CPEs): CPEs are a type of organic material that has been used as an electron injection material in QDLEDs. They possess a high work function and high electrical conductivity, which makes them suitable for electron injection into the QD layer. Additionally, CPEs have been found to improve the stability and lifetime of QDLEDs due to their ability to passivate surface defects in the QD layer.

Titanium dioxide (TiO_2): TiO_2 is another metal oxide that has been used as an electron injection material in QDLEDs. It has a high work function and can be easily processed. Additionally, TiO_2 has been found to improve the efficiency of QDLEDs due to its ability to improve charge balance and reduce electron leakage.

Lithium fluoride (LiF): LiF is a material that has been used as an electron injection material in QDLEDs. It has a very low work function, which allows for efficient electron injection into the QD layer. Additionally, LiF has been found to improve the efficiency and stability of QDLEDs due to its ability to reduce charge imbalance and improve electron confinement.

Zinc sulfide (ZnS): ZnS is another material that has been used as an electron injection material in QDLEDs. It has a high work function and can efficiently inject electrons into the QD layer. ZnS has been found to improve the efficiency and stability of QDLEDs due to its ability to reduce electron leakage and improve charge balance.

Conductive polymers: Conductive polymers such as poly(3-hexylthiophene) (P3HT) and poly(9,9-dioctylfluorene-co-benzothiadiazole) (F8BT) have also been investigated as electron injection materials in QDLEDs. These materials possess a high work function and can be easily processed. Additionally, conductive polymers have been found to improve the efficiency and stability of QDLEDs due to their ability to enhance charge injection and reduce electron leakage.

Graphene oxide (GO): GO is a modified form of graphene that has been investigated as an electron injection material in QDLEDs. GO possesses a high work function and can efficiently inject electrons into the QD layer. Additionally, GO has been found to improve the efficiency and stability of QDLEDs due to its ability to enhance charge injection and reduce electron leakage.

Overall, the choice of electron injection material depends on several factors, including the desired device performance, the compatibility with other device components, and the ease of processing.

Surface dipolar polymers such as PEIE, branched polyethylenimine (PEI), and self-assembled dipolar molecular monolayers (SAMs) are a class of stable EIL materials (Figure 12.3a) [15]. π-CPEs with covalently tethered quarternized ammonium cations or multivalent anions acting as powerful latent anion donors that are capable of electron doping organic semiconductors that have an EA as low as 2.3 eV, while remaining ambient solution processable (Figure 12.3b) [16].

Poly-[(9,9-bis(3′-[N,N-dimethylamino]propyl)-2,7-fluorene)-alt-2,7-(9,9-ioctylfluorene)] (PFN) is widely used electron injection layer in QDLED applications due to its interfacial dipole characteristic properties and is known to achieve a vacuum level shift over 0.5 eV. In addition, PFN possesses poor solubility against nonpolar solvents (such as hexane or toluene) and enables orthogonal processing to realize multilayered structures (i.e. ZnO nanocrystal layer/PFN thin layer/colloidal QD layer) [17]. Introducing a thin layer of PFN at the interface of ZnO with InP@ZnSeS QDs could reduce the electron injection barrier between ZnO and InP@ZnSeS QDs via the vacuum level shift, and facilitate the electron injection from ZnO to QDs across the PFN layer (tunneling process) (Figure 12.4.).

12.5 Hole Transport Materials for QDLED

Hole transport materials (HTMs) are a key component of QDLEDs, which are a type of LED that uses QDs to emit light. In general, organic HTMs exhibit slower mobility in QDLEDs than QDs or inorganic electrons, resulting in a charge imbalance in the device. Moreover, the energy barrier for hole injection from the hole transport layer to the QD is quite high. In particular, Cd-based QDs feature a deep valence energy layer (approximately −6 to −7 eV) compared to typical organic hole-transporting materials (around 5.5 eV) (Figure 12.1). Therefore, several approaches have been proposed to enhance the conductivity and control the energy level of the hole transport layer. The development of new and improved HTMs is a critical area of research in the field of QDLEDs, and there are several technological developments that are likely to play a role in this process.

1. Molecular design: The design of new HTMs with improved hole transport properties is a key area of research. This involves synthesizing new molecules and optimizing their chemical structure to improve their hole mobility, charge transport efficiency, and stability.
2. Interface engineering: The interface between the HTM and the QD layer is critical for efficient charge transfer and light emission in QDLEDs. Therefore, researchers are exploring ways to engineer the HTM/QD interface to improve the overall performance of the device.
3. Doping: Doping the HTM with suitable dopants can improve the conductivity and stability of the material, and also optimize the energy levels of the material

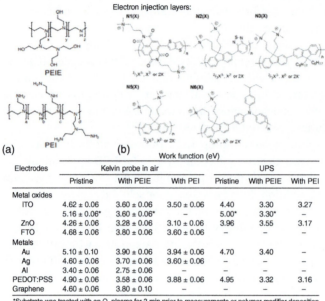

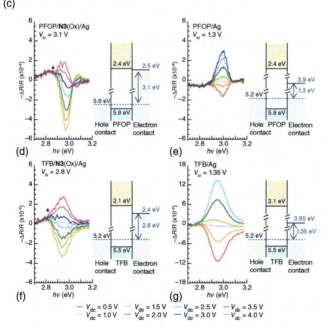

Figure 12.3 (a) and (b) Chemical structures of the materials; (c) Table. Work function of conducting materials with and without polymer modifiers, as independently measured by Kelvin probe in air and by UPS. Empty cells indicate no measurement for the corresponding sample; (d)–(g) In-phase electron absorption spectra (left panels) and energy level diagram (right panels) for ITO/PEDT:PSSH/ PFOP/10 nm $N_3(O_x)$/Ag (d); ITO/PEDT:PSSH/PFOP/Ag (e); ITO/PEDT: PSSH/TFB/10 nm $N3(O_x)$/Ag (f); and ITO/PEDT:PSSH/TFB/Ag (g).

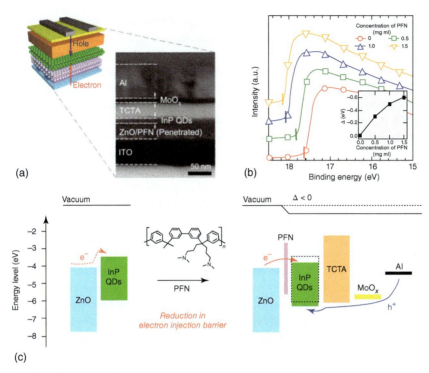

Figure 12.4 (a) Schematic illustration (left) and a cross-sectional TEM image (right) of QDLED in an inverted device structure. (b) Expanded UPS spectra near the binding energy cutoff region of a bare ZnO film and PFN-coated ZnO films on ITO substrates. The UPS spectra of PFN-coated ZnO films are vertically shifted for clarity. Inset: Change in the vacuum level shift (Δ) versus PFN concentration. (c) Flat-band energy level diagrams of QDLEDs illustrating the reduction in electron injection barrier between ZnO and QDs due to the presence of the PFN layer. Source: Lim et al. [18]/American Chemical Society.

to match those of the QDs. Researchers are exploring various dopants, including organic and inorganic molecules, to improve the performance of HTMs in QDLEDs.

4. Polymer-based HTMs: Polymeric HTMs offer several advantages, such as their ability to be processed from solution, which enables low-cost and large-scale manufacturing. Researchers are developing new and improved polymer-based HTMs with enhanced hole transport properties and improved stability.
5. Molecular self-assembly: The self-assembly of HTMs into ordered structures is another area of research that could improve their performance. Researchers are exploring methods such as Langmuir–Blodgett deposition and dip-pen nano-lithography to create ordered HTM monolayers with improved hole mobility.
6. Advanced characterization techniques: Advanced characterization techniques, such as X-ray diffraction, scanning tunneling microscopy, and atomic force microscopy, are critical for understanding the structure and properties of HTMs. These techniques enable researchers to study the morphology, crystal structure, and charge transport properties of HTMs, which can inform the design of new and improved materials.

The development of HTMs for QDLEDs has been focused on improving their charge transport properties, as well as their stability and compatibility with other layers in the device. One approach has been to modify the chemical structure of existing HTMs to enhance their electron affinity and charge mobility. For example, various derivatives of spirobifluorene, triarylamine, and carbazole have been investigated as potential HTMs for QDLEDs. Another approach has been to develop new materials specifically tailored for QDLEDs. For example, metal-organic frameworks (MOFs) have shown promise as HTMs due to their high surface area and tunable electronic properties. Other materials being investigated for use as HTMs in QDLEDs include graphene, carbon nanotubes, and conducting polymers. The development of HTMs for QDLEDs is an active area of research, and continued progress in this field is expected to lead to further improvements in the efficiency and performance of QDLED displays. In addition to improving the charge transport properties of HTMs, researchers are also investigating ways to improve their stability and compatibility with other layers in QDLEDs. One major challenge is the degradation of HTMs under the high-energy conditions present in QDLEDs, which can lead to reduced device performance and lifetime. To address this issue, researchers have developed various strategies such as introducing stabilizing functional groups, using crosslinking agents to improve the mechanical stability of the HTM layer, and developing self-assembling monolayers to enhance the adhesion between the HTM and other layers in the device.

Another challenge in the development of HTMs for QDLEDs is achieving high efficiency and color purity in the emitted light. This requires careful tuning of the energy levels of the HTM and the emissive layer to ensure efficient charge injection and recombination. Additionally, the HTM must have a high transparency in the visible light range to minimize absorption losses. Researchers are exploring various strategies such as incorporating dopants into the HTM layer to tune its energy levels, using multiple HTMs with complementary properties to optimize charge injection and transport, and developing hybrid HTM/emissive layer structures to enhance the light emission efficiency.

The development of HTMs for QDLEDs is a complex and multidisciplinary field that requires expertise in materials science, chemistry, and device engineering. However, the potential benefits of QDLED displays, such as high brightness, color purity, and energy efficiency, make this research area of great interest and importance.

QDLEDs typically have a multilayer structure where achieving a balance between electron and hole injection is crucial for their efficiency and lifetime. However, QDLEDs often suffer from severe charge imbalance issues due to the differences in charge injection and transport properties between holes and electrons. HTMs such as PEDOT:PSS, poly-TPD, poly[(9,9-dioctylfluorenyl-2,7- diyl)-alt-(4,4′-(N-[4-butylphenyl]))] (TFB), and poly(9-vinylcarbazole) (PVK), and small molecule materials such as 4,4-bis(carbazole-9-yl) biphenyl (CBP), 1,3-bis(carbazol-9-yl)benzene (mCP), 2,6-bis(3-[9H-carbazol-9-yl]phenyl)pyridine (26DCzPPy), 4,4′,4″-tris(carbazol-9-yl)triphenylamine (TCTA), 1,1-bis[4-[N,N′-di(p-tolyl)amino]phenyl]cyclohexane (TAPC), and N,N′-bis(naphthalen-1-yl)-N,N′-bis(phenyl) benzidine (NPB), are commonly used. Metal oxide materials such as n-type MoOx,

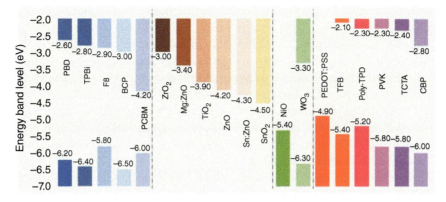

Figure 12.5 Charge-transporting materials commonly used in QDLEDs and their energy level diagrams.

VOx, WO$_3$, and p-type NiO and CuO can also be selected to enhance cavity injection, depending on the design of the front-loading and flip-chip QDLED device structures. PVK, TFB, and poly-TPD have all been successfully applied to red and green QDLED devices, but developing HTMs for blue and dark blue QDLED devices remains challenging. From an energy level alignment perspective, QD layers have the highest molecular orbitals (HOMO), making it difficult to inject holes into blue QDLEDs due to the large energy shift at the HTL/QD interface, resulting in carrier transport imbalances (Figure 12.5). In addition to developing new hole transport layer materials, some progress has been made in modifying existing materials to improve hole transport. Generally, two strategies are used to optimize the hole transport layer to increase hole injection density and achieve device carrier balance: doping and composition. The doping strategy optimizes the hole mobility and HOMO energy level of the hole transport layer through p-type doping, while the composite strategy uses a double or multilayer hole transport layer to form a stepped energy level and lower the hole injection barrier to enhance hole injection. Materials commonly used for doping and composition include organic small molecules, polymers, metal oxides, etc. Moreover, efforts to design and synthesize new material systems are ongoing. Studies have shown that achieving high efficiency in QDLED devices that rely on precise charge balance and exciton formation on QDs can be difficult and requires a hole transporting material with a lower HOMO level than the QD valence band and a band gap larger than that of QDs [19].

12.5.1 Doping of HTMs

PEDOT:PSS is a commonly used material in inverted structured full solution processing QDLEDs due to its high transparency, high conductivity, and high work function. However, the material also has some drawbacks that need to be addressed. One issue is the potential for interfacial defects between the PSS and the chalcogenide layer, which can lead to exciton bursts and affect device lifetime. Additionally, its acidity can corrode ITO substrates, and its hygroscopicity can impact the device stability of QDLEDs.

To mitigate these challenges, researchers have investigated physicochemical modifications of PEDOT:PSS and alternative organic HTMs. Strategies such as treating PEDOT:PSS with polar alcohols or incorporating surfactants have been shown to reduce the contact potential barrier and exciton burst. For example, Wang et al. treated PEDOT:PSS with methanol, which resulted in a reduced turn-on voltage and increased maximum brightness for $CH_3NH_3PbBr_3$ green light devices. Wu et al. also found that PEDOT:PSS films became more crystalline and had higher hole mobility when treated with more polar alcohols.

In another approach, Lee et al. incorporated a commercial fluorinated surfactant, Zonyl FS-300, into PEDOT:PSS to reduce the hole injection potential barrier and exciton burst. Pang et al. added a PSS-containing sodium salt (S-PSS) to the hole transport layer, which resulted in a PSS-rich layer between the PEDOT and chalcogenide layers and improved device performance. These strategies show promise for improving the performance and stability of QDLEDs using PEDOT:PSS and other organic HTMs.

Shi et al. demonstrated the use of a p-type semiconductor, MgNiO, as the hole transport layer in all-inorganic QD chalcogenide LEDs. This broad-band oxide semiconductor enables the modulation of hole injection performance by adjusting the forbidden band width through the Ni—Mg element ratios. The optimized device structure achieved a brightness of $3809\,cd\,m^{-2}$, a current efficiency of $2.25\,cd\,A^{-1}$, and an EQE of 2.39%. Remarkably, the device exhibited excellent air stability without encapsulation and retained 80% of its performance at 10 V for over 10 hours of continuous operation. Although inorganic oxides offer advantages as HTMs, additional passivation of surface defects is required due to the presence of metal vacancies that generate oxygen suspension bonds acting as exciton burst sites.

Doping with inorganic materials can overcome the inherent defects of organic HTMs such as PEDOT:PSS. Kim et al. employed MoO_3-doped PEDOT:PSS as the hole transport layer, which reduced the contact potential barrier with the light-emitting layer and improved hole injection. At a doping concentration of 0.3 wt%, the maximum device luminance reached $9200\,cd\,m^{-2}$. Furthermore, the unique surface physical properties of nanomaterials can enhance the properties of the hole transport layer and improve device performance. Xu et al. demonstrated the use of AuNPs doped into PEDOT:PSS, utilizing the far-field plasma effect to optimize the thickness of the luminescent layer to act at a similar distance to the plasma effect. This approach yielded excellent device performance with a maximum luminance of $1110\,cd\,m^{-2}$ and a maximum EQE of 1.64% in a blue chalcogenide device.

Xu et al. improved the performance of full-color QDLEDs by doping a monomer of poly-TPD as a small molecule into a PVK solution to obtain a composite hole transport layer. Compared to devices without dopants, the red, green, and blue (RGB) devices showed improved performance, indicating an overall enhancement of the QDLEDs based on this small molecule modification. CBP was used to dope TFB to achieve a small molecule/polymer composite hole transport layer for high-performance green devices with a maximum brightness of $90\,152\,cd\,m^{-2}$, much higher than CBP-only or polymer-only devices [20]. The homogeneity and wettability of the HTL surface were also optimized by adding the right amount of CBP to the TFB matrix. TCTA was used to dope TFB, and the blue QDLED achieved an EQE of 10.7% and a maximum luminance of 34 874 cd m^{-2} [21]. Theoretical simulations suggest that the small molecule doping strategy improves the carrier distribution balance in the device, increasing the compounding efficiency of electrons and holes and resulting in improved device performance. The lifetime of the device was also increased by a factor of approximately 3.5 due to more balanced charge injection [22]. Cross-linkable small molecules of CBP-V were added to TFB as dopants to reduce the injection barrier from the HTL to the QD layer, enhance the hole transport and compounding efficiency, and improve the thermal stability of the hole transport layer through cross-linking reactions. The presence of CBP-V also suppresses exciton bursts [23]. Heeyeop Chae et al. systematically investigated the effect of different small molecule and polymeric HTMs on the performance of QDLEDs and added different concentrations of small molecule materials as dopants to the devices as hole transport layers. Adding TCTA or CBP with high hole mobility to PVK significantly improves the electroluminescence performance of QDLEDs. Adding 20 wt% TCTA increased the maximum CE of PVK-based QDLEDs by 27%, reduced the on-state voltage, increased the current density, and improved the luminance, achieving a maximum luminance of 40 900 cd m^{-2} with a maximum C.E of 14.0 cd A^{-1} and a narrower full half-peak width (<35 nm) under optimal hole transport layer conditions [24].

Tian et al. devised a fast-carrier-mobility HTL by doping PVK TFB. The hole mobility of the doped TFB (P-TFB) HTL is increased from 1.08×10^{-3} to 2.09×10^{-3} cm^2 Vs^{-1}, which is attributed to the increased π–π stacking intensity (Figure 12.6). Meanwhile, P-TFB showed a small downshift of the HOMO level, resulting in a decreased barrier energy between the HTL and the QDs. QDLEDs based on the doped HTL presented a high EQE of 22.7%, a luminance of 133 100 cd cm^{-2}, a current efficiency of 29.3 cd A^{-1}, a power efficiency of 35.8 lm W^{-1}, and a long-term continuous operation stability T95 of 410 hours at 20 880 cd m^{-2} [25].

Inorganic metal oxides, with their deep HOMO energy levels (−5.5 to −6.3 eV), excellent moisture resistance, and air stability, are also widely used as cavity transport layers to help achieve high-efficiency QDLED devices. Li Guijun et al. demonstrated high-brightness green QDLED devices by doping with nickel oxide (NiO$_x$) and modifying the surface with wide-bandgap magnesium oxide (MgO). The luminescent devices using both strategies achieved a current efficiency of 6.08 cd A^{-1} and a maximum luminance of 40 000 cd m^{-2} at 10 V operating voltage [26]. Adding iron to nickel oxide can potentially reduce nickel vacancy defects and control the energy levels of NiO. QDLED devices with a hole injection layer of Fe-doped NiO demonstrate a high carrier concentration, which improves the device's EQE and lifespan [27]. Chen et al. utilized 11-mercapto-undecanoic acid (MUA)-modified NiO NPs as the hole transport layer in their study. The addition of MUA helped to decrease interfacial defects in the NiO layer and suppress exciton bursts. Furthermore, the reduction in valence band energy level of the modified

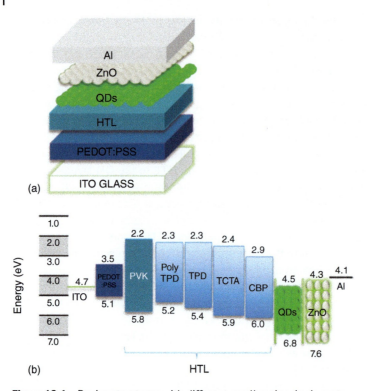

Figure 12.6 Device structures with different small molecule dopants.

NiO facilitated hole transport and improved charge balance. Surface engineering also improved the quality of the NiO film, leading to a decrease in leakage current. The resulting all-inorganic blue QDLED devices exhibited an EQE of 1.28% and a lifetime of up to 6350 hours. Importantly, the stability of the devices was significantly improved by more than 20 times after the modification of NiO with MUA [28]. Cl-passivated tungsten phosphate (Cl-TPA) was reported by Cao et al. They reported that TPA exhibits a relatively shallow HOMO energy level compared to NiO, which is favorable for hole injection, and that Cl effectively passivates the oxygen vacancies and suppresses the luminescence quenching [29]. Hong et al. employed an inorganic–organic hybridization approach to lower the HOMO energy level of PVK by incorporating metal oxides. They developed V_2O_5/PVK heterojunctions as hole transport layers, effectively reducing the injection barrier between QD and PVK. Inverted QDLEDs with this strategy showed improved electroluminescence performance compared to standard stacked structures, as the smoothly stepped hole conduction energy levels and significantly lower maximum Δh at the interface of the QD/PVK heterojunction achieved a more balanced charge carrier injection. These findings are detailed in their study (Figure 12.7) [30].

The incorporation of perfluorinated inomer (PFI, Nafion 117) PFI has been shown to effectively induce phase separation between the metal oxide and the polymer, resulting in energy band bending of the metal oxide and enhanced

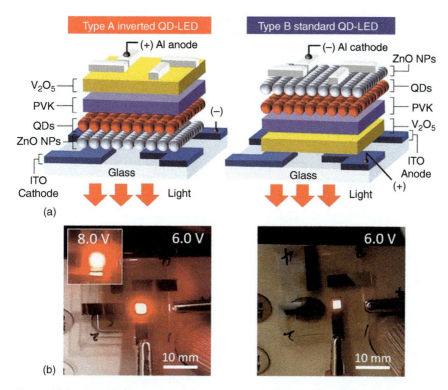

Figure 12.7 Organic/inorganic hybrid construction of V_2O_5/PVK heterojunctions as hole transport layers.

interfacial contact between the cavity transport layer and the luminescent layer. The performance of inorganic green QDLEDs was significantly improved as a result (Figure 12.8). The green QDLEDs with PFI and Cu–NiO hybrid HTL demonstrated a maximum current efficiency, power efficiency, and EQE of 7.3 cd A^{-1}, 2.1 lm W^{-1}, and 2.14%, respectively, which is four times higher than the maximum QE value of the QDLEDs that did not incorporate PFI and Cu–NiO [31].

12.5.2 Compositions of HTMS

Bilayer or cascade structures with CBP have been reported to suppress the accumulation of hole carriers at the QD interface and to enhance charge balance by blocking electron injection. A HTM of 2,2′,7,7′- tetrakis[N-naphthalenyl(phenyl)amino]-9,9-spirobifluorene (spiro-2NPB), with a HOMO of 5.5 eV, has been introduced into an inverted-structured Cd-based QDLED that exhibited 18% EQE. PVK is one of the most commonly used hole transport polymers owing to its deep HOMO level, but to overcome its slow hole mobility, the addition of conducting small molecules such as TCTA, CBP, or TPD dramatically improved the hole mobility of PVK. The bilayer of PVK/poly-TPD utilized the advantages of the deep HOMO of PVK to facilitate efficient hole injection into the QD EML and high mobility of TPD to reduce a turn-on voltage (Figure 12.9) [20].

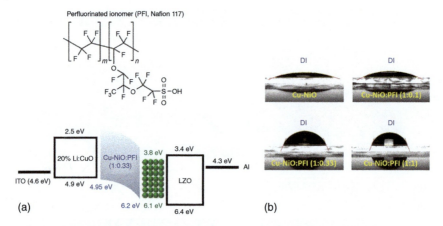

Figure 12.8 Molecular structure of PFI and the energy level alignment of all solution-processed conventional QDLEDs with the Cu–NiO:PFI HTL layer has the valance band bending by 1.25 eV from the interface of HIL/HTL (a); The contact angle images of DI droplet on Cu–NiO:PFI layer (b).

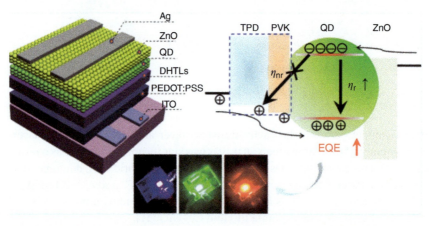

Figure 12.9 Device architecture and flat-band diagram of the QDLED. Source: Li et al. [20]/American Chemical Society.

Yang et al. fabricated a bilayer hole transport layer based on 4,4′-bis(9-carbazole) biphenyl (CBP)/1,3-bis(9H-pyrido[2,3-b]indol-9-yl)benzene (mCaP). By using the bilayer stepwise HTL, the HOMO energy level was reduced to −6.27 eV, resulting in an EQE of 12.6% in green QDLEDs. The results suggest that developing stepwise HTMs with high LOMO levels is an effective approach to improving hole transport/injection and carrier balancing in QDLEDs [32]. In another study, Wang et al. prepared red QDLEDs with a double-hole transport layer by thermally spin-coating a PVK layer on a poly TPD film. The thermal spin-coating technique allowed for the progressive preparation of high-quality HTLs, resulting in a reduced turn-on voltage and a much higher EQE of 15.3% compared to single-transport-layer devices. Thus, the thermal spin-coating strategy can be considered an effective technique for

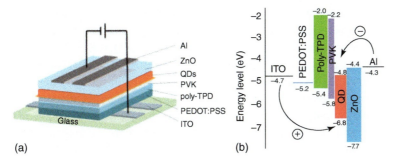

Figure 12.10 (a) The device structure of the multilayered QDLED device. (b) The energy level diagram for the various layers.

improving surface morphology and charge balance to achieve high-performance and low on-state voltage QDLEDs (Figure 12.10) [33].

12.5.3 New HTM Materials for QDLED

In addition to improving efficiency and stability, researchers are also focused on developing HTMs that can enable the production of flexible and transparent QDLEDs. To this end, materials such as conductive polymers and graphene-based materials are being explored as potential HTMs for flexible QDLEDs. One important consideration in HTM development is their energy level alignment with the emissive layer. This is because the energy level offset between the HTM and the emissive layer determines the efficiency of hole injection into the QDs. Therefore, researchers are also studying the electronic properties of new HTMs using computational modeling and spectroscopic techniques. Another important consideration in the development of new HTMs for QDLEDs is their compatibility with the other layers in the device, such as the electron transport layer (ETL), the emissive layer, and the cathode. This is because any chemical or electronic interactions between the layers can affect the devices performance.

Researchers are also investigating the use of multi-layered HTMs that can form a stable interface with the other layers in the device. For example, a recent study reported the use of a multi-layered HTM composed of a triazine-based material and a metal oxide layer, which resulted in improved device efficiency and stability.

In addition to improving the performance of QDLEDs, the development of new HTMs can also address the issues of toxicity and sustainability in the production and disposal of QDLEDs. Many of the currently used HTMs in QDLEDs contain heavy metals, such as lead and cadmium, which can be harmful to human health and the environment. Therefore, researchers are exploring the use of nontoxic and sustainable materials as HTMs for QDLEDs. A recent study reported the use of a bio-based HTM derived from cellulose, which demonstrated high efficiency and stability in QDLEDs. Furthermore, researchers are also investigating the use of earth-abundant materials, such as copper, as HTMs for QDLEDs. These materials offer low costs and high scalability for large-scale production of QDLEDs.

The use of solution-processed HTMs for large-scale production of QDLEDs is also explored. Solution processing methods such as spin-coating, inkjet printing, and spray coating offer low-cost and high-throughput fabrication of QDLEDs. However, the choice of HTM is critical for the success of these methods, as the HTM must have good solubility and film-forming properties. In summary, the development of new HTMs is a critical aspect of QDLED research, as it can significantly impact the efficiency, stability, and flexibility of these devices. The use of a wide range of organic, inorganic, and hybrid materials, coupled with computational and spectroscopic studies, is expected to lead to the discovery of new and improved HTMs for QDLEDs in the future.

In the following parts, we will discuss the very recent HTM development. To improve the surface morphology and hole charge mobility of HTMs, Zhang et al. developed a polymeric HTM, the poly(indenofuorene-co-triphenylamine) copolymer (PIF-TPA) to replace the poly(fuorene-co-triphenylamine) copolymer (PF-TPA) as the hole-transporting layer (HTL) in red QDLEDs. The introduction of alkyl chains enhanced the intermolecular π–π interactions, resulting in excellent hole mobility of up to (10^{-2} cm^2 V^{-1} s^{-1}), QDLEDs using PIF-TPA exhibited higher current efficiency, lower on-state voltage, and longer device lifetime compared to those using PF-TPA, demonstrating the potential of PIF-TPA as a promising HTM for QDLEDs (Figure 12.11) [34].

The cross-linked polymer provides a robust platform for the consecutive QD deposition. Huang et al. developed a novel cross-linkable small molecule HTM, VB-FNPD, for interfacial modification and hole transport between PEDOT:PSS and light-emitting layers. They reported on CdSe@ZnS light-emitting diodes (QDLEDs) showing both high efficiency and long operational lifetime by introducing thermally cross-linkable 2,7-disubstituted fluorene-based triaryldiamine (VB-FNPD) as the hole transporting layer and 1,4,5,8,9,11-hexaazatriphenylene hexacarbonitrile (HAT-CN) as the hole injection layer. This reduced the injection energy barrier while improving the morphology of chalcogenide films, resulting in green light devices with a maximum EQE of 8.0% and power efficiency (PE) of 32.1 lm W^{-1} with low-efficiency roll-offs (Figure 12.12) [35]. VB-FNPD HTLs have smoother and more uniform morphologies when compared to TFB. As a result using VB-FNPD instead of TFB doubles the electroluminescence half-life (LT50) of the devices, leading to an LT50 of 10 100 hours versus only 4900 hours for the TFB device at an initial luminance (L_0) of 1000 cd m^{-2}. This may help improve the quality of the HTL/QD interface and QD film uniformity, both of which are important for long-lived QDLEDs [36].

Efficient approach based on the photochemistry of benzophenone has been developed for the cross-linking of the polymer hole-transporting layer (C-TFB) (Figure 12.13). The spin-coated red QDLEDs based on the cross-linked HTLs showed the maximum current efficiency (CE), the maximum power efficiency (PE), and the peak EQE of 32.3 cd A^{-1}, 42.3 lm W^{-1}, and 21.4%, respectively. The inkjet-printed red QDLEDs with the cross-linked HTLs exhibited the CE, PE, and EQE of 26.5 cd A^{-1}, 37.8 lm W^{-1}, and 18.1%, respectively. The solution-processed

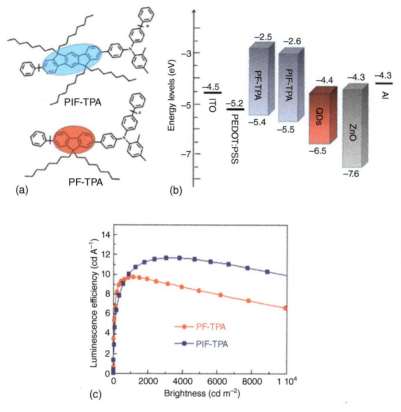

Figure 12.11 (a) Molecular structure of PIF-TPA and PF-TPA; (b) energy level diagram of the used materials; (c) luminescence efficiency versus brightness of red QDLEDs with PIF-TPA and PF-TPA as HTLs. Source: Gao et al. [34]/Springer Nature.

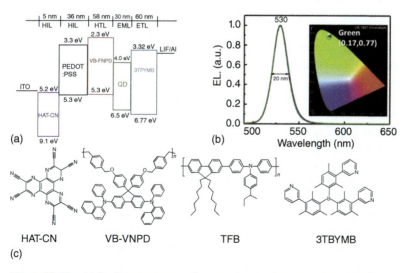

Figure 12.12 (a) Device structure and energy levels of the materials; (b) EL spectra, (c) the molecular structure of HIL(HAT-CN), cross-linked HTL(VB-FNPD), HTL(TFB), and ETL(3TPYMB).

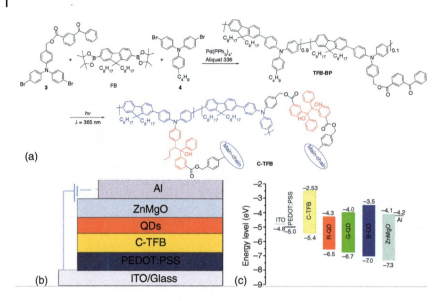

Figure 12.13 (a) Scheme of synthesis and photocross-linking of the TFB − BP polymer; (b) QDLED device structure; and (c) energy levels of the functional layers of red, green, and blue QDLEDs.

RGB QDLEDs based on the cross-linked HTLs showed much better performances than the corresponding devices based on the pristine TFB HTLs [37].

Recently, a new design strategy has been developed that uses a series of linear bis-benzophenone-based cross-linkers (BPO4, BPO6, BPO8) to enable the photothermal curing of a hole transport layer (HTL) in high-performance QDLEDs with ultralow efficiency roll-off. By blending these cross-linkers with TFB at an extremely low weight ratio and subjecting them to mild photothermal curing, a crosslinked compact HTL film can be generated. This study demonstrates that the addition of cross-linkers to traditional polymer HTMs can significantly improve the efficiency and stability of solution-processed QDLEDs. This technique also has potential applications in the manufacturing of large-area display panels using inkjet printing processes [38].

Recent studies on OLEDs and QDLEDs have indicated that double or blended HTL structures with a stepwise HOMO energy level alignment could benefit the hole transport at interfaces. A new cross-linkable small molecule 4,4′-bis(3-vinyl-9H-carbazol-9-yl)- 1,1′-biphenyl (CBP-V) (HOMO ∼ −6.2 eV) blending with polymer TFB (HOMO ∼ − 5.4 eV) is presented for red QDLEDs. Compared with the TFB-only devices, the EQE of devices with the blended HTL improved from 15.9 to 22.3% without the increase in turn-on voltage for spin-coating-fabricated devices. Furthermore, the blended HTL prolonged the T90 and T70 lifetimes from 5.4 and 31.1 to 39.4 and 148.9 hours, respectively (Figure 12.14). These enhancements in lifetime are attributed to the low hole-injection barrier at the HTL/QD interface and high thermal stability of the blended HTL after cross-linking. Moreover, the cross-linked blended HTL showed excellent solvent resistance after crosslinking, and the EQE of the inkjet-printed red QDLEDs reached 16.9%. The new

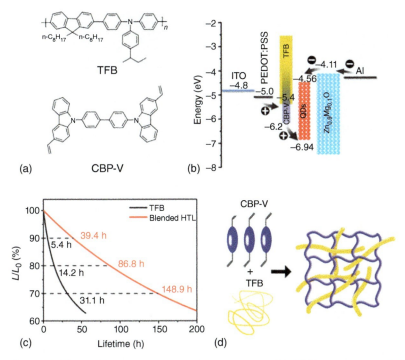

Figure 12.14 (a) Chemical structures of the polymer HTM TFB and cross-linkable small-molecule HTM CBP-V; (b) Schematic of the device structure, the energy level of the materials, and holes transporting from the hole injection layer (HIL) to QD layer d); (c) Schematic of the cross-linking process of the blended HTL. Source: Tang et al. [23]/American Chemical Society.

solvent-resistant blended HTL paves the way for further development of highly efficient and stable printed QDLEDs.

It has been identified that electron leakage from QDs to hole transport layers (HTLs) is a significant efficiency loss channel in green and blue QDLEDs, resulting from the energetic disorder of polymeric HTLs and geometric properties of the interfacial materials. Accordingly, Huang et al. designed the copolymer of poly((9,9-dioctylfluorenyl-2,7-diyl)-alt-(9-[2-ethylhexyl]carbazole-3,6-diyl)) (PF8Cz), which employs a HTM with rigid backbones and limited conjugation, resulting in a shallower LUMO level and reduced energetic disorder (Figure 12.15). This copolymer was developed to eliminate electron leakage, and the researchers demonstrated high efficiencies (peak external quantum efficiencies of 28.7% for green and 21.9% for blue). The best-performing solution-processed green and blue QDLEDs also exhibited a long lifetime (extrapolated T95 lifetime is 580 000 hours for green and 4400 hours for blue QDLEDs).

12.6 Hole Injection Materials for QDLED

Typically, QDLED devices have a hybrid structure that includes an organic hole injection/transport layer and an inorganic electron injection/transport layer.

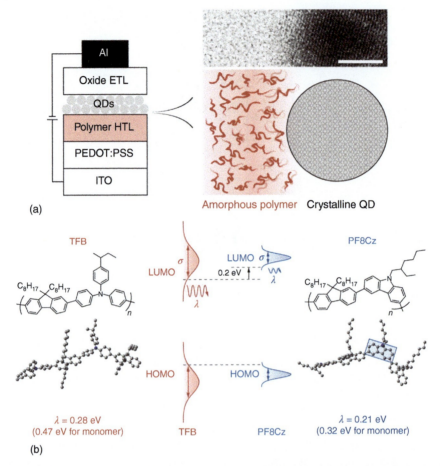

Figure 12.15 (a) Typical QDLED structure (left), HTL/QD interface (bottom right), and cross-sectional transmission electron microscopy image of the interface (top right; scale bar, 5 nm); (b) Theoretically optimized geometries and reorganization energies (λ) for TFB (left) and PF8Cz (right) dimers. Middle, a comparison of the electronic structures. Source: Deng et al. [39]/Springer Nature.

Poly(3,4-Poly (3,4-ethylenedioxythiophene):poly(styrene-sulfonate) (PEDOT:PSS) is widely used as the hole injection layer (HIL) for QDLED. It is widely used in QDLED devices because of its moderate work function, high electrical conductivity, transparency, and solution processability. However, the long-term stability of PEDOT:PSS remains problematic because of its acidity and hygroscopicity, which may corrode the indium tin oxide (ITO) substrates, and may further accelerate the degradation of the subsequently deposited QDs layer.

In replacing the PEDOT:PSS hole injection polymer, interfacial layers incorporating exfoliated molybdenum disulfide (MoS_2) flakes into self-doped conductive CPEs were proposed. The pH-neutral and self-doped conductive CPE, poly[2,6-(4,4-bis-potassiumbutanylsulfonate4H-cyclopenta-[2,1-b;3,4-b']-dithiophene)-alt-4,7-(2,1,3-benzo-thiadiazole)] (P1, Figure 12.16.), was designed to modulate the electrolyte

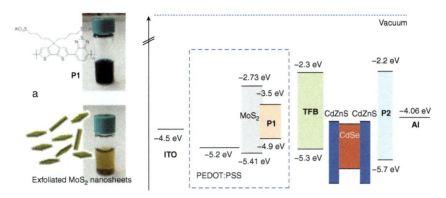

Figure 12.16 (a) Chemical structures of P1 and exfoliated MoS$_2$ nanosheets; (b) flat-band energy diagram of the QDLED device. Source: Lagonegro et al. [40]/American Chemical Society.

energy level state and charge transport capability by adjusting the polymer backbone monomer, alkyl side chain length, and cation species. This allowed for effective hole injection and electron blocking, and the optimized device structure achieved a maximum EQE of 5.66%.

Compared with the organic materials-based devices, solution-processed metal-oxide films as the interfacial buffer layers are often more efficient as compared to the metal oxide bulk films prepared by vacuum deposition because of their better interfacial phase compatibility with the QD layer. Transition metal oxides have been investigated with various thin film fabrication methods, such as NP dispersion, sol–gel method, and dissolution of metal oxide powders. In addition, MoO$_3$, CuO, NiO$_x$, VO$_x$, and WO$_3$ are commonly investigated as HILs for organic electronic and QDLED devices. However, more research is still needed to develop solution-based HILs to achieve high stability and efficiency. High-performance QDLEDs using NiO$_x$ hole-injection layers with a high and stable work function. Doping can play an important role in adjusting the electrical properties of semiconducting materials and improving their charge transport properties. Cu-doped NiO-based QDLED exhibits the optimal device performances, with a maximum luminance of 61 030 cd m^{-2}, a current efficiency of 45.7 cd A^{-1}, and an almost fourfold operating lifetime enhancement compared with PEDOT: PSS-based QDLED [41].

NiO$_x$ is an intrinsic p-type oxide semiconductor with work function normally lower than 5.2 eV, which cannot form low-resistance contact with the organic HTLs with high ionization potentials. Jin et al., by introducing a self-assembled monolayer using 4-(trifluoromethyl) benzoic acid, which creates strong interfacial dipoles to induce a shift of vacuum level achieved solution-processed NiO$_x$ films with high and stable work functions of up to ≈5.7 eV (Figure 12.17). The use of the NiO$_x$–BA–CF$_3$ films as HILs offers excellent hole injection into the polymeric HTLs, leading to efficient driving and long-lifetime QDLEDs [42].

SAMs containing a trifluoromethyl group exhibit an angled orientation relative to the NiO$_x$ surface, which activates the hole injection into the active layer without inducing luminescence quenching. By optimizing the energy level shift and

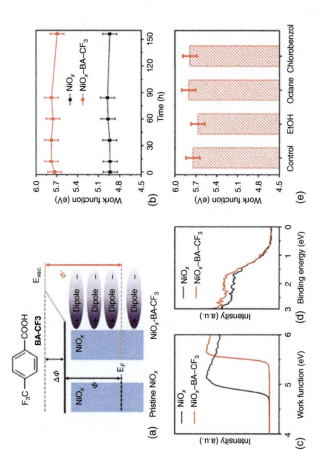

Figure 12.17 High and stable work functions of the NiO$_x$–BA–CF$_3$ films. (a) Schematic diagram showing the surface dipole-induced change of the work function of the NiO$_x$ film. (b) Time-dependent work functions of the pristine NiO$_x$ film and the NiO$_x$–BA–CF$_3$ film. (c) Work function and (d) valence spectra of the pristine NiO$_x$ film and the NiO$_x$–BA–CF$_3$ film measured by UPS. (e) Work functions of the NiO$_x$–BA–CF$_3$ films after solvent rinsing. Source: Lin et al. [42]/John Wiley & Sons.

passivating surface defect states of NiO$_x$, the Sargent group was able to achieve an EQE of 18.8% in InP QDLEDs using inorganic hole injection layers (HILs) based on the 4-CF$_3$-BA self-assembled monolayer (SAM) [43].

Recently, phosphomolybdic acid (PMA), which is a heteropolyacid containing MoO$_3$, has been reported to work as a HIL in QDLEDs and organic electronics. Because PMA is soluble in polar solvents, it has process orthogonality with nonpolar characteristics of organic materials [44].

The use of multi- or bi-layered HILs is an excellent approach to reducing the energy barrier of each step. A graphene oxide (GO)/PEDOT:PSS bi-layered HIL and obtained a maximum luminance three times higher than that of a single-HIL QDLED [45]. Vanadium oxide (V$_2$O$_x$) layer was deposited on a typical HIL, PEDOT:PSS layer to form bi-layer HIM. The combination of PEDOT:PSS and V$_2$O$_x$ as a bi-layered HIL resulted in significant improvements in QDLED performances [46].

The single-layered V$_2$O$_x$ HIL or PEDOT:PSS HIL has weaknesses such as a lower hole mobility and a higher density of deep defects, respectively. However, the bi-layered HIL alleviates the weakness of each single-layered HIL by compensating for the weakness of the other layer. In addition, the step-wise energy level structure in the bilayer of the HIL device enhances the hole current density and improves the charge balance. As a result, the bi-layered-HIL device exhibits a 60% higher EQE and more than twice the maximum luminance of the single-HIL devices (Figure 12.18) [47].

Here are some of the key breakthroughs in HTMs development for colloidal QDLED from 2000 to 2023:

1. Introduction of organic small molecule HTMs such as 4,4′,4″-tris(N-carbazolyl)-triphenylamine (TCTA), which improves the device efficiency and lifetime.
2. Development of inorganic HTMs such as CuSCN, CuI, and CuO, which exhibit high hole mobility and stability.
3. Hybrid organic–inorganic HTMs such as PEDOT:PSS with added inorganic NPs such as ZnO and TiO$_2$, which enhance the hole transport and injection efficiency.
4. Development of novel hole transport polymers such as PTP, PTAA, and PFN, which exhibit high hole mobility and stability, as well as good compatibility with colloidal QDs.

The device performance data varies depending on the specific HTMs used, as well as the device structure and fabrication process. However, here are some general performance data for colloidal QDLED using different HTMs:

- TCTA: CIE value ~0.24, maximum EQE ~4.8%, and lifetime >1000 hours
- CuSCN: CIE value ~0.18, maximum EQE ~3.5%, and lifetime >500 hours
- PEDOT:PSS with ZnO NPs: CIE value ~0.22, maximum EQE ~5.6%, and lifetime >500 hours
- PTP: CIE value ~0.22, maximum EQE ~6.7%, and lifetime >1000 hours

In addition to facilitating hole injection, the HTM can also serve as a barrier layer to prevent exciton quenching or charge recombination at the anode/QD interface. This can help to improve device stability and reduce device degradation over time.

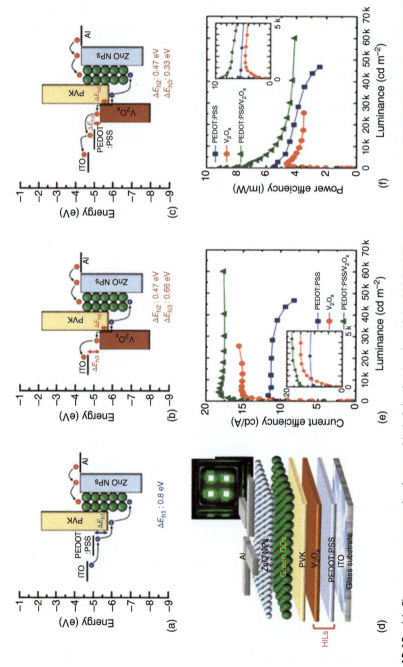

Figure 12.18 (a). Charge transport mechanisms and hole injection barriers of the PEDOT:PSS-; (b) V_2O_x-; and (c) PEDOT:PSS/V_2O_x-bi-layered-HIL devices. (d) Schematic of the multi-layered QDLEDs. (inset) Photograph of the QDLED under bias; (e) CE, and (f) PE versus luminance for the QDLEDs. The insets present efficiencies up to 5 kcd m^{-2}.

12.7 Summary and Outlook

The choice of HIMs, HTMs, EIMs, and ETMs is critical for achieving high efficiency and stability of QDLEDs. Different materials have different properties, such as energy levels, charge carrier mobility, and chemical stability, that affect their performance in QDLEDs. Several types of materials have been developed and studied for use as HIMs, HTMs, EIMs, and ETMs in QDLEDs.

In conclusion, the development of high-performance HIMs, HTMs, EIMs, and ETMs is essential for advancing the technology of QDLEDs, which have the potential to revolutionize display and lighting applications with their high efficiency, color purity, and tunability. In recent years, there has been a growing interest in developing hybrid materials that combine the advantages of both organic and inorganic materials. For instance, a hybrid ETM based on a blend of organic and inorganic materials was found to have higher electron mobility and stability than traditional inorganic ETMs.

ZnO has been widely used as an ETM, but there are limitations to balancing charge transport to improve the efficiency of QDLED devices. Various attempts have been made with other metal oxide NPs, such as metal doping, metal oxide doping, metal salt doping, etc., and composite material strategies using organic polymers and doped metal oxide composites. These strategies can improve efficiency to some extent, while also reducing exciton bursts, offering the possibility of efficient QDLEDs in the future. The design of electron transport layers for QDLEDs offers possibilities, and the composite of multilayer electron transport layers may become the mainstream development trend. Organic small molecules can also be considered to further enhance carrier mobility and injection characteristics.

Although progress has been made in modifying hole transport layers using organic small molecules, polymers, and inorganic metal oxides, there is still a lack of deep HOMO energy level HTM systems suitable for blue QDLED devices. New materials play a key role in improving the performance of blue and other light-colored QDLED devices, helping to address the core issues of QDLED devices for displays and lighting.

One of the main challenges in the development of HIMs, HTMs, EIMs, and ETMs for QDLEDs is achieving a balance between charge injection/transport and preventing quenching of the luminescence in the QDs. This requires careful optimization of the energy levels of the materials and their interfaces with the QDs. Another important consideration is the chemical and thermal stability of these materials, as QDLEDs can be subject to degradation over time due to exposure to oxygen and moisture, as well as high temperatures during operation. This has led to the development of materials with enhanced stability, such as encapsulation layers and passivation coatings. Furthermore, the performance of HIMs, HTMs, EIMs, and ETMs can also be affected by the device architecture and processing conditions, such as the deposition technique, substrate material, and the choice of electrodes.

Another area of research is the development of environmentally friendly and low-cost materials for use as HIMs, HTMs, EIMs, and ETMs in QDLEDs. This includes materials such as carbon nanotubes and graphene, which have shown promising properties as charge injection and transport materials. Furthermore, the integration of QDLEDs with other technologies, such as flexible and transparent substrates, is also an active area of research. This could enable the development of new types of flexible and transparent displays and lighting systems. In addition, there is a need for further research into the mechanisms that govern the charge injection and transport processes in QDLEDs, as well as the interactions between the charge carriers and the QDs. This could lead to a better understanding of the factors that affect the performance and stability of these devices and help guide the development of new materials and device architectures.

In summary, the development of high-performance HIMs, HTMs, EIMs, and ETMs for QDLEDs is an active and interdisciplinary field of research. Progress in this area has led to significant improvements in the efficiency and stability of QDLEDs, making them a promising technology for future display and lighting applications. However, further research is needed to optimize the performance of these materials and to address remaining challenges in device fabrication and integration. The development of HIMs, HTMs, EIMs, and ETMs in QDLEDs is an exciting and rapidly evolving field of research with many potential applications. The continued progress in this area will require the collaboration of researchers from various disciplines, including chemistry, physics, materials science, and engineering. With continued research and development, QDLEDs have the potential to revolutionize these industries, offering new levels of efficiency, color purity, and tunability.

References

1 Qian, L., Zheng, Y., Xue, J., and Holloway, P.H. (2011). Stable and efficient quantum-dot light-emitting diodes based on solution-processed multilayer structures. *Nature Photonics* 5 (9): 543–548.
2 Cao, S., Zheng, J., Zhao, J. et al. (2017). Enhancing the performance of quantum dot light-emitting diodes using room-temperature-processed Ga-doped ZnO nanoparticles as the electron transport layer. *ACS Applied Materials & Interfaces* 9 (18): 15605–15614.
3 Wang, L., Pan, J., Qian, J. et al. (2018). A highly efficient white quantum dot light-emitting diode employing magnesium doped zinc oxide as the electron transport layer based on bilayered quantum dot layers. *Journal of Materials Chemistry C* 6 (30): 8099–8104.
4 Sun, Y., Wang, W., Zhang, H. et al. (2018). High-performance quantum dot light-emitting diodes based on Al-doped ZnO nanoparticles electron transport layer. *ACS Applied Materials & Interfaces* 10 (22): 18902–18909.
5 Alexandrov, A., Zvaigzne, M., Lypenko, D. et al. (2020). Al-, Ga-, Mg-, or Li-doped zinc oxide nanoparticles as electron transport layers for quantum dot light-emitting diodes. *Scientific Reports* 10 (1).
6 Lee, Y., Kim, H.-M., Kim, J., and Jang, J. (2019). Remarkable lifetime improvement of quantum-dot light emitting diodes by incorporating rubidium carbonate

in metal-oxide electron transport layers. *Journal of Materials Chemistry C* 7 (32): 10082–10091.

7 Park, M., Roh, J., Lim, J. et al. (2020). Double metal oxide electron transport layers for colloidal quantum dot light-emitting diodes. *Nanomaterials* 10 (4).

8 Zhao, J., Zhang, X.Y., Zhang, Y. et al. (2013). Quantum dot array LED research with ZnO as an electron transport layer. *Applied Mechanics and Materials* 333–335: 1895–1898.

9 Mueller, A.H., Petruska, M.A., Achermann, M. et al. (2005). Multicolor light-emitting diodes based on semiconductor nanocrystals encapsulated in GaN charge injection layers. *Nano Letters* 5 (6): 1039–1044.

10 Hikmet, R.A.M., Talapin, D.V., and Weller, H. (2003). Study of conduction mechanism and electroluminescence in CdSe/ZnS quantum dot composites. *Journal of Applied Physics* 93 (6): 3509–3514.

11 Qasim, K., Chen, J., Zhou, Y., and Lei, W. (2012). Enhanced electrical efficiency of quantum dot based LEDs with TiO_2 as the electron transport layer fabricated under the optimized annealing-time conditions. *Journal of Nanoscience and Nanotechnology* 12 (10): 7879–7884.

12 Zhang, Q., Gu, X., Zhang, Q. et al. (2018). ZnMgO:ZnO composite films for fast electron transport and high charge balance in quantum dot light emitting diodes. *Optical Materials Express* 8 (4): 909–918.

13 Yuan, Q., Guan, X., Xue, X. et al. (2019). Efficient CuInS2/ZnS quantum dots light-emitting diodes in deep red region using PEIE modified ZnO electron transport layer. *Physica Status Solidi (RRL)–Rapid Research Letters* 13 (5): 1800575.

14 Alsharafi, R., Zhu, Y., Li, F. et al. (2019). Boosting the performance of quantum dot light-emitting diodes with mg and PVP co-doped ZnO as electron transport layer. *Organic Electronics* 75: 105411.

15 Zhou, Y., Fuentes-Hernandez, C., Shim, J. et al. (2012). A universal method to produce low&-work function electrodes for organic electronics. *Science* 336 (6079): 327–332.

16 Tang, C.G., Syafiqah, M.N., Koh, Q.-M. et al. (2019). Multivalent anions as universal latent electron donors. *Nature* 573 (7775): 519–525.

17 Chen, L., Lee, M.-H., Wang, Y. et al. (2018). Interface dipole for remarkable efficiency enhancement in all-solution-processable transparent inverted quantum dot light-emitting diodes. *Journal of Materials Chemistry C* 6 (10): 2596–2603.

18 Lim, J., Park, M., Bae, W.K. et al. (2013). Highly efficient cadmium-free quantum dot light-emitting diodes enabled by the direct formation of excitons within InP@ZnSeS quantum dots. *ACS Nano* 7 (10): 9019–9026.

19 Anikeeva, P.O., Madigan, C.F., Halpert, J.E. et al. (2008). Electronic and excitonic processes in light-emitting devices based on organic materials and colloidal quantum dots. *Physical Review B* 78 (8): 085434.

20 Li, J., Liang, Z., Su, Q. et al. (2018). Small molecule-modified hole transport layer targeting low turn-on-voltage, bright, and efficient full-color quantum dot light emitting diodes. *ACS Applied Materials & Interfaces* 10 (4): 3865–3873.

21 Zhao, Y., Chen, L., Wu, J. et al. (2020). Composite hole transport layer consisting of high-mobility polymer and small molecule with deep-lying HOMO level for

efficient quantum dot light-emitting diodes. *IEEE Electron Device Letters* 41 (1): 80–83.

22 Wang, F., Sun, W., Liu, P. et al. (2019). Achieving balanced charge injection of blue quantum dot light-emitting diodes through transport layer doping strategies. *The Journal of Physical Chemistry Letters* 10 (5): 960–965.

23 Tang, P., Xie, L., Xiong, X. et al. (2020). Realizing 22.3% EQE and 7-fold lifetime enhancement in QLEDs via blending polymer TFB and cross-linkable small molecules for a solvent-resistant hole transport layer. *ACS Applied Materials & Interfaces* 12 (11): 13087–13095.

24 Ho, M.D., Kim, D., Kim, N. et al. (2013). Polymer and small molecule mixture for organic hole transport layers in quantum dot light-emitting diodes. *ACS Applied Materials & Interfaces* 5 (23): 12369–12374.

25 Cheng, C., Liu, A., Ba, G. et al. (2022). High-efficiency quantum-dot light-emitting diodes enabled by boosting the hole injection. *Journal of Materials Chemistry C* 10 (40): 15200–15206.

26 Jiang, Y., Jiang, L., Yan Yeung, F.S. et al. (2019). All-inorganic quantum-dot light-emitting diodes with reduced exciton quenching by a MgO decorated inorganic hole transport layer. *ACS Applied Materials & Interfaces* 11 (12): 11119–11124.

27 Zhang, Y., Wang, X., Chen, Y., and Gao, Y. (2020). Improved electroluminescence performance of quantum dot light-emitting diodes: a promising hole injection layer of Fe-doped NiO nanocrystals. *Optical Materials* 107: 110158.

28 Yoon, S.-Y., Kim, J.-H., Kim, K.-H. et al. (2019). High-efficiency blue and white electroluminescent devices based on non-cd I–III–VI quantum dots. *Nano Energy* 63: 103869.

29 Cao, F., Wu, Q., Sui, Y. et al. (2021). All-inorganic quantum dot light-emitting diodes with suppressed luminance quenching enabled by chloride passivated tungsten phosphate hole transport layers. *Small* 17 (19): 2100030.

30 Park, Y.R., Choi, W.K., and Hong, Y.J. (2018). Hole barrier height reduction in inverted quantum-dot light-emitting diodes with vanadium(V) oxide/poly(N-vinylcarbazole) hole transport layer. *Applied Physics Letters* 113 (4): 043301.

31 Kim, H.-M., Kim, J., and Jang, J. (2018). Quantum-dot light-emitting diodes with a perfluorinated ionomer-doped copper-nickel oxide hole transporting layer. *Nanoscale* 10 (15): 7281–7290.

32 Wang, X., Shen, P., Cao, F. et al. (2019). Stepwise bi-layer hole-transport interlayers with deep highest occupied molecular orbital level for efficient green quantum dot light-emitting diodes. *IEEE Electron Device Letters* 40 (7): 1139–1142.

33 Chen, H., Ding, K., Fan, L. et al. (2018). All-solution-processed quantum dot light emitting diodes based on double hole transport layers by hot spin-coating with highly efficient and low turn-on voltage. *ACS Applied Materials & Interfaces* 10 (34): 29076–29082.

34 Gao, P., Lan, X., Sun, J. et al. (2020). Enhancing performance of quantum-dot light-emitting diodes based on poly(indenofluorene-co-triphenylamine) copolymer as hole-transporting layer. *Journal of Materials Science: Materials in Electronics* 31 (3): 2551–2556.

35 Chao, S.-W., Chen, W.-S., Hung, W.-Y. et al. (2019). Cross-linkable hole transporting layers boost operational stability of high-performance quantum dot light-emitting device. *Organic Electronics* 71: 206–211.

36 Ghorbani, A., Chen, J., Samaeifar, F. et al. (2022). Stability improvement in quantum-dot light-emitting devices via a new robust hole transport layer. *The Journal of Physical Chemistry C* 126 (42): 18144–18151.

37 Sun, W., Xie, L., Guo, X. et al. (2020). Photocross-linkable hole transport materials for inkjet-printed high-efficient quantum dot light-emitting diodes. *ACS Applied Materials & Interfaces* 12 (52): 58369–58377.

38 Yi, Y.-Q.-Q., Yang, J., Xie, L. et al. (2022). Linear cross-linkers enabling photothermally cured hole transport layer for high-performance quantum dots light-emitting diodes with ultralow efficiency roll-off. *Chemical Engineering Journal* 439: 135702.

39 Deng, Y., Peng, F., Lu, Y. et al. (2022). Solution-processed green and blue quantum-dot light-emitting diodes with eliminated charge leakage. *Nature Photonics* 16 (7): 505–511.

40 Lagonegro, P., Martella, C., Squeo, B.M. et al. (2020). Prolonged lifetime in nanocrystal light-emitting diodes incorporating MoS2-based conjugated polyelectrolyte interfacial layer as an alternative to PEDOT:PSS. *ACS Applied Electronic Materials* 2 (5): 1186–1192.

41 Cao, F., Wang, H., Shen, P. et al. (2017). High-efficiency and stable quantum dot light-emitting diodes enabled by a solution-processed metal-doped nickel oxide hole injection interfacial layer. *Advanced Functional Materials* 27 (42): 1704278.

42 Lin, J., Dai, X., Liang, X. et al. (2020). High-performance quantum-dot light-emitting diodes using NiO_x hole-injection layers with a high and stable work function. *Advanced Functional Materials* 30 (5): 1907265.

43 Lee, S., Park, S.M., Jung, E.D. et al. (2022). Dipole engineering through the orientation of Interface molecules for efficient InP quantum dot light-emitting diodes. *Journal of the American Chemical Society* 144 (45): 20923–20930.

44 Hwang, J.H., Seo, E., Park, S. et al. (2023). All-solution-processed quantum dot light-emitting diode using phosphomolybdic acid as hole injection layer. *Materials* 16 (4): 1371.

45 Song, D.-H., Song, S.-H., Shen, T.-Z. et al. (2017). Quantum dot light-emitting diodes using a graphene oxide/PEDOT:PSS bilayer as hole injection layer. *RSC Advances* 7 (69): 43396–43402.

46 Jiang, X., Ma, Y., Tian, Y. et al. (2020). High-efficiency and stable quantum dot light-emitting diodes with staircase V2O5/PEDOT:PSS hole injection layer interface barrier. *Organic Electronics* 78: 105589.

47 Song, S.-H., Yoo, J.-I., Kim, H.-B. et al. (2022). Hole injection improvement in quantum-dot light-emitting diodes using bi-layered hole injection layer of PEDOT:PSS and V2O. *Optics & Laser Technology* 149: 107864.

13

Quantum Dot Industrial Development and Patent Layout

13.1 Introduction

Quantum dots are inorganic semiconductor nanocrystals with a size between 1–100 nm that exhibit quantum size effects and are solution-processable. They have excellent luminescence properties, including a continuously adjustable emission peak, high color purity, good luminescence stability, high fluorescence quantum yield (QY), and long lifetime. Quantum dots are widely used in various fields, such as solid-state lighting, displays, solar cells, biomedical labeling, and photocatalysis. They possess great market potential and value. The main application of quantum dot materials is in the display industry, with two primary technical development directions: quantum dot photoluminescence and quantum dot electroluminescence. Quantum dot photoluminescent display products (quantum dot backlight, quantum dot-enhanced liquid crystal display (QD-LCD)) have already entered the consumer market, with terminal annual sales at around 100 billion RMB, while quantum dot electroluminescent (AM-QDLED) has seen rapid technological development since 2014 and is widely regarded as a strong candidate for the next generation of display technology. Currently, China, South Korea, and United States are in competition with each other in the fields of quantum dot materials and quantum dot display technology. Given the scale and importance of the new display industry, quantum dot display technology plays a crucial role in China's pursuit of industrial transformation and upgrading. Domestic and foreign enterprises and research institutions, including SAMSUNG, Beijing Oriental Electronics Group Co., Ltd. (BOE), and China Star Optoelectronics Technology (CSOT), have carried out patent layout in the field of quantum dot display to consolidate and improve their technological strengths and advantages. These patents cover a broad scope, including innovative methods for the synthesis of colloidal quantum dots, surface modification, and the design and preparation of QDLED. This analysis utilizes the free and open-source patent search and analysis websites www.lens.org and http://pss-system.cnipa.gov.cn/ from China National Intellectual Property Administration to search for patents related to "quantum dots," in order to track the current trends in quantum dot industrialization and patent layout. This chapter provides a useful reference for government policymakers, enterprises, and researchers.

Colloidal Quantum Dot Light Emitting Diodes: Materials and Devices, First Edition. Hong Meng.
© 2024 WILEY-VCH GmbH. Published 2024 by WILEY-VCH GmbH.

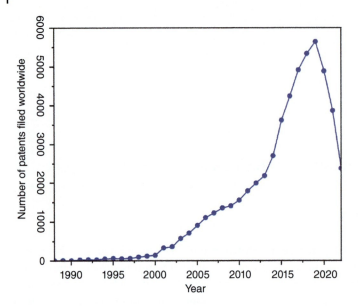

Figure 13.1 Number of patents published worldwide search by keywords in Title: ("quantum dots") OR (Abstract: ("quantum dots") OR Claims: ("quantum dots")). Source: https://www.lens.org/.

Figure 13.1 illustrates the number of global patents filed related to quantum dot materials and devices from 1988 to 2022, sourced from www.lens.org. The development can be roughly divided into four periods in terms of the number of patents filed. From 1988 to 2000, this field experienced a slow and steady growth with less than 150 patents filed per year. Between 2001 and 2013, the number started growing faster with over 300 patents filed each year, indicating growing interest in this area. From 2014 to 2019, the number of global patents filed surged sharply with more than 2500 patents filed per year, reaching the peak in 2019 and suggesting rapid industrialization of this technology. The fourth period (from 2020 to present) witnesses the number dropping quickly from 5643 to 2378, indicating that the investment is withdrawing and the commercialization of this technology is approaching maturity. This trend is closely linked to the quantum dot display industry, as main display panel giants such as SAMSUNG, BOE, and TCL have invested tremendously in quantum dot display research and development, and researchers are actively promoting its industrialization.

We analyzed and sorted out the roadmap of QDLED technology development by selecting important patents in the field of quantum dot optoelectronic devices. Before 2010, the patents filed in this field mainly related to synthetic methods of making quantum dots. The main topics in QD synthetic methods include thermal injection, cation exchange, ligand exchange, surface passivation, and core–shell structure design. With the maturity of synthetic methods, patents filed and granted after 2010 are mainly related to quantum dots-based optoelectronic devices such as QLED, QD-LCD, and solar cells. QDLED technology has the advantages of wide color gamut, long lifespan, high color saturation, and external quantum efficiency

(EQE), which can provide better visual experience and is the development direction of the new generation of display technology. The EQE of QDLED is still unsatisfactory, mainly because of the high hole injection barrier and low charge carrier recombination efficiency of quantum dots.

In the field of displays, quantum dot technology has four main development directions: QD-LCD, quantum dot-enhanced organic light-emitting diode (QD-OLED), AMQDLED, and quantum dot-enhanced MicroLED. Among them, QD-LCD combines quantum dot photoluminescence technology with traditional liquid crystal display (LCD) technology and is also known as quantum dot backlight technology or the first generation of quantum dot display technology. This approach has been widely adopted in the LCD TV industry, with industry data showing that all high-end color TVs from SAMSUNG in 2017 and 2018 used quantum dot backlighting products. In China, Najing Tech is currently the dominant supplier of quantum dot backlighting products and has established stable partnerships with well-known display panel manufacturers such as TCL and BOE. Many high-end smart TV products have already introduced large-sized 4K UHD quantum dot TVs featuring QD-LCD photoluminescence display technology to the market, which have gained popularity among consumers. The second development direction is the combination of organic light-emitting diode (OLED) display and quantum dot technology, known as QD-OLED. It is essentially an enhanced version of OLED technology that utilizes quantum dot technology in the light-emitting material, making it more efficient than pure OLED while also avoiding the issue of screen burning. SAMSUNG is currently leading this technology, and mass production of OLED panels with QD-OLED technology commenced in 2021. The third direction is active matrix QDLED (AMQDLED), which is an electroluminescent technology that is considered to be a second-generation quantum dot display technology that can rival AMOLED. Currently, this technology is still under development and has not yet been industrialized. The fourth direction is the quantum dot-enhanced MicroLED display technology. Although MicroLED display technology has not reached mass production yet, it has attracted the attention of the industry due to its ability to provide very high pixel density and ultra-high brightness on a very small area, making it suitable for specific display requirements. However, it is challenging to achieve high-density triple primary color LED arrays on small areas. The stability of blue, green, and red MicroLEDs varies greatly, resulting in color differences in the triple primary color arrays. One possible technical solution is to cover the three-dimensional array formed by the blue MicroLED with a thin film of red, green, and red quantum dots. Currently, quantum dot-enhanced MicroLEDs are the most promising option in MicroLED technology, somewhat similar to QD-LCD display technology. However, MicroLED technology is limited by its own technology and cost, so it is only able to fulfill its potential in small displays and is unlikely to become a mainstream product in the next generation of display technology.

As shown in Figure 13.2, the quantum dot display industry chain consists of three sections. To analyze the patent layout of this industry chain, we focus on representative companies involved in quantum dots, QD-LCD and QD-OLED modules, and TVs & monitors as consumer products. For the first section, companies with core

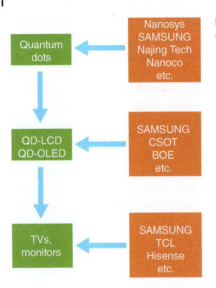

Figure 13.2 Key sections of the quantum dot display industry chain.

patents on quantum dot materials for display include Nanosys and SAMSUNG in the United States, Nanoco in the United Kingdom, and Najing Tech in China. The main players in QD-LCD and QD-OLED fields include SAMSUNG, BOE, and CSOT. For quantum dot-based consumer products such as QDLED TVs and monitors, many companies such as SAMSUNG, TCL, and Hisense sell related products.

13.2 Patent Layout

13.2.1 Nanosys

Nanosys, established in 2001, is a world leader in quantum dot materials with over 363 patents owned in the field of quantum dots. In early 2012, the company launched quantum dot enhancement film (QDEF) technology in collaboration with 3M. The QDEF technology extends the color gamut from 70% to 100% and increases EQE by around 50%, which is expressed as the ratio of the brightness of the LCD panel to the power of the backlight. Despite the high manufacturing cost of QDEF technology, it is gaining momentum as the quantum dot display technology shifts to the QD-film route. Nanosys boasts the world's largest production base for quantum dot materials, with an annual production capacity of over 50 tonnes. The company has established close partnerships with renowned display panel manufacturers such as SAMSUNG, Sharp, and LG.

Nanosys has a significant patent portfolio in the field of quantum dot technology, covering a broad range of areas including materials synthesis, device fabrication, and applications. We have analyzed and highlighted the key patents of Nanosys. Among them, patent US2019390109A1 describes a method for preparing cadmium-free $ZnSe_{1-x}Te_x/ZnSe/ZnS$ core–shell quantum dots and their QDLED devices, particularly for blue light quantum dot materials [1]. The QDLED device has a fluorescence emission peak at 455.9 nm, a fluorescence half-height width of 30 nm, an external QY of 2.0%, and CIE color coordinates of $x = 0.145$ and $y = 0.065$. In terms of green light quantum dot materials, the patent WO2019084135A1 discloses a synthetic method for producing cadmium-free InP/ZnSe/ZnS core–shell structured quantum dots [2]. By varying the size of the InP quantum dot core and the thickness of the shell layer, the fluorescence emission peak of the green light quantum dots can be tuned between 535 and 550 nm, with a fluorescence QY of up to 81%. As for red-light quantum dot materials, patent US2018155623A1 describes a synthetic method for producing cadmium-free InP/ZnSeS/ZnS core–shell quantum dots [3]. Through the adjustment of the InP quantum dot core size and the thickness of the shell layer, the fluorescence emission peak of the cadmium-free InP/ZnSeS/ZnS core–shell structured quantum dots located at 636 nm with a high fluorescence QY of up to 65%. In order to enhance the stability of quantum dots in thiol resin films, a patent (WO2018226925A1) discloses a method for modifying the surface of the quantum dots with thiol groups [4].

Nanosys holds several patents for quantum dot backlighting as well. Patent EP2946411B1 describes a method for manufacturing quantum dot films [5]. The disclosed design solution for a display device in patent US2020168673A1 includes a quantum dot layer sandwiched between a first blocking film and a second blocking film of the quantum dot film [6]. The quantum dot layer is composed of quantum dots dispersed in a polymeric material, which consists of a methacrylate polymer, an epoxy polymer, and a photoinitiator. The methacrylate polymer makes up 5–25% by weight of the quantum dots and is formed through the radiation of a methacrylate polymer precursor polymerization. The epoxy polymer, on the other hand, is formed through thermal polymerization of the epoxy polymer precursor. Patent US2020098951A1 describes a display device that includes quantum dot coated with a barrier layer as well as a method of manufacturing the device [7]. The display device includes a quantum dot layer that emits light of a desired color. The quantum dot layer is coated with a barrier layer to improve the stability and durability of the quantum dots. This design solution can be used in various types of displays, such as LCDs, organic light-emitting diode (OLED) displays, and MicroLED displays. The display device can produce vivid and accurate colors and can be controlled by independent control circuits for each light source. Each quantum dot that is coated with a barrier layer consists of a core–shell structure, with a hydrophobic barrier layer placed on top of it. The hydrophobic barrier layer is designed to keep a distance between the core–shell structure of one quantum dot and the core–shell structures of other quantum dots that are in close contact with it. A method for creating a quantum dot that is coated with a barrier layer involves forming a reverse micelle using a surfactant, and then incorporating the quantum dot into the reverse micelle. The process then entails coating the individual bound quantum dots with a barrier layer and separating the quantum dots that are coated with the barrier layer using a surfactant that is set on the reverse micelle of the barrier layer. Patent WO2020023583A1 describes a display device that incorporates quantum dots directly on the surface or substrate of a backlight unit, without the need for an intermediary layer to form a light transmission layer on top [8]. The quantum dot film in the design solution may contain numerous optical features, eliminating the need for additional films such as separate optical films. Optical features can be embedded in microspheres to enhance optical effects, improving the thickness uniformity of the quantum dot film, or both. Additionally, optical features may be embossed onto the quantum dot films, such as reflective and/or refractive features, prisms, notches, slotted prisms, biconvex lenses, microlenses, microspheres, or any other lenses, pitch, or suitable brightness enhancement and/or optical features. This enables the omission of separate optical films from the device structure altogether.

13.2.2 SAMSUNG

SAMSUNG Group, headquartered in Gyeonggi-do, Korea, has more than 3334 patents in the field of quantum dots, mainly in the field of quantum dots and QDLED displays. In the area of quantum dot display devices, SAMSUNG has explored improving display quality and simplifying the structure of display devices. SAMSUNG is also a major player in the QDLED field, with a strong patent portfolio in the area. SAMSUNG has been focusing on developing QDLED technology for several years and has been actively filing patents in the field. SAMSUNG also has a number of patents related to QDLED materials, including patents related to the synthesis of high-quality quantum dots and the development of new materials with improved properties, which were acquired from QD Vision.

In the field of quantum dot materials, SAMSUNG has pursued numerous technological innovations and patent strategies. Patent WO2013078252A1 discloses a composition comprising a quantum dot and an emission stabilizer, as well as a method for improving or enhancing the emission stability of quantum dots. The emission stabilizer may include a thiol compound, such as thioglycolic acid, as well as a surface passivating agent [9]. The composition may be used in a variety of applications, including as a component of an ink composition for use in printing quantum dots, or as a component in the production of a display device or a photovoltaic device. The disclosed method allows for improved stability of the quantum dots over time, thereby enhancing the performance and longevity of devices that incorporate them. Compositions that include an emission stabilizer can enhance the stability of at least one of the emission properties of quantum dots, compared to compositions that are otherwise identical. To mitigate the negative effects of variations in ambient temperature during the use on the solid-state photoluminescence efficiency of quantum dot nanocrystals, patent WO2006033396A1 discloses a composition that comprises a plurality of semiconductor nanocrystals [10].

In order to simplify the manufacturing process of backlight display devices, patent US10663648B2 discloses a design solution [11]. The display device comprises a display panel and a backlight source configured to supply light to the display panel. The backlight source includes a light source configured to emit light, a light guide provided on the rear side of the display panel and configured to direct the light emitted from the light source toward the display panel, and a plurality of optical sheets provided between the light guide and the display panel. These optical sheets are attached to the front surface of the light guide and may comprise: a quantum dot sheet configured to increase color reproduction; and a prism sheet attached to the front surface of the quantum dot sheet. Wherein the quantum dot sheet and the prism sheet are attached to each other by a plurality of portions of adhesive spaced apart from each other, such that an air layer is provided between the quantum dot sheet and the prism sheet. In order to remove, reduce, or minimize color defects generated on a display screen, patent CN106353921A discloses a reflector, a backlight unit, and a display device using the backlight unit [12]. The display device comprises: one or more light sources; a reflector plate, onto which light emitted from the one or more light sources is incident, the reflector plate having a reflective surface for reflecting the incident light; one or more selective light absorbing portions, provided on the reflective surface and configured to selectively absorb a portion of the incident light; and a quantum dot sheet, into which a portion of the light emitted and radiated from the light sources without being absorbed by the selective light absorbing portions, is incident. In order to improve the brightness of a display panel and reduce the losses caused by the color conversion of quantum dots, patent CN111048556A provides a display panel and display device [13]. The display panel comprises a light source configured to emit blue light and a quantum dot color filter layer. The quantum dot filter layer comprises: a red light converter comprising red quantum dots configured to convert blue light into red light; a green light converter comprising green quantum dots configured to convert blue light into green light; a light transmitting portion configured to emit blue light; and a white light generator comprising a first region and a second region, wherein the first region comprises a plurality of yellow quantum dots configured to convert blue light into yellow light, and wherein the second region emits blue light. To improve the color reproduction and light-gathering power of a backlight display device, patent US10914887B2 discloses a display device [14]. The display device comprises: a light source; a quantum dot sheet, on which a reflective region and a quantum dot region are alternately provided, the reflective region reflecting light emitted from the light source and the quantum dot region comprising quantum dots scattering light emitted from the light source; and a display panel configured to display an image using light provided from the quantum dot sheet. In order to improve the color reproduction and EQE of images, patent US10698255B2 discloses a photoluminescent device and a method of manufacturing the device [15]. The photoluminescent device comprises: a first substrate; a blue light blocking pattern, provided in a first color pixel region, a second color pixel region, and a first light blocking region between the first color pixel region and the second color pixel region on the first substrate; a blue color filter, provided in a blue pixel region, a first light blocking region and a second light blocking region between the blue pixel region and the second color pixel region on the first substrate in the second light-blocking region between the blue pixel region and the second color pixel region; a first color conversion pattern, provided on the blue light-blocking pattern, in the first color pixel region; and a second color conversion pattern, provided on the blue light-blocking pattern, in the second color pixel region. The first color conversion pattern may be a green conversion pattern, and the second color conversion pattern may be a red conversion pattern or vice versa.

The green conversion pattern includes green quantum dots or green phosphor, and the red conversion pattern includes red quantum dots or red phosphor. In order to reduce color bleeding in display devices, patent KR20200088532A discloses a display device [16]. The display device comprises: a light-emitting unit, which serves to generate and emit a first color light; an optical member, which serves to color convert the first color light incident from the light-emitting unit in the optical member and from which the color converted light is emitted; and a display panel, which serves to provide the color converted light emitted from the optical member to the display panel. The optical member comprises: a quantum dot member, which serves to transmit a portion of the first color light and to convert a portion of the first color light into a second color light and a third color light; a filter member, which is located between the light-emitting unit and the quantum dot member, and the filter member comprises a cholesteric liquid crystal layer, which serves to reflect the second color light or the third color light incident from the quantum dot member to the filter member.

In terms of QDLED devices, SAMSUNG has also actively engaged in patent layout. In order to ensure the heat dissipation as well as the EQE of quantum dot units, patent US10424691 provides a display device and its manufacturing method [17]. The display device comprises: a quantum dot unit or a quantum dot sheet capable of improving heat dissipation performance. A line with high thermal conductivity is provided in the QDLED unit or QDLED sheet, and the line is connected to a substrate of the display device to dissipate the heat generated in the QDLED. To improve the hole transport capability and surface properties of the emitting layer in QDLED devices to enhance their luminescence, patent US20200083450A1 discloses a design solution for an electroluminescent device and a display device [18]. The electroluminescent device comprises: a first electrode and a second electrode facing each other; an emitting layer provided between the first electrode and the second electrode and comprising quantum dots and a first hole transport material having a substituted or unsubstituted C4-C20 alkyl group attached to the backbone structure; a hole transport layer provided between the emitting layer and the first electrode and comprising a second hole transport material; and an electron transport layer provided between the emitting layer and the second electrode. The electron transport layer is provided between the emitting layer and the second electrode. In order to develop environmentally friendly quantum dots with good luminescence properties without heavy metals (e.g. cadmium, mercury, lead, or a combination thereof), patent US10851298B2 discloses quantum dots and an electroluminescent device thereof [19]. The electroluminescent device comprises a first electrode and a second electrode opposite each other, and an emitting layer disposed between the first electrode and the second electrode and comprises quantum dots. The quantum dots comprise: a core layer comprising InP nanocrystals; a first shell layer of ZnSe nanocrystals and a second shell layer of ZnS nanocrystals coated on the core layer. The QDLED device has an EQE of over 9% and a maximum luminance of over $10\,000\,\text{cd}\,\text{m}^{-2}$. In order to prevent current leakage from QDLED devices, patent US11233211B2 provides an electroluminescent device and a display device thereof [20]. The electroluminescent device comprises: a first electrode and a second electrode facing each other; an emitting layer provided between the first electrode and the second electrode and comprising light-emitting particles; an electron transport layer provided between the first electrode and the emitting layer; and a hole transport layer provided between the second electrode and the emitting layer, wherein the electron transport layer comprises inorganic oxide particles and a metal-organic compound, wherein the metal-organic compound or the metal thermal decomposition products of the organic compound are soluble in a nonpolar solvent. In order to improve the charge carrier balance and light extraction efficiency of the emitting layer while minimizing leakage current, patent US10879484B2 discloses an electroluminescent device and its display device [21]. The electroluminescent device comprises a first electrode and a second electrode opposite to each other, an emitting layer disposed between the first electrode and the second electrode and comprising a plurality of light-emitting quantum dots, a hole transport layer, and an electron transport layer. The electron transport layer comprises an inorganic layer provided on the emitting layer and comprising a plurality of inorganic nanoparticles, and an organic layer provided directly on at least a part of the upper surface of the inorganic layer on the side opposite to the emitting layer, the work function of the organic layer being higher than that of the inorganic layer. In order to improve the electron–hole balance in the light-emitting layer of quantum dot light-emitting devices and to enhance the hole transport properties and quantum efficiency, patent US10988685B2 discloses a quantum dot light-emitting device [22]. The light-emitting device comprises a first electrode and a second electrode opposite each other, and a light-emitting layer disposed between the first electrode and the second electrode and comprising quantum dots, wherein the quantum dots comprise semiconductor nanocrystals and a surface ligand bound to the semiconductor nanocrystals. The ligand comprises an organothiol ligand or a salt thereof and a multivalent metal compound comprising a metal.

13.2.3 Nanoco

Nanoco is a Manchester-based company that was established in 2001. It is a leading global producer of nanomaterials, including cadmium-free quantum dots (CFQDs). With over 223 patents in the field of quantum dots, the company's main products are CFQDs and CFQD quantum dot film technology. Although Nanoco's quantum dot materials are based on CFQD, their current photoluminescence linewidth (PL linewidth), QY, and chemical stability are not as good as those of CdSe quantum dots. However, due to the EU's ban on the use of toxic cadmium-containing materials, the development of CFQDs has become a crucial area of focus in the industrialization of quantum dots.

To be suitable for commercial applications in next-generation light-emitting devices, quantum dot materials must be incorporated into the LED encapsulant material while maintaining their monodispersity as much as possible and without significant loss of QY. However, current methods face several challenges, particularly due to the properties of LED encapsulants. The quantum dots may aggregate when formulated as an LED encapsulant, resulting in reduced optical properties. Additionally, after the quantum dots have been incorporated into the LED encapsulant, oxygen can migrate through the encapsulant to the surface of the quantum dots, causing photooxidation and a subsequent reduction in QY. To prevent oxygen from entering the quantum dots, one solution is to incorporate them into a medium with low oxygen permeability, such as a polymer. Quantum dot materials are crucial for the development of next-generation light-emitting devices, but incorporating them into LED encapsulant materials without aggregation or loss of QY poses significant challenges. LED encapsulants can cause quantum dots to aggregate, leading to reduced optical properties. Additionally, oxygen can penetrate the encapsulant and cause photooxidation, reducing QY. To address this, quantum dots can be incorporated into a low oxygen-permeable medium such as a polymer, which can be used to prepare films or beads for use in light-emitting devices. However, not all polymer systems are compatible with quantum dots, especially CFQDs, which are particularly difficult to match with suitable polymers. Incompatible polymer systems can react with quantum dots, leading to decreased QY. To date, polymers based on acrylate monomers, such as methacrylates, have been found to be the most compatible with quantum dots. However, most acrylate systems are slightly permeable to oxygen, and acrylate polymers can degrade under high temperatures, UV irradiation, and oxidation. To address this issue, patent WO2015140642A2 proposes a silicone-containing surface-modified ligand for the preparation of quantum dots that are more compatible with polysiloxanes [23]. In some embodiments, the polysiloxane film on the surface of the quantum dots is as flexible as films prepared using lauryl methacrylate (LMA) as a monomer and trimethylolpropane trimethacrylate (TMPTM) as a cross-linking agent, and the matrix of the polysiloxane film is more stable to heat, UV, corrosion, and oxidation. Compared to conventional polymeric materials obtained from methacrylates and polyurethanes, polysiloxanes have a high solids content, low volatile organic compounds, and low toxicity, which makes them less of an environmental and health concern.

Quantum dot nanocrystals can be passivated with protective organic groups on the "bare" surface atoms to partially overcome the instability problems associated with unliganded dangling bonds on the surface. Capping or passivation of quantum dot surfaces not only prevents aggregation but also protects the dots from the surrounding chemical environment and provides electrochemical stability to the dots. The capping agent is usually a Lewis base compound or other electron-donating compound bound to the surface metal atoms of the outermost inorganic layer of the quantum dot (e.g., to the outermost shell of the core–shell quantum dot). When performing quantum dot shell synthesis in an electron-giving solvent such as TOP/TOPO, the capping agent may be only a solvent molecule bound to the surface of the quantum dot. When using a non-electron-giving solvent, an electron-rich capping agent can be added to the shell synthesis. Although the electron-donating capping ligands provide some stability and surface passivation, these ligands only weakly adhere to the surface of the quantum dot nanocrystals. Desorption of the capping ligands leaves voids on the surface, which can lead to aggregation, precipitation, and reduced QYs of the quantum dots. One way to solve the problem of weakly bound capping ligands is to use capping ligands that contain functional groups with specific binding affinities for the surface atoms of the quantum dots. For example, the sulfur of thiol compounds has an affinity for many of the metal atoms (e.g., zinc) that are typically components of quantum dot shell semiconductor materials such as ZnS and ZnSe. Therefore, thiols have been widely used as capping ligands for quantum dots. However, thiol-capping ligands can also desorb, leaving voids on the surface of the quantum dots. One possible mechanism for thiol ligand desorption is through the formation of disulfide bonds between adjacent thiols on the surface of the quantum dot and the subsequent desorption of the disulfide. Another problem with thiol-capped ligands is that in some cases, the steric hindrance may prevent complete surface coverage. Therefore, to maximize the performance and stability of quantum dot materials, there is a need to develop a more efficient ligand for quantum dot capping. To meet this need, patent WO2015118346A1 discloses a dithiol ligand [24]. Examples of this dithiocarbon ligand include dithiocarbamate or salt ligands. These strongly bound ligands are able to coordinate with both cations and anions on the surface of the nanoparticles. The ligands are didentate, so they do not cause spatial dislocations as monodentate ligands do when approaching the surface of the quantum dots, thus completely passivating the quantum dot surface. In terms of cadmium-free green quantum dots, patent WO2015101779A1 provides a quantum dot nanocrystal with a quantum dot nucleus comprising at least one of the following: zinc ions, magnesium ions, calcium ions, and aluminum ions [25]. The method involves preparing a precursor solution by dissolving group III–V precursors and ZnX precursors in a solvent and then injecting the precursor solution into a high-temperature reaction zone in the presence of a capping agent and a reducing agent. The reaction mixture is then rapidly quenched to terminate the reaction, and the resulting quantum dots are isolated and purified. The capping agent used in the method is a dithiocarbamate compound, which provides efficient and stable surface passivation of the quantum dots. The reducing agent used in the method is a borohydride compound, which ensures high yields and monodispersity of the quantum dots. This scale-up synthesis method is capable of producing large quantities of high-quality group III–V/ZnX quantum dots with high photoluminescence efficiency and stability, making them suitable for use in various optoelectronic devices [26]. In this method, zinc acetate is reacted with thiol or selenol [R–Se–H] compounds to generate in situ ZnS or ZnSe "molecular seeds," which serve as templates for the growth of Group III-V semiconductor nuclei. The addition of zinc and sulfur elements to the reaction solution results in a partial alloying of ZnX (X = sulfur element) within the InP core quantum dots, leading to larger nanoparticle sizes compared to pure InP core quantum dots, while retaining smaller fluorescence emission peaks. In some embodiments, a combination of ZnS clusters with additional zinc salts and thiols is utilized to grow blue-shifted emission peaks relative to those of quantum dot nuclei prepared using only clusters. The alloying behavior of the quantum dot nuclei promotes enhanced absorption in the blue region of the electromagnetic spectrum. Furthermore, the reaction solution used in this synthetic method is substantially more concentrated than that in conventional synthetic methods, which facilitates commercial-scale production of InP quantum dots using this synthetic approach.

13.2.4 Najing Tech

Founded in 2009 and headquartered in Hangzhou, China, Najing Tech specializes in quantum dot materials as its core technology. Najing Tech has been actively involved in the development of quantum dot materials for display applications and has made significant contributions to the field. The company has expertise in the synthesis of high-quality colloidal quantum dots and has developed a series of high-performance quantum dot materials by controlling the growth process of quantum dots to regulate their exciton state properties. The company has a strong patent portfolio in the field of quantum dot technology, particularly in the area of colloidal quantum dot synthesis and its applications in display technology.

The company has more than 500 patents in the field of quantum dots and is led by its founder, Professor Xiaogang Peng from Zhejiang University. The invention of the quantum dot nanocrystal synthesis method by Professor Peng has become the standard technology in both academia and industry. The related patents have laid the foundation for the application of quantum dot materials. Najing Tech focuses on the precise regulation of the excitonic state of quantum dots through chemical synthesis to study their fluorescence emission properties. The company has also developed patents in other related fields, including electron transport layer, quantum dot post-processing, quantum dot backlight, and QDLED, achieving some innovative results. In this context, we have selected some of the core invention patents of Najing Tech for a brief introduction and analysis.

Patent CN111117595A discloses a blue-light core–shell quantum dot, a preparation method, and its application [27]. The quantum dot is a core–shell structure with a core of copper indium sulfide (CIS) and a shell of zinc sulfide (ZnS). The core–shell structure is prepared by a one-pot synthetic method using an amine-based solvent. The resulting quantum dot has a narrow size distribution, excellent photoluminescence properties, and good stability. The blue-light quantum dot is suitable for use in optoelectronic devices, such as displays, lighting, and sensors. The patent provides a detailed preparation method, including the materials used and the specific synthetic process, as well as the properties and applications of the resulting quantum dot. Regarding quantum dot materials, patent CN111117595A discloses a blue-light core–shell quantum dot, its preparation method, and its application. The blue-light core–shell quantum dots comprise a quantum dot core and a first shell layer encapsulated outside the core. The quantum dot core is a copper-doped II–II–VI ternary alloy quantum dot or a copper-doped II–II–VI–VI quaternary alloy quantum dot. The material of the first shell layer is an II–II–VI alloy material. The blue CdZnSe/ZnSeS/CdZnS core–shell quantum dots can be utilized in QDLED devices, which can significantly improve the life span of QDLED devices and enable blue QDLED that meet commercial requirements. The dots have a size of 18 nm, an EQE of 12.6%, and a T_{50} life of up to 19 200 hours. The InP quantum dot disclosed in patent CN106701059B is prepared by a high-temperature organic solution method using trioctylphosphine oxide as a phosphorus source and myristic acid as a solvent and stabilizer [28]. The resulting InP quantum dots have a narrow particle size distribution, high QY, and good stability. The method allows for the precise control of the reaction parameters to tune the size and properties of the InP quantum dots. The resulting quantum dots can be used in various applications, such as biological imaging, displays, and solar cells. The InP quantum dot comprises: a metal salt-modified InP core and a shell layer wrapped around the core. The shell layer is ZnSe/ZnS or $ZnSe_xS_{1-x}$, where $0 < x \leq 1$, the PL linewidth of the InP quantum dot is less than or equal to 50 nm, and the quantum efficiency is greater than or equal to 70%. As the modification effect of the metal salt makes the suspension bond on the surface of InP eliminated, it is able to reduce the loss of energy by its nonoptical nature and improve the fluorescence QY of InP quantum dots with this metal salt modification. In addition, since the difference between the lattice constants of InP and ZnSe is small, the shell layer with S and Se can be easily wrapped around the surface of the InP quantum dot nucleus, thus making the InP quantum dot with high quantum efficiency and stability; meanwhile, since the PL linewidth of the InP quantum dot is less than or equal to 50 nm, it can be concluded that the particle size distribution of the quantum dot is more uniform. In order to reduce the PL linewidth of CFQDs, improve the stability and enhance the photoluminescence quantum yield (PLQY), patent CN106479481A discloses a ZnSe/III–V group/$ZnSe_xS_{1-x}$ or ZnSe/III–V group/ZnSe/ZnS quantum dot and its synthesis method [29]. The process of the reaction consists mainly of the mixing of various solutions in which the ZnSe nuclei and their outer shell layers are gradually formed, and all the various raw materials and reaction conditions that can be applied to this system are covered in the patent. For the InP/ZnS structure of quantum dots, it has been shown that the lattice parameters of the two do not match, resulting in irregular structures of quantum dots with variable sizes. Patent CN106497546A proposes a white light quantum dot composition and its preparation method to solve the problem of short life span of white light diodes in the prior art [30]. The light color of the white light quantum dots is determined by adjusting the ratio of Zn/Cd and the amount of Se and S added. The white light quantum dots in the invention contain a core and a shell layer with different compositions. The general formula for the chemical structure of the core is $Cd_xZn_{1-x}Se_yS_{1-y}$ and the general formula for the shell layer is $Cd_zZn_{1-z}S$. These two quantum dots work together to emit white light. Likewise, the patent covers the possible structures and compositions of quantum dots inside and outside and the process of their preparation. The patent CN110317609A discloses a cadmium-free green light InZnPS quantum dot, preparation method, and optoelectronic device [31]. The method of preparing the III–IV–VI quantum dots comprises: formulating a first nanocluster, a second nanocluster, and a third nanocluster; mixing the first nanocluster with a non-coordinating solvent to form a first quantum dot solution; mixing the second nanocluster with the first nanocluster is mixed with the non-ligand solvent to form the first quantum dot solution; the second nanocluster is mixed with the first quantum dot solution and heated to form the second quantum dot solution; the third nanocluster is mixed with the second quantum dot solution and heated to form the third quantum dot solution, and the third quantum dot in the third quantum dot solution is III–IV–VI quantum dot. The nanoclusters are mixed one by one to improve the energy band structure, and reduce surface defects and dangling bonds, resulting in a tunable composition and uniform quantum dot size distribution. The fluorescence emission peak is located at 520 nm, the PL linewidth is 36 nm, and the QY can reach 48%. Patent CN108841373A discloses a red-light quantum dot, a synthesis method, and QDLED [32]. The synthesis method comprises: providing a solution containing a CdSe quantum dot core, and then epitaxially growing a plurality of $Zn_xCd_{1-x}S$ monolayer shells on the CdSe quantum dot core, with the x of each $Zn_xCd_{1-x}S$ monolayer shell gradually increasing from 0 and the maximum value taking a value between 0.5 and 0.8 in the direction away from the CdSe quantum dot core to obtain a system containing $CdSe/Zn_xCd_{1-x}S$ quantum dots; or in $CdSe/Zn_xCd_{1-x}S$ quantum dots and then continue to cover them with a ZnS shell layer. By controlling the gradual transition of the composition in the shell layer from CdS to CdZnS or ZnS, a better lattice match is ensured, and quantum dots of high optical quality, photobleaching resistance, and high air stability can be obtained.

In terms of electron transport layers, Najing Tech achieves electronic properties, chemical stability, and dispersion stability of electron transport layer materials by modifying ZnO nanocrystals. Patent CN109628082A discloses a method for the preparation of ZnOS/ZnO nanocrystals [33]. The method first obtains ZnOS cores by sulfiding the ZnO nanocrystals to a certain extent, then grows a ZnO shell layer outside the ZnOS cores to obtain ZnOS/ZnO nanocrystals. The ZnOS/ZnO nanocrystals are stable and not easy to agglomerate and ripen. In addition, the S in the ZnOS core can play a role in regulating the energy band, thus making it easier to obtain nanocrystals with a wide forbidden band. Patent CN110144139A discloses a method for the preparation of ZnO-based nanoparticle ink [34]. The ZnO-based nanoparticle ink comprises ZnO nanoparticles as well as a solvent. The first solvent is an alcohol, ether, or alcohol–ether that does not include a benzene ring, and the first solvent has a carbon number of not more than 7, and the second solvent is an alcohol, ether, or alcohol–ether that includes a benzene ring, and the second solvent has a carbon number of not less than 7. The device prepares the surface of the ZnO-based nanoparticle ink with better dispersion stability, which is conducive to avoiding the formation of coffee ring effect. It can effectively improve the uniformity of the film thickness. Patent CN110484233A discloses a method for the preparation of ZnO nanocrystals [35]. The ZnO nanocrystals comprise a ZnO nanocrystal body and a surface ligand, the surface ligand comprising one or more mercaptan ligands, the mercaptan ligands having the following structural formula: –S–R–OH, wherein R represents a straight chain alkyl chain or a branched chain alkyl chain. Because the coordination ability of element S with element Zn is stronger than that of element O, the mercaptan alcohol is selected to exchange ligands for the initial carboxylate and hydroxide roots of element Zn on the surface of ZnO nanocrystals, ultimately forming ZnO nanocrystals with a Zn–S–ROH surface, which can effectively solve the surface defect luminescence caused by the easy shedding or incomplete coordination of carboxylate and hydroxide roots, and improve the electrical properties of the ZnO nanocrystal film. The electrical properties of the ZnO nanocrystal film are improved, thus solving the problem of delayed luminescence in electroluminescent devices. Patent CN111115679A provides a method for the preparation of ZnO nanocrystals [36]. The method of preparing ZnO nanocrystals comprises the following steps: a first solution system of a cationic salt, the cationic salt comprising an organozinc salt, the organozinc salt comprising an alcohol-soluble group on its carbon chain; and adding a base to the first solution system for reaction to produce a solution containing ZnO nanocrystals. As the preparation method uses zinc salts with good solubility in the alcohol phase as the synthetic raw material for the synthesis of ZnO nanocrystals, it successfully increases the shelf life of the ZnO nanocrystal solution itself and facilitates the application of ZnO nanocrystals in industrial production. Patent CN111326661A discloses a method for the preparation of ZnO-doped nanocrystals [37]. The preparation method includes step A: mixing zinc precursors, doped ionic precursors, fatty alcohols, and solvents to obtain a first mixture, maintaining the temperature of the first mixture at 200–300 °C for the method, and obtaining ZnO-doped nanocrystals after the reaction, with the doped ionic precursors including magnesium precursors and indium precursors. The microscopic morphology of the ZnO-doped nanocrystals produced by this method is good and facilitates the formation of dense thin films. The electroluminescent devices demonstrate that the ZnO-doped nanocrystals as an electron transport layer can significantly improve the EQE of the devices.

In terms of luminescent devices, Najing Tech has mainly conducted patent layouts for new electroluminescent devices and light conversion films in photoluminescence with respect to light output efficiency, service life, heat dissipation, and cost reduction. In order to improve the EQE of electroluminescent devices, patent CN106058072A provides a technical solution for using scattering particle layers to achieve the best luminous effect [38]. The invention relates to a light-emitting device consisting of a first electrode, a light-emitting layer, and a second electrode, with a scattering film layer incorporated between either electrode and the light-emitting layer. The structure of the scattering film layer includes the main body with metal/nonmetal semiconductor heterojunction particles to make the film electrically conductive. The patent covers several structures of substrate and heterojunction particles for a variety of electrode materials that can obtain applications with the scattering film layer. In QDLED, direct contact between the metal oxide electron transport layer and the quantum dot introduces non-radiative transition channels, leading to quenching of excitons in the quantum dot and reducing the efficiency of QDLED. Patent CN111326664A discloses a QDLED and an ink for making it [39]. The QDLED includes an anode, a hole transport layer, a quantum dot light-emitting layer, an electron transport layer, and a cathode set in sequence adjacent to each other. The QDLED also includes an interface layer, which is set between the quantum dot light-emitting layer and the electron transport layer, and the material forming the interface layer is a conductive metal chelate. The use of the interface layer set up with conductive metal chelate can not only avoid the exciton quenching at the interface between the electron transport layer and the light-emitting layer, but also has good electron transport characteristics, which in turn enables the QDLED to maintain the proper light-emitting efficiency, so that the EQE of the QDLED is improved, and at the same time reduces the energy consumption of the QDLED. In order to achieve a display substrate with thermally conductive members to facilitate effective thermal conductivity of the pixel definition layer, reduce the coffee ring effect, facilitate the formation of a uniform film layer as well as reduce the device temperature, and improve the device life and stability, patent CN111354869A discloses a display substrate and a method of making a display device [40]. The display substrate includes a substrate, a pixel definition layer, a thermally conductive member, and a smoothing part. The pixel definition layer includes a plurality of pixel isolation structures, and the pixel isolation structures form a plurality of mutually isolated sub-pixel regions. The thermally conductive member is provided between the substrate and the pixel definition layer and includes a dominant thermal section and a secondary thermal section. The secondary thermal section is connected to each dominant thermal section; the upward projection of the dominant thermal section is located at the inner periphery of the sub-pixel region; the upward projection of the secondary thermal section is located at the outer periphery of the sub-pixel region; the projection area of the dominant thermal section in the sub-pixel region is larger than the projection area of the secondary thermal section in the sub-pixel region; the smoothing section is insulatively provided above the thermally conductive member; and the pixel definition layer is insulatively located above the thermally conductive member. The smoothed section is insulated above the heat-conducting member, and the pixel definition layer is located on the upper surface of the smoothed section. This design solution not only helps to reduce the coffee ring effect during the drying process and facilitates the formation of a homogeneous film layer, but also helps to increase the heat dissipation area of the device. To solve the problem of complicated and expensive process of pixel isolation structure made by yellow light process in the prior art, patent CN106158916A discloses a quantum dot film, its fabrication method, and display device [41]. The method of making the quantum dot film includes forming a hydrophilic region and a hydrophobic region on a first surface of a light-transmitting substrate, then setting a surface-modified mask with a plurality of hollow sections on the first surface, and making the hollow sections in the surface-modified mask corresponding to the hydrophilic region or the hydrophobic region, the surface-modified mask having a first modified surface and a second modified surface, and the first modified surface and the second modified surface having a hydrophilic and a hydrophobic surface, respectively. The first modified surface is a hydrophobic surface, so that the hydrophobic quantum dot ink enters the hydrophobic region through the hollow section, or the first modified surface is a hydrophilic surface, so that the hydrophilic quantum dot ink enters the hydrophilic region through the hollow section, and finally the quantum dot ink in the hydrophilic region or the hydrophobic region is dried. This production method can effectively reduce the production cost of quantum dot film. The liquid quantum dot adhesive has a certain degree of fluidity before it is cured, and if the force applied to the second barrier film is uneven, the quantum dot layer will be unevenly distributed, resulting in a large color difference that affects the display effect and low yield of large-size products. In addition, when the light curing is carried out by UV light, the initiator in the liquid quantum dot adhesive is often less compatible with the quantum dots, resulting in poor dispersion of the quantum dots in the quantum dot adhesive. To solve the above problems, patent CN110518112A provides an aerogel quantum dot film, a preparation method, and a display device thereof [42]. The preparation method of the quantum dot film comprises the following steps: (i) setting a pixel isolation structure on one side surface of the first barrier film, thereby defining a plurality of light-emitting regions; the light-emitting regions include a first sub-pixel region, a second sub-pixel region, and a third sub-pixel region; (ii) setting a wet gel or aerogel slurry inside and later drying it to obtain an aerogel layer; (iii) placing a quantum dot ink on at least two of the light-emitting regions of the quantum dot ink will penetrate and adsorb into the pores of the aerogel layer due to the capillary phenomenon; (iv) a hydroxide barrier layer is provided on the aerogel layer. Compared with the prior art, the method of preparing the aerogel quantum dot film provided by the invention can easily control the number of quantum dots in each luminescent region and ensure the thickness uniformity of the quantum dot luminescent layer. In addition, by increasing the density of the quantum dots stored in the aerogel layer and reducing the thickness of the aerogel layer, the aerogel quantum dot film can effectively reduce the brightness of the backlight when used in combination with an LED backlight or an electroluminescent backlight, thereby increasing the service life of the backlight. To effectively reduce the blue light hazard in the quantum dot film, patent CN109616577A provides a quantum dot film and its preparation method [43]. The quantum dot film comprises: a first quantum dot encapsulation structure comprising encapsulated red quantum dots and green quantum dots; and a second quantum dot encapsulation structure provided on one side of the first quantum dot encapsulation structure, the second quantum dot encapsulation structure comprising a second encapsulation material and a plurality of quantum dot portions spaced in the second encapsulation material, the gap between the quantum dot portions allowing light to be directed from one side of the second quantum dot encapsulation structure to the other side, the gap between the quantum dot portions allowing light to be directed from one side of the second quantum dot encapsulation structure to the other side. The quantum dot section comprises a third encapsulating material and blue quantum dots dispersed in the third encapsulating material. The invention reduces the blue light hazard by providing a quantum dot section that converts harmful blue light. In order to reduce the production cost of quantum dot films in the prior art, patent CN107680900A provides a quantum dot film, its production method, and a quantum dot device [44]. The fabrication method comprises: step one, preparing a cover plate and a plurality of microcapsules, setting each microcapsule on the surface of one side of the cover plate, each microcapsule comprising a capsule wall and a quantum dot dispersion capable of emitting a first light located in the capsule wall; step two, orienting the side of the cover plate having the microcapsules toward a substrate, the substrate comprising a first pixel region, and the first pixel region comprising a plurality of first subpixel regions such that at least one microcapsule corresponds to a first sub-pixel region so that the capsule wall is in contact with the surface of the substrate; step 3, applying a predetermined condition on the side of the cover plate away from the microcapsules so that the capsule wall ruptures and releases the quantum dot dispersion on the surface of the substrate corresponding to the first sub-pixel region; step 4, removing the cover plate with the ruptured capsule wall and curing the released quantum dot dispersion to obtain the first quantum dot film. The technical solution is simple to operate, requires low equipment, and does not require complex lithography process and expensive lithography equipment, which solves the problem of high cost of quantum dot film production and greatly reduces the cost of quantum dot film production. In addition, the quantum dot dispersion in the microcapsules is precisely delivered, and the technical solution can significantly improve the material utilization rate compared to the material waste in the photolithography method.

13.2.5 CSOT

CSOT has over 1093 patents in the field of quantum dots, mainly in the field of backlight modules and display panels. In the field of backlight modules, CSOT carries out patent layout for the design and manufacturing process of high-performance backlight modules. It is worthwhile to analyze some of the key patents owned by CSOT.

In order to realize the screen control of quantum dot natural light, patent US20230123638A1 discloses a quantum dot display panel, quantum dot display device, and its preparation method [45]. The quantum dot display panel includes: an array substrate, a color film substrate, and a liquid crystal layer set between the array substrate and the color film substrate, wherein the color film substrate includes: a cover plate, a cut-off layer, a quantum dot pixel layer, a blocking layer, a reflective layer, a coating layer, a built-in polarizing layer, an isolation column, and a PI layer. The interplay of the external polarizer, the built-in polarizing layer, and the liquid crystal layer allows the screen to control the passage of natural light from the quantum dots; at the same time, the built-in polarizing layer can be prepared by applying the polarizing solution directly inside the box, simplifying the manufacturing process and saving production costs. To improve contrast and display uniformity, patent CN111338129A discloses a backlight module and display device [46]. The backlight module includes a number of backlight units comprising a backplane, a light-emitting part, a reflective lens, a quantum dot film, and an encapsulated bezel. The back panel has a first side and a second side set opposite to each other, the light-emitting part is set on the first side, the reflective lens is set on the first side and corresponds to the light-emitting part; the reflective lens is used to reflect light from the light-emitting part, and the quantum dot film is set on the side of the reflective lens away from the back panel. The backplane, the light-emitting part, the reflective lens, and the quantum dot film are encapsulated in the encapsulation frame. The backlight module has a plurality of independent backlight units, each of which can form the backlight source. By controlling the backlight units independently, backlight crosstalk between adjacent units can be eliminated. In order to solve the problems of high water and oxygen permeability to meet the working requirements of quantum dots under existing packaging methods and poor heat dissipation leading to low EQE of quantum dots, patent WO2019148599A1 provides a quantum dot LED, backlight module, and display device [47]. The quantum dot LED includes a holder, a LED chip, at least one optical fiber layer, and an encapsulation layer. The LED chip is fixed to the holder and is connected to the holder. The optical fiber layer is provided above the LED chip and consists of an optical fiber hermetically sealed with quantum dots. The encapsulation layer encapsulates at least one layer of the optical fiber layer and the LED chip in the holder. The design solution ensures a high EQE of the quantum dots in this case by encapsulating the quantum dots with silica optical fibers with low oxygen and water permeability and high thermal conductivity. In order to improve the problem of large visual role bias of display devices, patent WO2019140711A1 provides a backlight module design solution [48]. The backlight module includes a backlight source, a light guide plate, and a quantum dot layer set in successive layers. The quantum dot layer includes a first side and a second side set opposite to each other, and the second side is laminated to the light guide plate. The quantum dot layer is provided with a plurality of first regions and a plurality of second regions formed between the first side and the second side, each first region is provided between two adjacent second regions to form an alternating arrangement of the first region and the second region. The concentration of quantum dots in the first region is less than the concentration of quantum dots in the second region, and the first face forming the first region comprises a first curved surface, wherein the first curved surface is used for scattering light passing through the first region. In order to improve the photoluminescence utilization rate and backlight brightness of the quantum dot layer and to avoid the bluish backlight color, patent CN107688255A discloses a method of making a quantum dot film and a backlight module [49]. The quantum dot film includes a quantum dot layer and two protective layers in which the quantum dot layer is sandwiched; the upper layer is a micro-hole structure with bubbles inside. By forming a bubble inside the protective layer on one side of the quantum dot layer, light entering the bubble inside the protective layer from the quantum dot layer will be totally reflected and reenter the quantum dot layer, thus increasing the photoluminescence efficiency and backlight brightness. To avoid the influence of water vapor and oxygen on the quantum dot material, patent WO2019019253A1 announces a backlight module that facilitates the display of LCDs by using a glass plate to seal and encapsulate the quantum dot material [50]. The backlight module comprises: a transparent light guide with light-intake sides and a hermetically sealed cavity. A number of quantum dots are housed in the cavity, and a light source is provided adjacent to the light-entry side.

In the field of display panels, CSOT mainly focuses on improving the EQE, color gamut, and color purity of display panels and other related patent layouts. In order to solve the problem that the color purity and color gamut of LCD devices are reduced due to the RGB color barrier not being able to fully absorb other wavelengths of light in the existing technology, which affects the visual effect, patent WO2021189590A1 discloses a quantum dot material structure and LCD device [51]. The quantum dot material structure is applied to the LCD device. The quantum dot core includes cadmium arsenide nanoclusters for absorbing green light at a predetermined wavelength; the quantum dot shell is used to protect the quantum dot core; and the ligand layer is used to facilitate the dispersion of the quantum dot material structure. To enhance the color gamut, patent CN111077698A provides a design solution for a backlight module and a LCD device [52]. The backlight module includes a backplane, a light source, a quantum dot film, and a purification film. The backplane is formed with a holding cavity and a light source fixing member. The light source is fixed in the cavity by the light source fixing member. The quantum dot film is provided facing the light source. The photopurification film is provided facing the quantum dot film and is provided with a color light absorption factor for absorbing a specific color light band. The design solution can further enhance the color gamut by providing a light purification film to absorb the yellow–orange and cyan–green light output from the quantum dot film. To solve the optical brightness loss and uneven brightness caused by the direct lamination of quantum dot layer to polarizer and light guide, patent CN110888254A discloses a quantum dot substrate, LCD panel, and double-sided LCD panel [53]. The quantum dot substrate comprises a substrate, a quantum dot layer, and a frame adhesive, wherein the quantum dot layer is formed on the substrate and the frame adhesive is coated on and around the quantum dot layer to support the film layer and form an air layer. In order to reduce the light loss during the photoluminescence process and to improve light utilization, patent WO2021088139A1 provides a display device and a method for making it [54]. The display device comprises: an array substrate and MicroLED. An array of MicroLEDs is arranged on the array substrate. The surface of one side of the array substrate is provided with a plurality of micro-cavity structures recessed toward the interior of the MicroLED device, and the plurality of micro-cavity structures containing quantum dots are filled into the micro-cavity structures by a quantum dot film layer provided on the MicroLED device to disperse quantum dots into the various micro-cavity structures on the surface of the MicroLED device, preventing the aggregation of quantum dots, while at the same time confining the quantum dots to the interior of the MicroLED device. The energy transfer effect between the MicroLED device and the quantum dots can be enhanced. In order to maintain the quantum dot display panel in the dark state display, to prevent the excitation of quantum dots by ambient light, and also to eliminate the reflection of ambient light by metal electrodes in the backlight, patent WO2021088148A1 provides a design solution for quantum dot display panel, including a pixel layer set in a stack, a color filter layer, a reflective filter layer, and a circular polarizer [55]. Two of every three sub-pixels in the pixel layer are filled with red quantum dots and green quantum dots. The color filter layer consists of a color filter, a red filter corresponding to the red quantum dots, and a green filter corresponding to the green quantum dots. The reflective filter layer comprises a reflective filter, at least one first film, and at least one-second film. In order to effectively improve the contrast and penetration rate of the display panel, patent WO2021093043A1 discloses a design solution for a display panel and a display device [56]. In particular, the display panel comprises a first display panel, a quantum dot layer, and a second display panel. The first display panel is provided opposite to the second display panel, and the quantum dot layer is provided on the side of the first display panel facing the second display panel. By setting the first display panel and the second display panel and enabling one of them to achieve dynamic backlighting and the other to display a picture, and by setting the quantum dot layer between the two display panels, the display panel is thus provided with a high color gamut, a wide viewing angle, a high contrast ratio, and a high penetration rate.

In terms of quantum dot ink and inkjet printing, CSOT has actively carried out relevant patent layout. In order to suppress the coffee ring effect of quantum dot ink in the inkjet printing process, patent WO2021109246A1 provides a production method of quantum dot ink and display panel [57]. The quantum dot ink includes an organic solvent, and quantum dots are dispersed in the organic solvent. The quantum dots include luminescent quantum dots and blocking quantum dots. By adding blocking quantum dots to the quantum dot ink, the diffusion of quantum dot ink during the inkjet printing process can be suppressed, and the flatness and uniformity of the quantum dot film surface can be improved, so that the display panel can exhibit excellent display quality. To avoid the impact of quantum dot ink printing on the electrical conductivity and luminescence performance of the film, patent WO2017161629A1 provides a method for making quantum dot printing ink [58]. The method of making this quantum dot printing ink is to adjust the viscosity of the quantum dot printing ink to a desired range by mixing a first solvent with a second solvent, and later to adjust the surface tension of the quantum dot printing ink to a desired range by mixing a third solvent with the first and second solvents while maintaining the quantum dot printing ink with a predetermined range of viscosity. By adding a fourth solvent to the ink, the vapor pressure can be adjusted to a reasonable range. By mixing the above solvents, a quantum dot ink with suitable viscosity, surface tension, drying conditions, and other performance indicators for inkjet printing can be formulated, thus avoiding the addition of surfactants to the ink and achieving the goal of not affecting the electrical conductivity and luminescence of the film by the quantum dot ink. The goal is to avoid the addition of surfactants to the ink and to achieve the goal that the quantum dot ink does not affect the conductivity and luminescence of the film.

13.2.6 BOE

BOE, one of the world's leading display panel manufacturers, has also been actively involved in the research and development of QDLED technology. BOE has over 1531 patents in the field of quantum dots, mainly focusing on the design of high-performance QDLED display panel structures.

In order to reduce the radiation-free transition between the interface of the quantum dot core and the interface of the quantum dot shell and to improve the luminescence of quantum dots, patent WO2020228403A1 discloses a quantum dot and its production method, QDLED and display panel [59], wherein the quantum dots comprise a quantum dot core, a charge transition layer encapsulated outside the quantum dot core, and a quantum dot shell encapsulated outside the charge transition layer. The charge transition layer is doped with metal ions within the body material, the metal ions being charge valence variable metal ions, the charge-valence of the metal ions comprising the charge valence of the cations in the quantum dot core and the charge valence of the cations in the quantum dot shell. In order to improve the light output efficiency and reduce the thickness of the display panel, patent US11567403B2 provides a quantum dot color filter layer and its production method, display panel, and device [60]. The quantum dot color filter layer is divided into a plurality of filter zones spaced apart, and a quantum dot functional layer is provided in the filter zones, and the quantum dot functional layer emits light under the excitation of blue light. There are tiny patterns in the quantum dot functional layer, and the tiny patterns are hollow patterns. When blue light irradiates into the quantum dot functional layer, it can irradiate to the side of the quantum dot functional layer through the gap of the hollow pattern, thus increasing the contact area between blue light and quantum dot functional layer, improving the excitation rate of blue light to quantum dots and the EQE of quantum dot functional layer. To effectively solve the problem of short service life of existing quantum dot films, patent CN110703498A provides a quantum dot film and its preparation method, backlight, and display device [61]. The quantum dot film comprises a first substrate and a second substrate set opposite to each other, and a quantum dot layer disposed between the first substrate and the second substrate, the material of the first substrate or second substrate comprising a thermally triggered self-healing material. By absorbing the heat generated during the operation of the LED and the light conversion of the quantum dots through the substrate of the thermally triggered self-healing material, the operating temperature of the quantum dots is not only reduced but also the aging or damage of the substrate can be repaired, maximizing the performance and service life of the quantum dot film. To solve the problem of low EQE of quantum dot display devices, patent CN110646977A provides a quantum dot display panel and display device. The quantum dot display panel has a plurality of sub-pixels [62]. The quantum dot display panel includes: a quantum dot array, a first lens array, and a dimming layer. The quantum dot array comprises a plurality of quantum dot light-emitting layers spaced apart, with one quantum dot light-emitting layer disposed within a sub-pixel. The first lens array is located on the light-emitting side of the quantum dot array and includes a plurality of spaced first lenses. One lens is located within a sub-pixel and the first lens is used to emit parallel light. A dimming layer, located on the outgoing side of the first lens array, and having a plurality of dimming zones with adjustable transmittance, one dimming zone being located within a sub-pixel. In order to avoid the undesirable phenomenon that the blue MiniLED module will show a blue glow all around in the lit state, patent CN109765726A discloses a quantum dot film, preparation method, backlight module, driving method, and display device [63]. By setting the quantum dot structures within each grid of the isolation structure, each quantum dot structure is surrounded by the isolation structure, so that in the process of using the quantum dot film, for example, when the quantum dot film is cut and applied to the backlight module, the cut quantum dot film is protected by the isolation structure, so that moisture and oxygen in the surrounding environment do not easily enter the intermediate quantum dot structures surrounded by the isolation structure. The middle quantum dot structure will not be damaged by the surrounding environment, and only a small part of the quantum dot structure around the cut is damaged, which does not affect the effectiveness of the quantum dot film. In order to be able to effectively improve the EQE of the quantum dot layer, patent US11322706B2 discloses a quantum dot film, a quantum dot luminous assembly, and a display device [64]. The quantum dot film comprises: a quantum dot layer and a conductive layer disposed on at least one side of the quantum dot layer along the thickness direction, wherein the conductive layer comprises a plurality of nanoscale metal particles, a portion of the plurality of nanoscale metal particles being configured to produce surface plasmon resonance under the action of electromagnetic radiation. In order to prevent quantum dots from generating cluster failure problems, minimize the effect of the operating temperature of the LED chip on the quantum dots, and maintain the high efficiency of the quantum dots, so that the color rendering and saturation of the LED light source can be improved, the color gamut can be widened, and the lighting and display effects can then be improved, patent CN109301056B discloses an LED light source, backlight, and display device [65]. Wherein, the LED light source comprises: an LED chip; at least two quantum dot excitation layers laminated on the light-emitting surface of the LED chip; and wherein the concentration of quantum dots in the quantum dot excitation layer away from the LED chip is greater than the concentration of quantum dots in the quantum dot excitation layer close to the LED chip. In order to reduce the fluorescence quenching phenomenon and improve the EQE, patent WO2019062365A1 provides a QDEF [66]. The QDEF comprises a substrate layer and a functional layer, the functional layer being formed on the substrate layer, wherein the functional layer is provided on at least one side of the thickness direction of the substrate layer, the functional layer comprising a limiting layer and a quantum dot material layer alternately laminated along the thickness direction of the QDEF, the limiting layer comprising a hydrotalcite material or a hydrotalcite-like material, and the quantum dot material layer comprising quantum dots. In order to improve the color gamut to some extent and to improve the problem of large viewing angle bias, patent US10330849B2 discloses a quantum dot film and a backlight module thereof [67]. The quantum dot film comprises: a quantum dot layer and an optical waveguide layer, with the quantum dot layer over the optical waveguide layer. The optical waveguide layer is a laminated structure composed of a number of sub-layers, the laminated structure in which the refractive index of the number of sub-layers becomes larger layer by layer, starting from a sub-layer close to the quantum dot layer. The method comprises: preparing an optical waveguide layer; providing a quantum dot layer; and laminating the quantum dot layer with the optical waveguide layer to obtain a quantum dot film. The backlight module comprises: a light guide plate, a quantum dot film, and a prism film, the quantum dot film being sandwiched between the light guide plate and the prism film, and the quantum dot film being the aforementioned quantum dot film, the quantum dot layer being located between the light waveguide layer and the prism film.

In the field of QDLED, BOE has also made patent layouts. In order to improve the electron injection ability and thus the EQE and lifetime of QDLED devices, patent CN111341926A provides a QDLED device and its production method, display panel, and display device [68]. The QDLED device comprises: a quantum dot light-emitting layer and an electron transport layer provided on one side of the quantum dot light-emitting layer. The QDLED layer comprises a quantum dot material, and the electron transport layer comprises a heterogeneous multimer, wherein the difference in the conduction band positions of the heterogeneous multimers is used to build a multi-energy gradient so that the electron injection barrier can be lowered when electrons are jumped. The heteropolymer is a nanoparticle comprising at least a first electron transport material and a second electron transport material. The first electron transport material and the second electron transport material are connected by van der Waals forces, the conduction band energy level of the first electron transport material is lower than the conduction band energy level of the quantum dot material, and the conduction band energy level of the second electron transport material is higher than the conduction band energy level of the first electron transport material. In order to significantly improve the adhesion of the quantum dot layer in the quantum dot device backplane, make it firmly bonded, not easy to fall off, ensure good display effect, improve the processing yield, and reduce costs, patent US11258028B2 provides a quantum dot device backplane and its production method [69]. The quantum dot device backplane comprises: a substrate; a cathode, the cathode being provided on a first surface of the substrate; an electron transport layer, the electron transport layer being provided on a surface of the cathode away from the substrate; a connection layer, the connection layer being provided on a surface of the electron transport layer away from the substrate and bonded to the electron transport layer by chemical bonding; a quantum dot layer, the quantum dot layer being provided on a surface of the connecting layer on a surface of the substrate away from the substrate and bonded to the connecting layer by chemical bonding. In order to effectively promote electron–hole injection balance and improve the efficiency and lifetime of QDLED devices, patent WO2020233293A1 discloses a quantum dot, its preparation method, and quantum dot light-emitting devices [70]. The invention provides the outermost shell layer of quantum dots as hole transport material, which makes the quantum dots of this structure applied to the preparation of QDLED devices. On the one hand, the outermost hole transport material of the quantum dots can be used as the hole transport layer in QDLED devices, reducing the process of making a separate layer of hole transport layer and effectively simplifying the device structure and process; on the other hand, the outermost hole transport material of the quantum dots. On the other hand, the outermost hole transport material of the quantum dot is in contact with the electron transport layer in the QDLED device, which acts as an electron-blocking layer and can block part of the electron transport, solving the problem of electrons becoming multiplets in the QDLED device due to more efficient electron transport in the prior art. In order to improve the QDLED of the prior art, there will be the problem that the light-emitting efficiency will drop when the voltage rises again after a certain level, patent WO2020224334A1 discloses a quantum dot electroluminescent device, display panel, and display device [71]. The quantum dot electroluminescent device includes a stacked anode layer, a composite light-emitting layer, and a cathode layer, wherein the composite light-emitting layer comprises at least two stacked quantum dot light-emitting layers, with an intermediate layer between each of the two adjacent quantum dot light-emitting layers; the intermediate layer is configured to transport holes and block electrons. In order to solve the problem that excitons formed by electrons and holes are easily quenched at the interface between the light-emitting layer and the electron transport layer and improve the performance of QDLED, patent WO2020233358A1 provides a quantum dot complex and its preparation method, a light-emitting device, and its preparation method [72]. By forming an electron transport material layer on the outer side of part of the core–shell quantum dots, the electron transport material layer wraps part of the core–shell quantum dots and exposes the other part of the core–shell quantum dots. The quantum dot complex is used to form an interface layer located between the light-emitting layer and the electron transport layer, which can solve the problem that excitons formed by electrons and holes are easily quenched at the interface between the light-emitting layer and the electron transport layer, and enable the interface between the light-emitting layer and the electron transport layer to form a non-heterogeneous structure, thereby extending the service life and performance of QDLED.

13.2.7 TCL

TCL has more than 1440 patents in the field of quantum dots, mainly focusing on high-performance QDLED and QDLED display panels. TCL is a multinational electronics company with a significant presence in the quantum dot display industry and has a number of patents related to QDLED technology.

In the area of QD materials synthesis, TCL has patented various methods for synthesizing and preparing quantum dot materials, including core–shell quantum dots, metal chalcogenide quantum dots, and other semiconductor quantum dots. These methods involve the use of different precursors, solvents, and reaction conditions to control the size, shape, and properties of the quantum dots.

In terms of ink formulation and printing techniques, TCL has developed several ink formulations that can be used for the inkjet printing of quantum dot materials. These formulations include quantum dot inks with high stability, high efficiency, and high concentration, as well as surface-modified quantum dot inks that can be used for printing on various substrates.

TCL has also developed several device engineering and display application patents related to QDLED technology. For example, they have patented methods for manufacturing QDLED with improved efficiency and stability, including the use of various device structures and encapsulation methods. Additionally, they have developed techniques for producing QD-LCDs and QD-OLEDs, as well as methods for producing high-brightness, high-color gamut displays using quantum dot technology.

In the field of QDLED devices, TCL mainly focuses on improving the EQE and lifetime of the devices to carry out the relevant patent layout. In order to solve the problem of low carrier narrow-band shift rate of existing QDLED devices, which affects the EQE of the devices, the thin thickness of quantum dot luminous layer, and the service life, patent CN106328822A provides a QDLED device structure design [73]. The device includes a first electrode, a hole injection layer, a hole transport layer, a light-emitting layer, an electron transport layer, and a second electrode in that order. The light-emitting layer is made of a quantum dot light-emitting material and a hybrid transport material. The hybrid transport material is a hole transport material and an electron transport material, and the hole transport material and the electron transport material form a bi-continuous network structure in the light-emitting layer, while the quantum dot light-emitting material is dispersed in the bi-continuous network structure. The hole transport material and the electron transport material form a bi-continuous network structure, and the quantum dot light-emitting material is dispersed in this bi-continuous network to form the light-emitting system. By replacing the quantum dots with a bi-continuous network as the transport medium for the injected carriers, the holes injected from the electrodes can be transported in the network of the hole transport material, and the electrons can be transported in the network of the electron transport material. Since the hole transport material and the electron transport material are optimized for the transport of holes and electrons, respectively, the bi-continuous network structure described can significantly increase and balance the electron and hole migration rates in the light-emitting layer, thereby increasing the EQE of the QDLED device. As the carrier mobility and conductivity increase, the driving voltage decreases and the thickness of the light-emitting layer can be increased, thus providing a new dimension for optimizing the light output efficiency. The increase in device thickness reduces the effective electric field of the light-emitting layer, thus increasing the QDLED device lifetime. In addition, the double-continuous network structure formed by the hole transport material and electron transport material makes the quantum dot light-emitting materials no longer closely aligned with each other, which can reduce the quenching of excitons due to concentration and the intermittent compounding caused by carriers, ultimately improving the EQE and brightness of the light-emitting single component. Currently, metal oxides are widely used as electron transport and hole transport layers in QDLED devices; compared to organic compounds, metal oxides have a high carrier narrow-band shift rate, tunable energy level structure, and excellent stability. However, most metal oxides are synthesized in air and therefore contain a large number of hydroxyl groups on their surface, which can act as exciton quenching groups and lead to weakened device performance. To solve the problem that existing metal oxide surface treatment methods make the device structure complex and cannot be used in devices on a large scale, patent CN106450042A discloses a metal oxide, QDLED, and a preparation method [74]. The metal oxide preparation method includes the following steps: dissolving the precursor of the metal oxide in a solvent, then mixing it with PVP, and then reacting it at 100–300 °C for 0.5–2 hours to obtain PVP-coated metal oxide, or dissolving the precursor of the metal oxide in a solvent, reacting it at 100–300 °C for 0.5–2 hours, and then reacting it with PVP mixing to obtain PVP-coated metal oxide; or mixing the nanoparticle solution of metal oxide with PVP to obtain PVP-coated metal oxide. By coating the metal oxide surface with PVP, the defects on the metal oxide surface can be reduced and the metal oxide can be passivated, thus effectively covering the defects that may arise on the metal oxide surface as well as the exciton compounding center, and thus increasing the ratio of exciton compounding and improving the QDLED efficiency. The method is a simple process, easy to implement, and can be applied to device preparation on a large scale. In QDLED devices, the light on the backlight side of the quantum dot light-emitting layer is often not effectively utilized, and the metal electrodes, although reflective, also have a large absorption loss, so the existing QDLED have yet to improve their light-emitting efficiency. In order to solve the problem that the EQE of QDLED in the prior art needs to be improved, patent CN106206976A discloses a QDLED based on a photonic crystal structure and a preparation method [75]. The QDLED device includes, from bottom to top, a substrate, an electron transport layer, a quantum dot light-emitting layer, a hole transport layer, a hole injection layer, and a top electrode. The hole injection layer is a hole injection layer with a photonic crystal structure. By making a photonic crystal structure in the hole injection layer, the surface effect of the photonic crystal, i.e. total reflection and coupling of the light emitted from the quantum dots with the surface state of the photonic crystal, can be effectively used to improve the light output efficiency of QDLED by shooting light from the quantum dots to the metal electrode side. In order to solve the problem that the existing QDLEDs, especially short-wavelength QDLEDs, have low performance due to the imbalance of carriers (electrons and holes) caused by the difficulty of hole injection, patent CN105280829A provides a QDLED design scheme and its preparation method [76]. The QDLED includes a cathode, a quantum dot light-emitting layer, a hole transport layer, and an anode set in a sequential lamination. The cavity transport layer is made of a deep blue light body material. By using a deep blue light body material with a deep HOMO (highest occupied orbital) energy level and a high T1 (triplet state) energy level as the hole transport layer material for the QDLED. On the one hand, the deeper HOMO energy level (around 7eV) can effectively reduce the hole injection barrier between the hole transport layer and the quantum dot emitting layer; at the same time, due to the good hole transport properties of the deep blue light body material, thus ensuring the effective hole transport. On the other hand, the higher T1 energy level effectively prevents the quenching of excitons at the interface of the hole transport layer due to energy reversal, thus effectively improving the QDLED device's performance. Current QDLED devices all have high leakage currents in the low current region, resulting in low carrier complex efficiency and, hence, low efficiency and short lifetime of QDLED devices. The main reason for the leakage current is that the QDLED films are not dense, and even densely stacked QDLED films still have gaps. In order to solve the above problem, patent CN106450013A provides a QDLED device design solution [77]. On the one hand, the QDLED device does not have a separate quantum dot layer, thus avoiding the impact of the introduction of quantum dot films on the EQE and lifetime of the device. At the same time, the doping of quantum dots into the transport material to form a hole transport layer and an electron transport layer makes the QDLED device have only a heterojunction structure between the interfaces of different functional layers, which can improve the electrical characteristics of the QDLED device, enhance the stability of the QDLED device, and contribute to the EQE. On the other hand, the

QDLED device described in the invention can employ an insulating material as a carrier for the hole transport layer or the electron transport layer of the quantum dot doping, thus allowing for better confinement of electrons and holes and facilitating the formation of quantum potential wells. The insulating material can better protect the quantum dots, fill the gaps between quantum dots, and reduce the occurrence of leakage currents, thus improving the EQE and lifetime of the QDLED device.

In terms of quantum dot display panels, TCL has also carried out some patent layouts. Quantum dots are applied to quantum dot display panels, which can only be prepared by wet method, so it is necessary to make pixel banks. After the pixel banks are made, a very thin layer of bank film will remain on the surface of the pixel electrodes. In order to solve the problems of removing the residual pixel bank layer by existing methods, which will affect the performance of the device at a later stage, and the process of removing the residual pixel bank layer will increase the production cost and be easy to mix colors, patent CN106601922A discloses a quantum dot display panel and its production method [78]. The quantum dot display panel includes a substrate, a pixel electrode located in the pixel electrode pattern area of the substrate, a pixel bank layer in the peripheral area of the pixel electrode, a residual pixel bank layer on the pixel electrode, an electron transport layer, a quantum dot light-emitting layer, a hole transport layer, a hole injection layer, and an anode layer in order from the bottom up. By using the bank film left on the pixel electrode during pixel bank fabrication as an electron-blocking layer to produce QDLED with inverted structures, the invention will reduce the charge injection from the pixel electrode to the QDLED device due to the insulating properties of the bank film, thus balancing the electron and hole injection balance in the QDLED and improving device performance, while also avoiding the process of removing the residual bank film and the ease of color mixing caused by this process. The QDLED is generally composed of a first electrode, a hole transport layer, a quantum dot emitting layer, an electron transport layer, and a second electrode; as each layer has a different energy level, i.e. an energy polarity difference, during the operation of the QDLED, charges will accumulate at the interface with the energy polarity difference, especially at the interface in contact with the quantum dot emitting layer. These defects can also limit the carriers. As the operating time of the QDLED increases, more and more charges are confined to the interface and become the center of burst photons, thus greatly reducing the luminous intensity and shortening the lifetime of the QDLED. During the driving process of QDLED display panels, the long-time accumulation of electric charges in the QDLED interface layer will likewise affect the service life of QDLED display panels as well as the luminous intensity. To solve the above problem, patent CN108932927A discloses a driving scheme for quantum dot display panels [79], wherein step A: a reverse trigger signal is added in advance to the driving circuit for driving the pixel dots to light up; step B: when the reverse trigger signal is excited, a reverse drive signal is applied to the pixel dots, thereby eliminating the electric charge confined in the pixel dots; the reverse drive signal changes the potential barrier of the defect potential well, eliminating the charge confined and collected in the potential well and reducing the density of the confined charge, thereby extending the service life and increasing the display brightness of the quantum dot display panel.

13.3 Summary and Outlook

Quantum dot material is the core material in the next generation of new display technology, which directly determines the performance of quantum dot display devices. At present, China is in the first echelon in the field of quantum dot display industry with a first-mover advantage. Whether in the field of quantum dot materials or in the field of QDLED, Chinese enterprises have their own patent layout. In the field of quantum dot materials, Najing Tech is leading the world in the field of colloidal quantum dot synthesis. By controlling the growth process of colloidal quantum dots to regulate the exciton state properties and optical properties of quantum dots, Najing Tech has developed a series of high-performance quantum dot materials. In the quantum dot backlight and QDLED fields, Najing Tech and CSOT have carried out a series of technological innovations and patent layouts focusing mainly on the light output efficiency, service life, heat dissipation, EQE, color gamut, and color purity. In the display panel field, BOE and TCL mainly focus on the technological innovation of display panel structure design and preparation processes, so as to achieve the goal of reducing the manufacturing cost of display panels, improving the display effect and EQE, and other quantum dot display technology industrialization developments.

In addition to Chinese enterprises, there are other major players in the quantum dot display technology industry. SAMSUNG is a major player in the display industry and has been actively developing quantum dot displays. The company has already released QDLED TVs, which use quantum dots to enhance color accuracy and brightness. In addition, SAMSUNG acquired QD Vision to expand the patent layout in the field. Nanosys is a US-based company that has developed a technology known as QDEF, which is designed to improve color accuracy and brightness in displays. These companies are key players in the field of quantum dot display technology. They have made significant contributions to the development and commercialization of quantum dot displays.

At present, the performance indicators of red and green QDLED are close to the requirements for commercialization, and the breakthrough in EQE and service life of blue QDLED is the key to the industrialization of QDLED display technology. With ongoing research and innovation, the quantum dot display technology industry is poised for continued growth and advancement. Despite these challenges, many experts believe that QDLED technology has the potential to become a mainstream display technology in the near future, particularly as the demand for high-quality, energy-efficient displays continues to grow.

References

1 Ippen, C., Truskier, J., and Manders, J. Nanosys Inc., Assignee (2019). Method for synthesis of blue-emitting $ZnSe_{1-x}Te_x$ alloy nanocrystals. Patent US20190390109A1.

2 Ippen, C., Ilan, J.-L.P., Kan, S. et al. (2017). Stable InP quantum dots with thick shell coating and method of producing the same. patent US20170306227A1.
3 Guo, W., Chen, J., Dubrow, R., and Freeman, W.P. (2015). Highly luminescent nanostructures and methods of producing same. Patent US9884993B2.
4 Plante, I.J.-L. and Wang, C. (2021). Thiolated hydrophilic ligands for improved quantum dot reliability in resin films. Patent US11021651B2.
5 Nelson, E.W., Eckert, K.L., Kolb, W.B. et al. (2014). Quantum dot film. Patent EP2946411B1.
6 Manders, J.R. and Berkeley, B.H. (2019). Display devices with different light sources in pixel structures. Patent US20200168673A1.
7 Hartlove, J., Hardev, V., Kan, S. et al. (2016). Quantum dot based color conversion layer in display devices. Patent US20200098951A1.
8 Lee, E.C.-W. (2019). Methods of improving efficiency of displays using quantum dots with integrated optical elements. Patent WO2020023583A1.
9 Nick, R.J. and Breen, C. (2012). Quantum dot-containing compositions including an emission stabilizer, products including same, and method. Patent WO/2013/078252.
10 Jin, T., Kinjo, M., Tamura, M. et al. (2005). Water-Soluble Fluorescent and Production Process Thereof. Patent WO2006033396A1.
11 Yoo, J.-M., Ahn, J.S., and LEE, K.H. (2018). Display apparatus. Patent US10663648B2.
12 Park, C.S., Lee, Y.C., Chae, S.H. et al. (2015). Reflecting plate, backlight unit, and display device. Patent US9857631B2.
13 Lee, K.H. (2019). Display panel and display apparatus having the same. Patent US10930713B2.
14 Cho, B., Lee, Y., Roh, N., and Min, K. (2019). Display apparatus. Patent US10914887B2.
15 Park, K., Kim, Y., Park, H., and Yoon, S.-T. (2017). Photoluminescence device, method of manufacturing the same and display apparatus having the same. Patent US10698255B2.
16 Lee, S.-G., Park, J., Son, D. et al. (2019). Display device. Patent KR20200088532A.
17 Jang, N.-w., Hur, J., Park, T.S., and Yoo, J.-M. (2017). Display apparatus having quantum dot unit or quantum dot sheet and method for manufacturing quantum dot unit. Patent US10424691B2.
18 Han, M.G., Chung, D.Y., Kim, K. et al. (2019). Electroluminescent device, and display device comprising thereof. Patent US20200083450A1.
19 Won, Y., Kwon, H.I., Jang, E.J. et al. (2019). Electronic device including quantum dots. Patent US10851298B2.
20 Kim, S.W., Kim, C.S., Kim, T.H. et al. (2019). Electroluminescent device, manufacturing method thereof, and display device comprising the same. Patent US11233211B2.
21 Kim, T.H., Kim, S.W., Jang, E.J., and Chung, D.Y. (2019). Electroluminescent device, and display device comprising the same. Patent US10879484B2.

22 Ahn, J., Kwon, H.I., Park, S.H. et al. (2017). Quantum dots, a composition or composite including the same, and an electronic device including the same. Patent US10988685B2.

23 Narrainen, A.P., Vo, C.-D., Quang, K.D. et al. (2015). Quantum dot compositions. Patent WO2015140642A2.

24 Daniels, S. and Narayanaswamy, A. (2015). Quantum dot nanoparticles having enhanced stability and luminescence efficiency. Patent WO2015118346A1.

25 Glarvey, P.A., Harris, J., Daniels, S. et al. (2014). Cadmium-free quantum dot nanoparticles. Patent WO2015101779A1.

26 Daniels, S., Harris, J., Glarvey, P. et al. (2014). Group iii-v/zinc chalcogenide alloyed semiconductor quantum dots. Patent WO2014162208A2.

27 Hu, B. and Mao, Y. (2019). Blue light core-shell quantum dot, and preparation method and application thereof. Patent CN103210576A.

28 Gao, J., Wang, J., Chen, C. et al. (2016). InP quantum dot and preparation method thereof. Patent CN106701059B.

29 Xie, S., Wang, J., Tu, L., and Gao, J. (2016). ZnSe/III V race/ZnSexS1 x or ZnSe/III V race/ZnSe/ZnS quantum dot and preparation method thereof. Patent CN106479481A.

30 Chen, X. and Su, Y. (2016). White light quanta point composition and preparation method thereof. Patent CN106497546A.

31 Qiao, P., Wu, H., and Gao, J. (2019). Quantum dot, preparation method and photoelectric device. Patent CN110317609A.

32 Chen, X., Shao, L., and Xie, Y. (2018). A kind of red light quantum point, its synthetic method and light emitting diode with quantum dots. Patent CN108841373A.

33 Gao, Y. and Xie, S. (2018). A kind of ZnOS/ZnO is nanocrystalline and preparation method thereof, luminescent device. Patent CN109628082A.

34 Peng, J. and Guo, H. (2019). A kind of zinc oxide base nano particle ink and photoelectric device. Patent CN110144139A.

35 Jing, Y., Zhang, Z., and Chen, X. (2018). Zinc oxide nano-crystal, zinc oxide nano-crystal composition, preparation method and electroluminescent device. Patent CN110484233A.

36 Zhang, Z. (2019). Preparation method of zinc oxide nanocrystal and photoelectric device. Patent CN111115679A.

37 Zhang, Z. (2018). Doped zinc oxide nanocrystal and preparation method thereof, quantum dot light-emitting device and preparation method thereof. Patent CN111326661A.

38 Chen, C. and Zhen, C. (2016). Electroluminescent device, and display apparatus and illumination apparatus provided with same. Patent CN106058072A

39 Jin, Y. and Li, Y. (2018). Quantum dot light-emitting diode device and ink for manufacturing same. Patent CN111326664A.

40 Hu, B. (2020). Display substrate, manufacturing method thereof and display device. Patent CN111354869A.

41 Gu, X. and Zhen, C. (2016). Quantum dot film, its manufacture method and display device. Patent CN106158916A.

42 Wang, B. (2019). A kind of aeroge quantum dot film, preparation method and the display device comprising it. Patent CN110518112A.
43 Jin, G. (2018). A kind of quantum dot film and preparation method thereof. Patent CN109616577A.
44 Du, Y. (2017). Quantum dot film and preparation method thereof, quantum dot device. Patent CN107680900A.
45 Yang, C., Huang, C., SHI, Y., and Cheng, W. (2020). Quantum dot display panel, quantum dot display device, and preparation method thereof. Patent US20230123638A1.
46 Xiang, C. (2020). Backlight module and display device. Patent CN111338129A.
47 Fan, Y. (2018). Quantum dot led, backlight module and display apparatus. Patent WO2019148599A1.
48 Chang, J., Li, Y., Xiao, Y., and Zhang, J. (2018). Backlight module and display device. Patent WO2019140711A1.
49 He, X. and Fu, L. (2017). A kind of backlight module, quantum dot diaphragm and preparation method thereof. Patent CN107688255A.
50 Han, M. and Guo, Q. (2017). Backlight module and liquid crystal display. Patent WO2019019253A1.
51 Zhou, M. (2020). Quantum dot material structure, liquid crystal display apparatus and electronic device. Patent WO2021189590A1.
52 Zha, B., Tang, M., and Chen, X. (2019). Backlight module and liquid crystal display device. Patent CN111077698A.
53 Zhou, M. (2019). Quantum dot substrate, liquid crystal display panel and double-sided liquid crystal display panel. Patent CN110888254A.
54 Hu, Z. (2019). Display apparatus and method for manufacturing display apparatus. Patent WO2021088139A1.
55 Pan, S. (2019). Quantum dot display panel filter. Patent WO2021088148A1.
56 Zhou, M. and Chen, L. (2019). Display panel and display device. Patent WO2021093043A1.
57 Zhang, S. (2019). Quantum dot ink, display panel manufacturing method and display panel. Patent WO2021109246A1.
58 Liu, Y. (2016). Method for preparing quantum dot printing ink, and prepared quantum dot printing ink. Patent WO2017161629A1.
59 Yu, G. and Zhang, A. (2020). Quantum dot and method for manufacture thereof, quantum dot light-emitting diode, and display panel. Patent WO2020228403A1.
60 Shi, G., Zhu, M., Fang, Z. et al. (2020). Quantum dot color filter, fabrication method thereof, display panel and device. Patent US11567403B2.
61 Zhang, Q., Zhou, J., Li, Q. et al. (2019). Quantum dot film, preparation method thereof, backlight source and display device. Patent CN110703498A.
62 Yang, S., Zhu, M., Liu, Y. et al. (2019). Quantum dot display panel and display device. Patent CN110646977A.
63 Zeng, W., Zhang, B., Li, H. et al. (2019). Quantum dot film, preparation method, backlight module, driving method and display device. Patent CN109765726A.
64 Zhu, W., Li, Z., Li, X. et al. (2019). Quantum dot film, quantum dot light-emitting assembly and display device. Patent US11322706B2.

65 Qu, L., You, Y., Yang, R. et al. (2018). LED light source, preparation method thereof, backlight source and display device. Patent CN109301056B.

66 Fan, G. (2018). Quantum dot enhancement film and manufacturing method thereof, backlight source, and display device. Patent WO2019062365A1.

67 Wang, B., Ma, Z., Xuan, M. (2016). Quantum dot film, method for manufacturing the same and backlight module. Patent US10330849B2.

68 Feng, J. (2020). QLED device, manufacturing method thereof, display panel and display device. Patent CN111341926A.

69 Zhang, A. (2020). Quantum dot device baseplate, manufacture method therefor and quantum dot device. Patent US11258028B2.

70 Feng, J. (2020). Quantum dot and manufacturing method therefor, quantum dot light-emitting device, related apparatus. Patent WO2020233293A1.

71 Kristal, B. (2020). Quantum dot electroluminescent device, display panel, and display device. Patent WO2020224334A1.

72 Mei, W. (2020). Quantum dot composition and preparation method therefor, and light-emitting device and preparation method therefor. Patent WO2020233358A1.

73 Chen, S., Qian, L., Yang, X. et al. (2016). QLED and fabrication method thereof. Patent CN106328822A.

74 Cao, W. and Wang, Y. (2016). Metal oxide, QLED. and preparation method. Patent CN106450042A.

75 Li, L., Cao, W., and Qian, L. (2016). A kind of QLED based on photon crystal structure and preparation method. Patent CN106206976A.

76 Chen, Y., Fu, D., and Yan, X. (2015). QLED and preparation method thereof. Patent CN105280829A.

77 Xiang, C., Yang, H., Yang, Y. et al. (2016). QLED device. Patent CN106450013A.

78 Chen, Y. (2016). Quantum dot display panel and manufacturing method thereof. Patent CN106601922A.

79 Xiang, C., Li, L., Qian, L. et al. (2017). A kind of driving method of quantum dot display panel. Patent CN108932927A.

14

Patterning Techniques for Quantum Dot Light-Emitting Diodes (QDLED)

14.1 Introduction

Quantum dot light-emitting diodes (QDLEDs) are a promising technology for next-generation display and lighting applications due to their high color purity, efficiency, and tunable emission spectra. Patterning and printing of QDLED are crucial for their practical implementation in various devices. In this chapter, we will discuss the different patterning and printing technologies that have been developed for QDLED. Current methods for patterning QDs contain three main types: (i) photolithography [1–4]; (ii) micro-contact transfer (MCT) [5–8]; and (iii) inkjet printing (IJP) [9–15].

14.2 Photolithography

Photolithography is a common technique for patterning QDLED. In this process, a photosensitive polymer is deposited onto a substrate and exposed to a patterned light source. This process selectively hardens the polymer, which can then be etched away to leave behind the desired pattern. Photolithography can produce high-resolution patterns with high accuracy, but it is a complex and expensive process that requires specialized equipment.

Conventional photolithography involves several steps, namely, photoresist (PR) coating, exposure, PR development, pattern transfer, and PR removal [16]. As shown in Figure 14.1, PR is a light-sensitive polymer that can be classified as positive or negative. For the positive-tone PR, the photochemical reaction leads to dissociation of the bonding between polymers; therefore, the regime of the PR layer, which is selectively exposed to light, can be easily dissolved using a developing solvent while selectively retaining the unexposed regime of the layer. For the negative-tone PR, on the other hand, the photochemical reaction yields chemical crosslinking between the polymer chains; therefore, the regime of the PR layer, which is selectively exposed to light (the crosslinked regime), remains structurally robust against the development process. Since the final pattern of the substance relies on the features of the PR pattern, the uniform application of the PR layer on a substrate is critical and is

Colloidal Quantum Dot Light Emitting Diodes: Materials and Devices, First Edition. Hong Meng.
© 2024 WILEY-VCH GmbH. Published 2024 by WILEY-VCH GmbH.

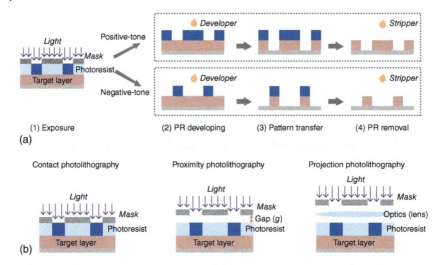

Figure 14.1 Basics of photolithography. (a) Schematic description of the photolithography process using positive-tone and negative-tone PRs. (b) Classification of the exposure systems used in photolithography [16].

typically achieved by spin-coating a PR solution (a mixture of resin, binder, photoactive compound [PAC], and solvent).

In 2016, Park et al. achieved the precise preparation of red, green, and blue QD patterns through conventional photolithography [1]. The preparation process is shown in Figure 14.2a, which consists of five key steps: (1) the PR material is spin-coated onto the substrate, and then the resulted wet PR film layer is pre-baked; (2) the substrate with PR is transferred to an exposure machine set with a suitable photomask, and then the photoresist film is exposed to UV light and developed to obtain the PR pattern. Here, the PR irradiated by UV light undergoes a cross-linking reaction and is removed during the development process; (3) the surface of the developed PR film layer is treated with oxygen plasma to endow negative charge and to improve the adhesion of quantum dots to the PR layer; (4) the substrate is then immersed in poly(diallyldimethylammonium chloride) (PDDA) solution and rinsed in deionized (D.I.) water to remove residual PDDA that is not bonded to the substrate, and it is dried with nitrogen gun. The surface-adhered PDDA layer temporarily provides positively charged surface state. After the PDDA treatment, conventional water-soluble II–VI core/shell-type red QDs terminated with carboxylic acid are diluted in water to approximately 0.01 wt% and pipetted onto the substrate so as to cover the entire patterned area. After the QD deposition, the substrate is rinsed in D.I. water to remove excess QDs, which are not strongly bonded to the PDDA layer, and it is dried with nitrogen gun. This results in assembly of QD monolayer on the entire substrate, and by repeating the process of immersing in PDDA solution and depositing QDs, one can achieve thicker QD layers with increasing luminescence with respect to the number of layer-by-layer cycles [17]. (5) The substrate is treated with a PR stripping solution, and the remaining PR is stripped off, together with the quantum dot layer above it, leaving the quantum dot film on the surface of the PR, thus patterning the quantum dot film [1]. The red, green, and blue quantum dot films were

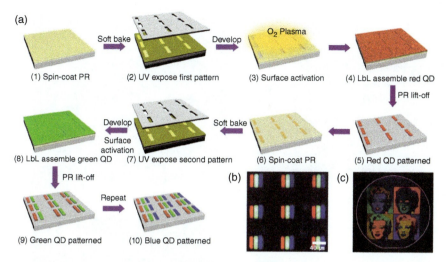

Figure 14.2 Fabrication of QD patterns by photolithography. (a) Illustration of the QD patterning process. (b) PL image of patterned QDs on a substrate with 405 nm laser as excitation source. (c) Large-scale QD pattern demonstration with pop art of Marilyn Monroe by Andy Warhol (1967) on 4-in. quartz wafer under UV lamp excitation [1].

patterned in three cycles, and as an example, the authors prepared a pattern showing the figure shown in Figure 14.2b under UV light and applied it to the patterning of Figure 14.2c [18].

Using the demonstrated method, they fabricated an inverted, multicolored, and active QDLED on a single substrate through solution-process. The device structure is shown in Figure 14.3a: ITO is used as cathode, sol–gel processed ZnO layer as electron transport layer (ETL), poly[N,N'-bis(4-butylphenyl)-N,N'-bis(phenyl)-benzidine] (poly-TPD) as hole injection layer (HIL) and electron blocking layer (EBL), and MoO_x/Ag layer as hole transport layer (HTL) and anode. The device fabrication and measurement are performed in ambient clean-room environment except for MoO_x/Ag electrode deposition and encapsulation processes, which are performed in vacuum and nitrogen glovebox, respectively. To confirm that the device fabrication has been performed as desired, cross-sectional transmission electron microscope (TEM) image of resulting device is provided in Figure 14.3b. The well-defined boundaries between each constituting layer allow effective carrier interactions between each layer without short-circuiting when bias voltage is applied. Figure 14.3c shows the energy levels of the device constituting materials with respect to the vacuum energy as 0 eV, determined from UPS and PL measurements. Figure 14.3d shows the EL photograph of the active RGCW QDLED under a common bias voltage of 6 V.

14.3 Micro-Contact Transfer

MCT is an alternative high-resolution patterning technique that involves the use of a patterned mold to transfer patterns onto a substrate. This technique can achieve sub-micron resolution and can be used with a wide range of materials and substrates.

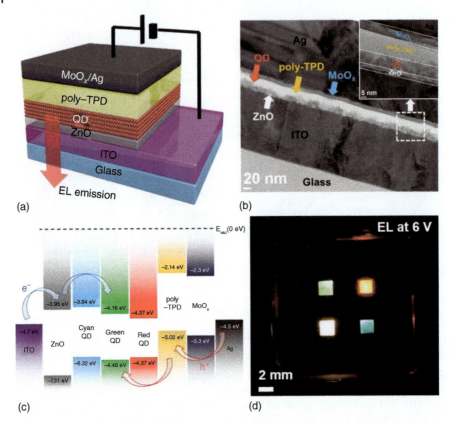

Figure 14.3 Multicolored, active QDLED demonstration. (a) Inverted QDLED structure. (b) Cross-sectional TEM image of the fabricated device. (c) Energy diagram of our inverted QDLED structure. (d) EL image of RGCW QDLED on single device at common bias voltage of 6 V [1].

However, nanoimprint lithography requires a specialized mold, which can be expensive to produce, and the process can be sensitive to defects and contamination.

In 2008, Kim et al. successfully applied MCT for deposition of patterned colloidal QD thin films as the electroluminescent layers within hybrid organic-QDLED [5]. The QD printing process follows the steps schematically shown in Figure 14.4a: (1) poly(dimethylsiloxane) (PDMS) is molded using a silicon master; (2) resulting PDMS stamp is conformally coated with a thin film of parylene-C, a chemical-vapor deposited (CVD) aromatic organic polymer; (3) parylene-C coated stamp is inked via spin-casting of a solution of colloidal QDs suspended in an organic solvent; (4) after the solvent evaporates, the formed QD monolayer is transferred onto the substrate by contact printing. Based on this approach, QDLED arrays as shown in Figure 14.4c were successfully demonstrated following the device structure as shown in Figure 14.4b.

In 2011, scholars from South Korea also achieved effective transfer of QD patterns in a similar way. The transfer process is shown in Figure 14.5a. Unlike the previous method, the quantum dots are first spin-coated onto a master substrate, and an additional octadecyltrichlorosilane self-assembled monolayer (ODTS SAM)

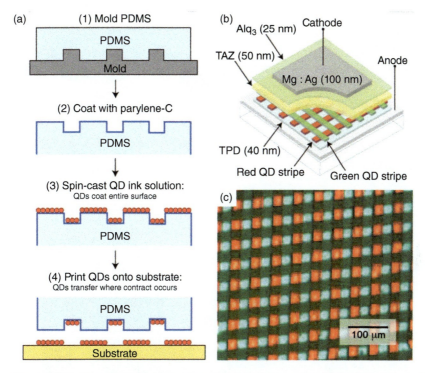

Figure 14.4 Transfer of QD pattern based on PDMS template and its application in QLEDs. (a) Schematic shows the four-step contact printing process. (b) Schematic diagram shows the structure of a QDLED with an emissive layer consisting of 25 μm wide stripes of green and red QD monolayers. (c) Electroluminescence of the structure shown in (b) at 7 V of applied bias. Blue emission is due to TPD hole-transporting underlayer [5].

is deposited on the surface of the master substrate to reduce the adhesion between the quantum dots and the master substrate and to avoid the quantum dot pattern being adsorbed away by the PDMS template during the transfer process. After the introduction of ODTS SAM, the surface energy of the master substrate decreased rapidly from 1140 to 21 mJ m^{-2}. On the other hand, the surface energy of PDMS itself is also low (19.8 mJ m^{-2}), which in turn makes it easier to dislodge the quantum dots when they adhere to the PDMS bumps. Although the surface energy of PDMS is also slightly lower than that of the ODTS SAM-modified master substrate, the transfer of quantum dots onto PDMS bumps is mainly based on their viscoelastic properties [19, 20]. By increasing the peel-off speed of the PDMS stencil to achieve effective adsorption of the quantum dot film, when the peel-off speed was increased to 60 mm s^{-1}, the quantum dot film was fully adsorbed on the entire PDMS bump. Based on the same process, the red, green, and blue quantum dot patterns were sequentially transferred onto the same substrate. The characteristics of the transferred QD patterns under UV lamp irradiation are shown in Figure 14.5b, and it can be found that the film layer transferred based on this method is continuous and uniform. The QD films were later integrated into a TFT substrate with a suitable device structure, as shown in Figure 14.5c, and a full-color QDLED flexible display was successfully prepared, as shown in Figure 14.5d [6]. In 2015, Kim and coworkers

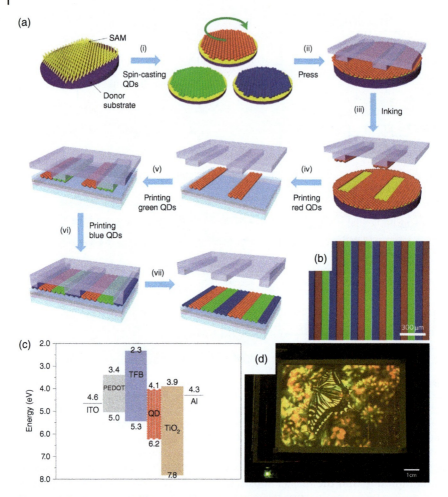

Figure 14.5 Transfer of QD pattern based on PDMS template and its application in full-color display. (a) Schematic of transfer printing process for patterning of quantum dots. (b) Fluorescence micrograph of the transfer-printed RGB QD stripes onto the glass substrate, excited by 365 nm UV radiation. (c) Energy band diagram of crosslinked QDLED using TiO_2 and TFB as the ETL and HTL, as well as quantitative energy levels. (d) Electroluminescence image of a 4-in. full-color QD display using a TFT backplane with a 320 × 240 pixel array [6].

also obtained high-resolution QD film patterns by gravure transfer. Differently, the QD was firstly spin-coated onto a PDMS template and then microcontacted through a gravure model with patterns to transfer the QD film to target substrate. Based on this approach, QDLED with a maximum brightness of 14 000 cd m^{-2} was successfully demonstrated [8].

14.4 Inkjet Printing

IJP is a simple and cost-effective technique for patterning QDLED [9–15]. In this method, a solution containing QDLED materials is loaded into an inkjet printer,

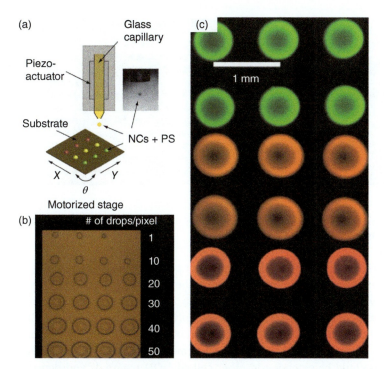

Figure 14.6 Fabrication of QD patterns by IJP technique. (a) Schematic representation of the inkjet setup. (b) Optical microscopic images of inkjet-printed pixels with different numbers of drops per lines. (c) Optical microscopy images of part of a 9 × 9 single-color array [15].

and droplets are ejected through a tiny nozzle onto the substrate in a specific pattern. The deposited material then solidifies to form the desired pattern. IJP can produce high-resolution patterns with low waste and low material consumption. However, it can be challenging to achieve uniform deposition and control over the thickness and morphology of the printed layers.

In 2009, Kim et al. successfully achieved the preparation of CdSe quantum dot patterns by IJP [15]. In this chapter, they first formulated quantum dot inks based on CdSe quantum dots, polystyrene, and chloroform, and set different concentrations of polystyrene from 1% to 5% mass ratio in order to achieve better printing results. Based on the scheme shown in Figure 14.6a, the process was carried out with a 70 μm diameter printhead to produce quantum dot patterns in a variety of forms. By adjusting the number of droplets, a variety of sizes of quantum dot patterns could be printed, as shown in Figure 14.6b, and by controlling the number of droplets, red, yellow, and green quantum dot patterns were successfully printed, as shown in Figure 14.6c.

In 2020, Xiang et al. successfully demonstrated QDLED with an high external quantum efficiency (EQE) and high lifetimes through IJP method [21]. The device structure and EL characteristics are shown in Figure 14.7. As they stated, passivating the trap states caused by both anion and cation under-coordinated sites on the

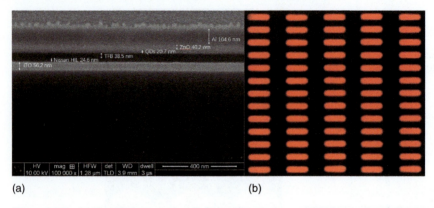

Figure 14.7 (a) The SEM of a typical Zn(OA)$_2$ QDs device and (b) the pixel array image of a typical Zn(OA)$_2$ QDs device [21].

quantum dot surface with proper ligands for ink-jet printing processing reminds a problem. Based on the requirements of IJP process, they proposed a dual ionic passivation method by liquid phase ligand exchange using Zn(OA)$_2$ organic salt. By successfully suppressing the mid-gap trap states, the Zn(OA)$_2$-QDs-based IJP QDLED shows high EQE of 16.6% and long lifetime of 1 721 000 hours. The IJP QLED performances could somewhat meet the basic requirements of industry applications and the methods could be extended to other colloidal QD materials and devices as well.

To analyze and summarize the above three patterning techniques. The photolithography process is similar to the conventional semiconductor manufacturing process. The light-emitting material and PR material are mixed together. The patterning of QDs relies on the characteristics of PR. The material utilization rate of this method is relatively low. In addition, the efficiency of light-emitting material in the exposure and development process is also greatly affected. MCT is mainly based on the pre-made precision template. The main problem with this method is its high production accuracy, and large-area fabrication is also more difficult. Compared to photolithography and MCT processes, IJP is a contact-free, print-on-demand, and photomask-free technique. The materials utilized in IJP are soluble, which makes them quite compatible with QDs. Clearly, the IJP is more suitable for use as a patterning method for QD materials.

14.5 Other Patterning Techniques

Besides abovementioned three techniques, there are also some other patterning approaches, such as: laser direct writing [22], self-assembly [17], and electrohydrodynamic-jet (E-jet) printing [23]. Laser direct writing is a promising approach for patterning QDLED, where a laser beam is used to selectively pattern QDLED materials onto a substrate. This technique is maskless, meaning that it does not require a physical mask to create the pattern, allowing for the creation of complex patterns with high resolution and accuracy. However, laser-direct writing can be a slow and expensive process, and it requires careful optimization to achieve the desired patterning results. Self-assembly methods use the intrinsic properties of QDLED materials to spontaneously form ordered patterns on a substrate, offering a simple and cost-effective approach for patterning. This technique has the potential to achieve high-resolution patterns, but it requires careful control over the properties of the QDLED materials and the substrate. E-jet printing is a precise and versatile printing technology that uses an electric field to control the deposition of tiny droplets of fluid onto a substrate. This technique has been increasingly used in the field of organic electronics, including the printing of QDLED. The E-jet printing process involves the use of a conductive nozzle, which is connected to a high-voltage power supply. The nozzle is filled with a solution containing quantum dots, which are tiny semiconductor nanocrystals that emit light when excited by an electric current. As the electric field is applied to the nozzle, the fluid in the nozzle is ionized and ejected as a highly charged jet. The electric field then guides the charged droplets of the solution toward the substrate, where they land and solidify to form a pattern. One of the key advantages of E-jet printing is its ability to print highly precise and uniform patterns with a resolution down to the submicron level. This level of precision makes it ideal for printing QDLED, which requires highly uniform and controlled deposition of quantum dots onto a substrate. Moreover, E-jet printing is a non-contact printing technology, which eliminates the risk of damaging delicate substrates. In addition to its precision and versatility, E-jet printing also offers a high throughput, making it suitable for large-scale production. Furthermore, the technology is compatible with a wide range of materials and substrates, including flexible substrates such as plastics, which are becoming increasingly popular in the field of organic electronics. E-jet printing is a promising technology for the fabrication of QDLEDs and other organic electronic devices. Its ability to print highly precise and uniform patterns with a high throughput and compatibility with various substrates makes it an attractive alternative to traditional printing technologies.

14.6 Conclusion

Patterning and printing technologies play a critical role in the practical implementation of QDLED in various devices, including displays, lighting, and sensors. The development of different patterning and printing techniques, such as photolithography, IJP, soft lithography, laser direct writing, nanoimprint lithography, and self-assembly, has enabled the fabrication of complex patterns with high resolution and accuracy. Continued development of these techniques is expected to further enhance the performance and application of QDLED.

References

1 Park, J.-S., Kyhm, J., Kim, H.H. et al. (2016). Alternative patterning process for realization of large-area, full-color, active quantum dot display. *Nano Letters* 16 (11): 6946–6953.
2 Huang, B.L., Guo, T.L., Xu, S. et al. (2019). Color converting film with quantum-dots for the liquid crystal displays based on inkjet printing. *IEEE Photonics Journal* 11 (3): 9.
3 Kim, H.J., Shin, M.H., Hong, H.G. et al. (2016). Enhancement of optical efficiency in white OLED display using the patterned photoresist film dispersed with quantum dot nanocrystals. *Journal of Display Technology* 12 (6): 526–531.
4 Bai, X., Yang, H., Zhao, B. et al. (2019). 4-4: flexible quantum dot color converter film for micro-LED applications. *SID Symposium Digest of Technical Papers* 50 (1): 30–33.
5 Kim, L., Anikeeva, P.O., Coe-Sullivan, S.A. et al. (2008). Contact printing of quantum dot light-emitting devices. *Nano Letters* 8 (12): 4513–4517.
6 Kim, T.H., Cho, K.S., Lee, E.K. et al. (2011). Full-colour quantum dot displays fabricated by transfer printing. *Nature Photonics* 5 (3): 176–182.
7 Lee, K.H., Han, C.Y., Kang, H.D. et al. (2015). Highly efficient, color-reproducible full-color electroluminescent devices based on red/green/blue quantum dot-mixed multilayer. *ACS Nano* 9 (11): 10941–10949.
8 Choi, M.K., Yang, J., Kang, K. et al. (2015). Wearable red–green–blue quantum dot light-emitting diode array using high-resolution intaglio transfer printing. *Nature Communications* 6: 8.
9 Torrisi, F., Hasan, T., Wu, W.P. et al. (2012). Inkjet-printed graphene electronics. *ACS Nano* 6 (4): 2992–3006.
10 Singh, M., Haverinen, H.M., Dhagat, P., and Jabbour, G.E. (2010). Inkjet printing-process and its applications. *Advanced Materials* 22 (6): 673–685.
11 Calvert, P. (2001). Inkjet printing for materials and devices. *Chemistry of Materials* 13 (10): 3299–3305.
12 Bao, B., Li, M.Z., Li, Y. et al. (2015). Patterning fluorescent quantum dot nanocomposites by reactive inkjet printing. *Small* 11 (14): 1649–1654.
13 de Gans, B.J., Duineveld, P.C., and Schubert, U.S. (2004). Inkjet printing of polymers: state of the art and future developments. *Advanced Materials* 16 (3): 203–213.
14 Jiang, C.B., Zhong, Z.M., Liu, B.Q. et al. (2016). Coffee-ring-free quantum dot thin film using inkjet printing from a mixed-solvent system on modified ZnO transport layer for light-emitting devices. *ACS Applied Materials and Interfaces* 8 (39): 26162–26168.
15 Kim, J.Y., Ingrosso, C., Fakhfouri, V. et al. (2009). Inkjet-printed multicolor arrays of highly luminescent nanocrystal-based nanocomposites. *Small* 5 (9): 1051–1057.
16 Park, S.Y., Lee, S., Yang, J., and Kang, M.S. (2023). Patterning quantum dots via photolithography: a review. *Advanced Materials* 2300546.

17 Bae, W.K., Kwak, J., Lim, J. et al. (2010). Multicolored light-emitting diodes based on all-quantum-dot multilayer films using layer-by-layer assembly method. *Nano Letters* 10 (7): 2368–2373.

18 Harwell, J., Burch, J., Fikouras, A. et al. (2019). Patterning multicolor hybrid perovskite films *via* top–down lithography. *ACS Nano* 13 (4): 3823–3829.

19 Kim, T.-H., Carlson, A., Ahn, J.-H. et al. (2009). Kinetically controlled, adhesiveless transfer printing using microstructured stamps. *Applied Physics Letters* 94 (11): 113502.

20 Feng, X., Meitl, M.A., Bowen, A.M. et al. (2007). Competing fracture in kinetically controlled transfer printing. *Langmuir* 23 (25): 12555–12560.

21 Xiang, C., Wu, L., Lu, Z. et al. (2020). High efficiency and stability of ink-jet printed quantum dot light emitting diodes. *Nature Communications* 11 (1): 1646.

22 Lee, J.Y., Kim, E.A., Han, J. et al. (2022). Nondestructive direct photolithography for patterning quantum dot films by atomic layer deposition of ZnO. *Advanced Materials Interfaces* 9 (22): 2200835.

23 Kim, B.H., Onses, M.S., Lim, J.B. et al. (2015). High-resolution patterns of quantum dots formed by electrohydrodynamic jet printing for light-emitting diodes. *Nano Letters* 15 (2): 969–973.

Index

a

AC/DC dual drive mode QDLED
  advantages 245
  disadvantages 245
  energy band diagram 250
  optimization
    carrier balance 256
    charge injection 255–256
    device structure 255
    electrical and optical properties 256
    QD materials 255
  structure 249, 250
  working mechanism 244
AC-driven QDLEDs, future research directions for 256
AC electroluminescence (ACEL) 237
active matrix QDLED (AM-QDLED) 19, 134, 143, 329
aerogel quantum dot film 343
Ag–Au–Se (AgAuSe) quantum dots 176
aging-induced enhancement 272
aging mechanism in QDLEDs 271–277
all-inorganic QDLED devices 25, 98, 122
allowed bands 37
alloy core–shell quantum dots 64–65
alloyed ZnCdSSe quantum dots 112
alternating current (AC) driven QDLEDs 235
  advantages 235, 246
  double-emission tandem structure mechanism 239–245
  drawbacks 236
  dual dielectric layer structure 236
  luminescence principle 236–239
  optimization strategies
    AC signal optimization 246
    device engineering 245–246
    hybridization with other materials 246
    QD material optimization 245
    structural modification 246
    surface modification 246
  structure and operating mechanisms 236–238
angle-resolved XPS 74
anharmonic oscillator model 74
asymmetric AC QDLEDs 244
atomic force microscopy (AFM) 278, 287, 303
Auger effect 64
Auger recombination 19–21, 270, 271

b

band engineering strategies, in InAs-based QDs 64
Beijing Oriental Electronics Group Co., Ltd (BOE) 347–351
binding rules, nanocrystals *vs.* ligands 66
bis-benzophenone based cross-linkers 314
blue CdSe/ZnS quantum dots 26, 146
blue CdZnSe/ZnSeS/CdZnS core–shell quantum dots 338
blue QDLEDs 141
  bad/poor performance factors of 141, 142

Colloidal Quantum Dot Light Emitting Diodes: Materials and Devices, First Edition. Hong Meng.
© 2024 WILEY-VCH GmbH. Published 2024 by WILEY-VCH GmbH.

blue QDLEDs (contd.)
  based on AlSb 155
  based on Cu 153–155
  based on InP 149–151
  based on ZnSe 151–153
  with cadmium-free quantum dots 149–155
  challenges for 167
  charge transport layer optimization 155–164
  containing cadmium quantum dots 145–149
  current density and brightness 147
  devices based on cadmium-containing quantum dots
    advantages 148
    challenges 148–149
  device structure 164–166
  external quantum efficiency of 144
  luminescent materials 143–155
  problems of 166–167
  stability studies of 260
  technology development breakthroughs 144–145
  with ZnMgO/Al/HATCN/MoO$_3$ interconnection layer 147
blue ZnSeTe ternary quantum dot devices, luminescence spectra of 217
bottom-up method, QD preparation 3
Bragg's law 77, 79
Br$^-$ QDLED 127
brush process 47, 48
bulk semiconductor 20
  bandgap 16
  physical parameters 9
  physical properties 8

C

cadmium-based core/shell QDs 63
cadmium-containing yellow light quantum dots 189, 190
cadmium-free blue QDLED 144
cadmium-free InP/ZnSe/ZnS core–shell structured quantum dots 331
cadmium-free QDLEDs
  performance parameters of 207
  structure and performance of 209
cadmium-free quantum dots
  I-III-VI quantum dots 218–222
  indium phosphide 208, 215
  QDLED performance optimization methods 222
    device structure optimization 225–227
    ligand engineering 223–225
    shell engineering 224–227
  ZnSe 215, 217–218
cadmium-free yellow light quantum dots 189–193
cadmium-free ZnSe$_{1-x}$Te$_x$/ZnSe/ZnS core–shell quantum dots 330
cadmium-free ZnSe/ZnS core–shell quantum dots 151
calixarene CsPb(Br$_x$I$_{1-x}$)$_3$ quantum dots, high purity yellow light emission from 191
carbon dots 17
carbon-13 NMR 69
carbon quantum dots 17
cation exchange 180
CdCl$_2$-aliphatic amine complexes 59
CdSe bulk material 89
CdSe/CdS-Cd(RCOO)$_2$ quantum dots, photoluminescence and electroluminescence properties of 8
CdSe@CdS dot/sheet core–shell structured quantum dots, UV–Vis absorption/fluorescence emission spectra of 70
CdSe/CdS quantum rods, absorption and emission spectra of 10
CdSe/CdS-RNH$_2$ quantum dots, photoluminescence and electroluminescence properties of 8
CdSe/CdS/ZnS colloidal quantum dots 51
CdSe/CdS/ZnS core–multishell quantum dots 262

CdSe/Cd$_{1-x}$Zn$_x$Se$_{1-y}$S$_y$/ZnS alloy quantum dots, chemical composition and energy band structure of 97
CdSe/Cd$_{1-x}$Zn$_x$Se$_{1-y}$S$_y$/ZnS QDs 13
CdSe nanocrystals
  solubility of 66, 93
  TEM image of 9
  UV–Vis and PL spectra of 52
  XRD patterns of 56, 57
CdSe quantum dot 15
  original ligands for 90
  patterns, by inkjet printing 367
CdSe quantum sheets, absorption and emission spectra of 10
CdSe/ZnS/CdSZnS quantum dots 121
CdSe/ZnS core–shell quantum dots 95, 200
  type I and type II 11, 12
CdSe/ZnSe core/shell structure based QDLEDs 26
CdSe@ZnS light-emitting diodes 312
CdSe/ZnS quantum dots, absorption and emission spectra of 10
CdSe/ZnS thick-shell quantum dots 127
CdSeZnS/ZnS/ZnS quantum dots 118, 122
CdSe/Zn$_{1-x}$Cd$_x$S quantum dots
  photoluminescence/electroluminescence performance vs. shell thickness 262
CdTe bulk material 89
CdTe/CdSe II QD 178
CdTe quantum dots 15
  for white light-emitting diodes 189
Cd$_{1-x}$Zn$_x$Se QDs 112
4-CF3-BA self-assembled monolayer 319
chalcogenide cadmium quantum dots 177–178
chalcogenide lead quantum dots 176–177
chalcopyrite CIS quantum dots 221
charge transport layer
  optimization, blue QDLEDs 155–164
  in QDLED performance 125
charge transport mechanism, in QDLEDs 292–293
charge transport properties, of quantum dot films 94
China Star Optoelectronics Technology (CSOT) 344–347
CIE chromaticity diagram, boundary of 33
CIGS quantum dots 153
CIGS/ZnS QDLEDs 154
CIS quantum dots 221–223
CIS/ZnS/ZnS quantum dots 191
Cl-passivated tungsten phosphate (Cl-TPA) 308
CNPr-TFB hole transport polymers 158
colloidal CdSe/CdS core/shell nanosheets 123
colloidal QDLEDs 33
  color purity 33–34
  solution processability 34
  stability 35–36
colloidal quantum dots 53
  direct heating method 55–56
  future development challenges 80–81
  hot injection method 54–55
  ligating and non-ligating solvents 56–58
  liquid phase methods 53
  nucleation and growth mechanism 58–59
  physical vapor phase epitaxial growth methods 53
  precursor chemistry of 56, 57
  size distribution
    focusing principle 59
    scattering 60
  surface chemistry of
    covalent bond classification method 65–66
    entropic ligands 66
colloidal semiconductor nanocrystals 1
color purity 33–34
color-tunable CIS quantum dots 221
color-tunable QDLEDs 200

composite ZnMgO:ZnO electron transport layers   296
conducting carriers, in semiconductors   38
conduction band   37–39, 62, 91, 95, 120, 241
continuously graded CdSe/Cd$_x$Zn$_{1-x}$Se/ZnSe$_{0.5}$S$_{0.5}$ QDs   14
conventional vs. inverted QDLEDs   123
core–shell quantum dots   15
  alloy   64–65
  cadmium-free ZnSe/ZnS   151
  non-alloyed   63
  ZCGS/ZnS   222
core–shell structure   8–12
  NIR quantum dot control   180
covalent bond classification (CBC) method   65–66
crystal growth rate vs. size, Sugimoto model   59
Cu-based quantum dots   153–155
Cu-doped NiO-based QDLED   317
CuInS$_2$-ZnS/ZnSe/ZnS QD device, fluorescence emission spectra and electroluminescence spectra of   219
current–voltage (IV) measurements   278, 296

d

differential absorption spectroscopy   283–285
1,4-dioxane polar solvent   166
dip coating process   47, 48
direct current (DC) driven QDLEDs   235
  device structure   235, 239, 245, 250
    anode   42
    cathode   42
    electron buffer layer   42
    hole buffer layer   42
    light-emitting layer   42, 43
  luminescence principle   236–239
  structure and operating mechanisms   236, 238

direct heating method   55–56
directional attachment, of nanocrystals   60, 62
displacement current measurement (DCM) technique   285–286
DMATP-QDLED   128
doping   130, 295
  of HTMs   301, 303, 305–309
double heterojunction device, energy level diagram of   179
double-insulated AC-QDLED device   239, 240
double-shell InGaP/ZnSeS/ZnS QDs   151
double-stacked ETMs   297–299

e

efficiency decay mechanism in QDLEDs   268–271
efficiency roll-off, QDLEDs   22, 118, 135, 270, 271, 283, 296
electro-absorption (EA) spectroscopy   281–282
electrohydrodynamic jet (E-jet) printing   368, 369
electrohydrodynamic jet-sprayed process   47, 48
electroluminescence (EL) spectroscopy   278
electroluminescent device   334–335, 341
electroluminescent quantum dot devices (ELQDs)   240
electron communalization movement motion   37
electron diffraction   76
electron injection materials, for QDLED   299
  conductive polymers   300
  conjugated polyelectrolytes   300
  graphene   300
  graphene oxide   300
  lithium fluoride   300
  metal oxides   299–300
  nanoparticles   300
  organic materials   300
  surface dipolar polymers   301

titanium dioxide   300
zinc sulfide   300
electron spectroscopy for chemical analysis (ESCA)   74
electron transport layer (ETL), QDLEDs   161, 164
  ageing and degradation   274
    energy level alignment   275
    interface issues   274
    stability   274
electron transport materials (ETMs), QDLEDs   265, 266, 293
  composite material design   296
  development breakthroughs   297
  device performance data   297, 299
  double-stacked ETMs   297–299
  inorganic organic hybrid ETMs   296–297
  metal-doped   293–295
  metal salt doped   296
  polymer-modified ETMs   296
embryonic nuclei-induced alloying process   145
emission wavelength range, of NIR QDs   174
empty band   37
energy band diagram, of multi-layer QDLED device with poly-TPD   157
energy bands
  of an insulator   38
  of conductor   37, 38
  of semiconductor   38
energy design strategy, CdSe/Cd$_{1-x}$Zn$_x$Se/ZnSe quantum dot synthesis   97
energy level
  alignment   36, 40, 156, 275, 305, 310, 311, 314
  and bands   36, 37
energy level diagram
  of common CTLs for InP quantum dots   228
  of different electron transport layers   220
  of double heterojunction device   179
  of molecular, quantum dot and bulk semiconductor materials   1, 2
  quantum dots
    with different shell layer thickness   221
    Ga doping effect on   220
entropic ligands   34, 66, 93
equivalent circuit, of sandwich AC-driven EL device   239
European Union's Restriction of Hazardous Substances (RoHS) rules   207
eutectic indium–gallium (EGaIn)   114, 132
exciton decay   125
exciton formation in QDLEDs   43
external quantum efficiency (EQE)   44, 269
  of double-junction green QDLEDs   141

f

Fermi Dirac statistics   39
Fermi distribution function   39
Fermi energy level   4, 5, 39, 268
field-driven AC QDLEDs
  dielectric layer optimization   248–250
  quantum dot layer optimization   250–251
field-generated AC QDLEDs   240, 241
field-induced AC QDLEDs   247
  AC voltage optimization   247
  carrier balance control   247
  device fabrication and packaging process   247
  device structure optimization   247
  engineering dielectric layers   247
  quantum dot properties, optimization of   247
  surface passivation   247
fluorescence electrophoresis   22
fluorescence intermittence   21
fluorescence intermittency   5, 21
fluorescence lifetime   10, 67
  of CdSe/ZnS quantum dot film   201

f

fluorescence lifetime (*contd.*)
  of colloidal QDs 68
fluorescence quantum yield 10, 36, 45, 79, 112, 113, 135, 151, 225
fluorescence resonance energy transfer 19, 21–22
fluorescence spectroscopy 66–68
fluorescent carbon nanomaterials 17
forbidden band 37–39, 94, 95
Forster energy transfer 36
Förster resonance energy transfer (FRET) 21, 24, 96, 114, 125, 149, 196, 212, 224
Fourier transform infrared spectroscopy (FTIR) 71, 74
free electrons and holes, in semiconductors 38–39

g

giant quantum dots 12
graphene oxide (GO)/poly(3,4-ethylenedioxythiophene):poly(4-styrenesulfonate)(PEDOT:PSS) bi-layered HIL 319
green QDLEDs
  device doping with nickel oxide (NiO_x) 307
  device performance improving strategy 132–134
  device structure development 122–125
  EQE development 114, 118, 119
  lifetime and stability of 135
  performance affecting factors 125–134
    device structure optimization 130–132
    QD core/shell structure 129–130
    QD ligand effect 126–128
  with PFI and Cu–NiO hybrid HTL 309
  stability studies of 260
  technology development breakthroughs 118, 120
  works on 115

h

half-field driven half-injected AC QDLEDs
  charge generation layer optimization 254
  optimization strategies
    device structure optimization 252
    driving voltage and frequency 252
    electrode materials optimization 252
    injection layer optimization 252
    quantum dot size and composition 252
    P(VDF-TrFE-CFE) dielectric layer 252
    tandem structure 254–255
half-field to half-injection (HFHI) AC QDLEDs 242
  advantages 242
  vs. asymmetric AC QDLEDs 244
  device structure 242
  disadvantage 242
  working mechanism 242
half-life of QDLEDs 164
^1H-^{13}C DIPSHIFT results and opening angles 73
hexadecylamine (HDA) 90
high-efficiency orange–yellow QDLEDs 200
high-resolution transmission electron microscopy (HRTEM) 76
^2H NMR lineshapes and chain flexibility 72
hole injection materials (HIM), for QDLED 316
hole mobility 135, 167, 229, 263, 276, 296, 301, 303, 305–307, 309, 312, 319
hole transport layer (HTL), QDLED 156–161, 263, 265
  ageing and degradation 275–276
hole transport materials (HTMs), for QDLED 301–315
  advanced characterization techniques 303
  bio-based HTM 311

challenges 304
compositions 309–311
composition strategy 305
conductive polymers 311
copper 311
development breakthroughs 319
device performance data 319
doping 301, 303, 305–309
graphene-based materials 311
interface engineering 301
metal-organic frameworks 304
molecular design 301
molecular self-assembly 303
polymer-based 303
solution-processed HTMs 312
homogeneous transfer 22
hot-electron impact excitation principle 239
hot injection method 54–55
hot injection synthesis method 59
hybrid nanocrystal-organic light-emitting device (NC-OLED) 178
hybrid organic/colloidal QDLEDs 21
hybrid QDLEDs 122, 200, 275
hydroxyl-capped CIS quantum dots 223

i

ICL-ZnMgO/Al/HATCN/MoO$_3$ tandem QDLED 125
I-III-VI quantum dots 218–222
II–VI semiconductor QDs 16
II$_3$–V$_2$ semiconductor QDs 17
InAs/CdSe core/shell QDs 64
InAs QDs 15, 64, 65
indium phosphide (InP) quantum dots 208, 215
InGaN-based blue-emitting chips 195
InGaN LED chips, photoluminescence spectra of 190
injection luminescence principle 42
inkjet printing (IJP) 47, 48, 366–368
inorganic multilayer thin-film QDLEDs 250
inorganic organic hybrid ETMs 296–297
InP quantum dots 207

InP/ZnS core–shell quantum dots 151, 189
InP/ZnSeS/ZnS quantum dots 151, 196, 266
InP/ZnS small core–shell tetrahedral QDs 149
in-situ EL–PL measurement technique 282–283
interfaces energetics 36
interfacial corrosion phenomenon, in tandem QDLEDs 124
internal quantum efficiency (IQE) 44, 102
inverted blue QDLED preparation 166
inverted QDLED 131, 255, 308, 364
  vs. conventional 123
  current density EQE characteristic curve of 165
  device structure 102, 103, 364
  ITO/MoO$_x$/NiO$_x$/Al$_2$O$_3$/QDs/ZnO/Al device 123
  ITO/NiO/Al$_2$O$_3$/ZnCdSSe/ZnS QDs/ZnO/Al, all-inorganic green QDLEDs of 122
  ITO/PEDOT/PSS /TFB/QDs/ZnO/Al 113
  ITO/ZnO/CdSe@ZnS/ZnSQDs/PEIE/Poly-TPD/MoO$_x$/Al device structure 131
  ITO/ZnO/PEIE/QDs/CBP/MoO$_3$/Al inverted device structure 102
IV–VI semiconductor QDs 17

j

Janus quantum dots 15, 16

l

LaMer diagram 53, 58
laser direct writing 368–369
layer printing methods 47
lead chalcogenide IV–VI semiconductor QDs 17
lead oleate 177
lead sulfide (PbS) quantum dots 15, 177, 182

LiF interface layer   162
ligand engineering   262
   cadmium-free quantum dots   223–225
   NIR quantum dot regulation   179–180
ligand-induced surface dipoles   94
ligand selection, in QD conductivity   34
ligating and non-ligating solvents   56–58
light-emitting characteristics, of QDs
   optical property   4–8
   particle size and emission color   3
liquid dispersed quantum-dots ion-beam deposition (LIQUID)   250
liquid-type QDLEDs   193
luminescence spectra   198, 199, 211
   of blue ZnSeTe ternary quantum dot device   217
   of I-III-VI quantum dots   218
luminescent layer materials, in green QDLEDs   120
   alloyed core/shell quantum dots   121
   core/multilayer shell quantum dots   121–122
   discrete core/shell quantum dots   120–121
luminous life, defined   45
luminous lifetime   45–46, 48

m

magic nanoclusters   59, 60
material characterization techniques for quantum dots   66
   Fourier transform infrared spectroscopy   71–74
   nuclear magnetic resonance spectroscopy   69–71
   SAXS   76–77
   transmission electron microscopy   76
   UV–Vis absorption and fluorescence spectroscopy   67–69
   WAXS   76–77
   XAFS spectroscopy   78–79
   X-ray diffractometer   77
   X-ray photoelectron spectroscopy   74–75
mercaptopropionic acid (MPA)   90

3-mercaptopropyl trimethoxysilane (MPTMS)   90
11-mercapto-undecanoic acid (MUA)-modified NiO nanoparticles, as hole transport layer   307
metal-doped ETM   293–295
metal oxide ETLs   272, 275
metal salt doped ETMs   296
metal-sulfide nanocrystals, stoichiometry and ligand exchange of   65
Mg-doped ZnO electron transport materials   296
micro-contact transfer (MCT)   363–366
MicroLED device   346
MicroLED display technology   329
mixed nanocrystal–organic light emitting device, structure and emission spectrum of   179
molar extinction coefficients, of CdX nanocrystals   67
molecular energy calculations   60
molecular orbital energy level diagram   1
monodisperse Ag_2S QDs   55
monodisperse quantum dots   51
MoO_3-doped PEDOT:PSS hole transport layer   306
MoO_3/TFB hole-generating layer   254
multicolor fluorescent LEDs based on cesium-lead halide chalcogenide QDs   191
multilayered QDLED device structure   311
multilayer structure of InP/ZnS QDLED   150

n

Najing Tech   338–344
Nanoco   335–338
nanocrystals
   CdSe (see CdSe nanocrystals)
   colloidal semiconductor   1
   crystalline seed-mediated growth   60
   directional attachment growth of   60, 61

semiconductor, 1
silicon   178
two-dimensional   61
ZnOS/ZnO   340
nanostructure engineering   162
Nanosys   330–332
near infrared (NIR) quantum dots
  chalcogenide cadmium quantum dots   177–178
  chalcogenide lead quantum dots   176–177
  classification of   173
  control of   180
  developments and device performance data   176
  light-emitting diodes   27
  material optimization   179–182
  materials   174
  regulation of   179–180
  silicon quantum dots   178–179
negative ageing effect, QDLEDs   273, 274
NiO-based QDLED   123
NiO_x–BA–CF3 films   317, 318
non-alloyed core–shell quantum dots   63
non-cadmium quantum dot light-emitting materials and devices   207
  characterization and understanding   230
  environmental impact studies   230
  new material development   229
  research directions   229–230
  technology development   208
non-coordinating solvents   57
non-metallic inorganic ligands   34
non-passivated S-sites   91
nuclear magnetic resonance (NMR) spectroscopy   69–71
nuclear-shell quantum dots, synthesis methods and band gap regulation engineering of   61–65

o

1-octadecene (ODE)   58, 177
oil-soluble quantum dot   193
OLED technology   259
oleic acid (OA)   177
one-dimensional-quantum wires   2
one-pot colloidal synthesis method, for high-quality CdS nanocrystals   55
one-pot scheme, core/shell QD synthesis   63
orange-yellow QDLEDs   201
organic/inorganic hybrid construction, of V_2O_5/ PVK heterojunctions   309
organic/inorganic hybrid QDLED device structures   122
Osher composite process   224, 227
Ostwald ripening   60
ozone plasma treatment   268

p

passivation of quantum dots   4, 89, 255, 277, 336
patterning techniques, for QDLED   361
  electrohydrodynamic jet printing   369
  inkjet printing   366–368
  laser direct writing   368
  micro-contact transfer   363–366
  photolithography   361–363
  self-assembly methods   369
PbS/CdS quantum dots with core–shell structure   175, 180
PbS quantum dots   15
  fine band energy calibration of   94
PbX QDs based on NIR–QDLEDs   175
PEDOT:PSS/V_2O_x-bi-layered-HIL devices   319, 320
perfluorinated inomer (PFI)   308, 310
perovskite QDs   15, 174
phosphine-free method   113
phosphomolybdic acid (PMA)   166, 319
photoconductivity measurements   278
photoexcitation theory   236
photolithography   48, 361–363, 368
photoluminescent device   193, 333
poly(3,4-poly (3,4-ethylenedioxythiophene):poly(styrene-sulfonate) (PEDOT:PSS)   316
poly(9-vinylcarbazole) (PVK)   113, 118, 304

poly-[(9,9-bis(3'-(N,N-dimethylamino)
propyl)-2,7-fluorene)-*alt*-2,7-
(9,9-ioctylfluorene)] (PFN)   301
poly((9,9-dioctylfluorenyl-2,7-diyl)-
alt-(9-(2-ethylhexyl)carbazole-
3,6-diyl)) copolymer   315
poly(indenofuorene-*co*-triphenylamine)
copolymer (PIF-TPA)   312, 313
polyethoxy polyethyleneimine (PEIE)
modified conventional ZnO layers
296
polymer-mediated QD assembly strategy
197
polymer-modified ETMs   296
poly [2-methoxy-5-(2'-ethylxyloxy)-
1,4-phenyl vinyl] (MEH-PPV)
matrix material   181
poly(3,4-ethylenedioxythiophene):
poly(styrene sulfonate)
(PEDOT:PSS)   305
poly-TPD   146, 156, 157, 211, 215,
304–306
positive aging effect, in QDLED   272–273
precursor chemistry   55, 56
proton NMR   69
PVK/poly-TPD bilayer   309
pyridine-treated CdSe nanocrystals, NMR
spectra of   71

q

QDiP structure simulation   182
QDiP system   181, 182
quantum confinement effect   5, 7, 8, 12,
27, 68, 141, 149, 151, 153, 155
quantum dot display industry   328
  chain   329
  patent layout
    BOE   347–351, 354
    CSOT   344–347
    Najing Tech   338–344
    Nanoco   335–338
    Nanosys   330–332
    SAMSUNG Group   332–335
    TCL   351–354

quantum dot-enhanced liquid crystal
display (QD-LCD)   143, 327, 329,
330, 351
quantum dot enhancement film (QDEF)
330, 349, 355
quantum dot light-emitting diodes
(QDLEDs)   18, 259
  advantages   235
  ageing and degradation, QD layer
    impact   276–277
  aging mechanism   271–277
  anode-interface regulation layer   267
  basic structure of   18–19
  cathode-interface regulation layer   267
  characterization technologies for
    278–286
  charge transport mechanism   292–293
  commercialization challenge   19
  common materials for functional layer
    262
  with Cs_2CO_3/Al cathode and
    MoO_3/Alq_3 anode   267
  current density-voltage and luminous
    intensity-voltage curves   265
  development, history of   22–27
  device structure
    conventional   99, 102
    inverted   102
    and performance   100
    tandem   102
    Type I   98
    Type II   98
    Type III   98
    Type IV   98–99
  efficiency and stability of   291
  efficiency decay mechanism   268–271
  electronic transport layer   265–267
  EQE-current density curves   265
  factors affecting light emission   19
    Auger recombination   19–21
    fluorescence resonance energy
      transfer   21–22
    surface traps and field emission burst
      22
  hole transport layer   263–265

with InP/ZnSe/ZnS quantum dots 99
performance degradation causes of 125
performance with WO$_3$ nanoparticles vs. PEDOT:PSS as anode interface layer 268, 269
QD light-emitting layer 261–263
stability
  factors affecting 261–268
  structure 261
  types of 235
  working mechanism 261
quantum dots (QDs) 51
  application on display devices 18–27
  classification 111
  with direct white light emission 197–200
  display technology 327
  energy spectroscopy 197
  feature of 10
  fluorescence quantum yield measurement 79
  ink 347
  ligand engineering 92
  light-emitting characteristics of
    continuously gradated core–shell structure 12, 14
    core-shell structure 8–12
    optical property 4–8
    particle size and emission color 3
  materials and their properties 14–17
  in matrix 181
  photoluminescent display products 327
  preparation route of 3
  QD-OLED 329
  for white LEDs 188–200
quantum dots light-emitting diodes (QDLED)
  device fabrication process 48
  operating parameters
    current efficiency 45
    luminescence color 45
    luminous brightness 44
    luminous efficiency 44
    luminous lifetime 45–46, 48
    power efficiency 45
    quantum efficiency 44–45
    turn-on voltage 44
  structure of 41–43
  working principle of
    carrier injection 43
    carrier transport 43
    exciton formation 43–44
    radiative recombination of excitons 44
quantum optical properties 6, 8
quantum size effect 1, 4, 5
quantum surface effect 4
quantum tunnelling effect 5, 6

r

red QDLEDs
  ageing of 275
  alloy core–shell structure 96–97
  breakthrough technology developments 87, 88
  challenges 103, 104
  core–shell structure 94–96
  device architecture development 97
  with double-hole transport layer 310
  external quantum efficiency development 87, 88
  fluorescence emission range 89
  materials 88–97
  PIF-TPA as HTL in 312
  stability studies of 260
  structure design and optimization range 90–91
  surface ligands on 91–94
research directions, QDLED degradation and aging 287
RGB color space parameters 46

s

SAMSUNG Group 332–335
sandwich AC-driven thin film electroluminescent (AC-TFEL) device 239
Scherrer's formula 77

Schottky barrier   39, 40
scintillation phenomenon of quantum dots   21
selected area electron diffraction (SAED) technique   77
self-annihilation of quantum dots   181
self-assembly methods   369
self-quenching   21
semiconductor nanocrystals   1, 51, 57, 188, 332, 335
Se-ZnSe colloidal quantum dots, UV–Vis absorption spectrum of   69
shell engineering   114, 118, 130, 134
   cadmium-free quantum dots   224–227
SILAR (sequential ion layer adsorption and reaction), core/shell QD synthesis   63
silicon nanocrystals   178
silicon quantum dots   15, 17, 178–179
single-layer quantum dot light-emitting device   41
single light-emitting white light devices   188
SiO_2 CQDs in perovskite matrix   181
slow dropping, core/shell QD synthesis   63
small-angle X-ray scattering (SAXS)   67, 76–77
solid-state ligand exchange process   179, 180
solid-state nuclear magnetic resonance (SSNMR) techniques   71
solubility of CdSe nanocrystals   66, 93
solution ligand exchange process   179, 180
solution NMR spectroscopy   70
solution-prepared inverted QDLEDs   122
solution processability   33, 34
   of colloidal quantum dots   66
solution-processed method   48
solution-processed QDLEDs   25, 27, 48, 200, 314
solvent-resistant blended HTL   315
spin-coating process   47, 48

stability   35–36
surface-adsorbed H_2S 91
surface chemistry, of colloidal quantum dots
   covalent bond classification method   65–66
   entropic ligands   66
surface dipolar polymers   301
surface ligand modification   87, 90, 91
surface passivation   61, 176, 201, 247, 287, 328, 336
surface states, of quantum dots   36

t

tandem QDLED device structure   102
tandem WQDLEDs   196
TCL   351–354
TC-SP (thermocyclically coupled single precursor), core/shell QD synthesis   63
ternary I–III–VI$_2$ chalcopyrite semiconductor QDs   17
tetradecane   57, 58
2,2′,7,7′-tetrakis[N-naphthalenyl(phenyl)amino]-9,9-spirobifluorene (spiro-2NPB)   309
TFB-BP polymer synthesis and photocross-linking   314
thermal spin-coating technique   310
thick-shell structured $Zn_{1-x}Cd_xSe/ZnS$ core/shell quantum dots   130
thioglycolic acid (TGA)   90, 332
three base color quantum dot composite   193–197
3D cluster-based bcc single supercrystals   78
time-resolved fluorescence spectroscopy   68
time-resolved photoluminescence (TRPL)   278, 287
top-down method, QD preparation   3
transient electroluminescence (EL)   278–281
transient fluorescence spectra   68, 70

transmission electron microscopy (TEM) 7, 67, 76, 278
tricolor QDs@Psi powders 197
tri-*n*-octylphosphine (TOP) 127
tri-*n*-octylphosphine oxide (TOPO) solution 54
trioctylphosphine oxide 56, 90, 339
tris (4-carbazoyl 9-ylphenyl) amine (TCTA) 158
two-dimensional nanocrystals 59, 61
two-dimensional quantum sheets 2
type I and type II CdSe/ZnS core–shell quantum dots 11, 12
type I core–shell structure 95
type II core–shell QDs 95
type II core–shell structure 95

u

ultra HD quantum dot TV sets 51
ultra-thin 1,3,5-tris(1-phenyl-1*H*-benzimidazol-2-yl)benzene (TPBi) interface layer 264
UV–Vis absorption spectroscopy 66–69

v

vacuum evaporation 132, 165
valence band 1, 3, 37–39, 62, 95, 120, 158
V_2O_5/PVK heterojunctions, as hole transport layers 308

w

white LEDs based on quantum dot photoluminescence mechanism 188
white QDLED based on CIS class 222
white quantum dot light-emitting diodes (WQDLEDs) 188, 195, 201–203
wide-angle X-ray scattering (WAXS) 67, 76–77

x

X-ray absorption fine structure (XAFS) spectroscopy 78–79
X-ray diffraction (XRD) 77, 278
X-ray diffractometer 77–78
X-ray diffractometry 77
X-ray photoelectron spectroscopy (XPS) 74–75

y

YAG hybrid phosphors, for white light-emitting diodes 189
yellow–blue composite white light quantum dots 189–193
yellow–green emitting $CuInS_2$ colloidal quantum dots 190

z

ZCGS/ZnS core–shell quantum dots 222
zero-dimensional-quantum dots 2
$Zn_{0.95}Mg_{0.05}O$ as electron transport layer material 295
ZnO as ETM 293
ZnO ETLs fabrication, n-type dopants for 295
$Zn(OA)_2$-QDs based IJP QDLED 368
ZnSe/CdSe/ZnS core-bilayer shell QDs 114
ZnSe quantum dots 215, 217–218
ZnSe/ZnS-based blue QDLEDs 215
ZnS QDs/CBP/MoO_3/Al inverted device 124
ZnS QDs/CBP/MoO_3/Al QDLED 113
ZnS quantum dots with 1-octanethiol ligands 121
$Zn_xCd_{1-x}Se$ alloyed quantum dots 121
$Zn_{1-x}Cd_xSe$/Zn core/shell quantum dots 126
$Zn_{1-x}Cd_xSe$/ZnSe/$ZnSe_xS_{1-x}$/ZnS core/multi-shell QDs 113